REMAKING REALITY

In the face of political-economic, technological and environmental changes of unrivalled scope, the matter of "nature" is high on the agenda as the new millenium approaches. From wilderness to the composition of the human body, nature is increasingly "artifactual", a social product fashioned by economic, cultural and scientific practices.

Rejecting apocalyptic pronouncements that the end of the millenium represents the "end" of nature as well, *Remaking Reality* brings together contributors from across the human sciences who argue that a notion of "social nature" provides great hope for the future. Applying a variety of theoretical approaches to social nature, and engaging with debates in politics, science, technology and social movements surrounding race, gender and class, the contributors explore important and emerging sites where nature is now being remade with considerable social and ecological consequences. The essays are organized around two themes: "capitalising and envisioning nature" and "actors, networks and the politics of hybridity". An afterword reflects on the problems and possibilities of future natures.

For critics and activists alike, *Remaking Reality* provides essential theoretical and political tools to rethink environmentalism and progressive social natures for the twenty-first century.

Bruce Braun is a visiting scholar in the geography department at the University of California, Berkeley and **Noel Castree** is a lecturer in geography at the University of Liverpool.

REMAKING REALITY

Nature at the millenium

Edited by Bruce Braun
and Noel Castree

London and New York

First published 1998
by Routledge
11 New Fetter Lane, London EC4P 4EE

Simultaneously published in the USA and Canada
by Routledge
29 West 35th Street, New York, NY 10001

Typeset in Galliard by Keystroke, Jacaranda Lodge, Wolverhampton
Printed and bound in Great Britain by TJ International Ltd, Padstow, Cornwall

British Library Cataloguing in Publication Data
A catalogue record for this book is available from the British Library

Library of Congress Cataloging in Publication Data
Remaking reality : nature at the millenium / edited by Bruce Braun and
Noel Castree.
Includes bibliographical references and index.
1. Human ecology. 2. Nature—Effect of human beings on.
I. Braun, Bruce, 1964– . II. Castree, Noel, 1968– .
GF41.R45   1998
304.2—dc21   97–35301
CIP

ISBN 0–415–14493–0 (hbk)
ISBN 0–415–14494–9 (pbk)

CONTENTS

FIGURES

CONTRIBUTORS

Bruce Braun is a Visiting Scholar in the Department of Geography at the University of California, Berkeley. His work addresses the cultural politics of nature, social and cultural theory, and environmental and anti-colonial politics in Canada.

Noel Castree is Lecturer in Geography at the University of Liverpool. His work addresses contemporary social theory and environmental thinking, with a particular focus on Marxist and post-Marxist thought.

Charis Cussins is Visiting Assistant Professor in the Department of Science and Technology Studies at Cornell University. She has written on human reproductive technologies and the politics of conservation in both the developed and developing worlds.

David Demeritt was until recently Post-Doctoral Fellow with the Sustainable Development Research Institute at the University of British Columbia. He is now a Lecturer in Geography at the University of Bristol.

Margaret FitzSimmons is Professor of Environmental Studies at the University of California, Santa Cruz. She has written extensively on the political economy of modern agriculture.

David Goodman is Professor and Chair of Environmental Studies at the University of California, Santa Cruz. He is the author of *Refashioning Nature* (with Michael Redclift, 1991) and *From Farming to Biotechnology* (with B. Sorj and J. Wilkinson, 1987).

John Graham is a contract researcher for the Deutsches Institut für Urbanistik of Germany and planner with Berridge Lewinberg Greenberg Dark Gabor, an international urban design/planning firm based in Toronto. He is an associate editor of *Alphabet City*.

Cindi Katz is Chair of the Environmental Psychology Programme and on the Women's Studies and Cultural Studies Faculty at the Graduate School of the City University of New York. She is author of the forthcoming

Disintegrating Developments: Global Economic Restructuring and the Struggle Over Social Reproduction, and is editor (with Janice Monk) of *Full Circles: Geographies of Women Over the Life Course* (1993).

Roger Keil is an Associate Professor at the Faculty of Environmental Studies, York University, Toronto. His research has centred on world city formation, urban ecology and urban politics, particularly in Los Angeles, Frankfurt and Toronto. His most recent book is *Globalization, Urbanisation and Social Struggles* (1998).

Bruno Latour is a philosopher and sociologist and author of *We Have Never Been Modern* (1993). He lectures at the Centre de Sociologie de l'Innovation of the École des Mines de Paris and at the London School of Economics.

James McCarthy is a doctoral candidate in the Department of Geography at the University of California, Berkeley, working on the political ecology of resource exploitation and the Wise Use movement. He has published on issues specific to the American West as well as on more general questions of nature and capitalist modernity.

Emily Martin is Professor of Anthropology at Princeton University. She has written on ideology and power in Chinese society (*The Cult of the Dead in the Chinese Village* and *Chinese Ritual and Politics*) and more recently on the anthropology of science and reproduction in the USA in *The Woman in the Body: A Cultural Analysis of Reproduction* (1987) and *Flexible Bodies* (1994).

Allan Pred is Professor of Geography at the University of California, Berkeley. He is author of *Making Histories and Constructing Human Geographies* (1990), *Reworking Modernity* (with Michael Watts, 1992) and most recently Recognising *European Modernities* (1995).

Hilary Rose is Emerita Professor of Social Policy at the University of Bradford. Author of *Love, Power and Knowledge: Towards a Feminist Transformation of the Sciences* (1994), she has established herself as a leading critic of modern science. She is currently working on a book on human genetics.

Neil Smith is Professor of Geography and Fellow at the Centre for the Critical Analysis of Contemporary Culture at Rutgers University, New Jersey. He is author of *Uneven Development: Nature, Capital and the Production of Nature* (1984), co-editor of *Geography and Empire* with Anna Godlewska, (1994) and author of *New Urban Frontier: Gentrification and the Revanchist City* (1996).

Michael Watts is Director of the Institute of International Studies and Chancellor's Professor of Geography and Development Studies at the University of California, Berkeley. He is author of numerous works on the political economy of famine and development and is currently working on a book on photography.

FOREWORD

REMAKING REALITY:
NATURE AT THE MILLENIUM

At a time of political-economic transformations global in reach and of unrivaled dynamism, technological changes of astonishing power and unfathomable extent, and apocalyptic pronouncements of ecological catastrophe and cultural crisis, the matter of "nature" is on the agenda as never before. Today, whether it be on behalf of Amazon trees, arctic tundra or even the human body, one sees unprecedented efforts to "protect" and "save" nature from the destruction wrought by accelerating socio-economic change. But is there a "nature" to be defended and preserved? Is it really possible to decry the "end of nature?" Or is nature, more than ever before, *social nature*, a nature ordered up, manipulated and constructed, as well as animate, unpredictable and consequential?

Remaking Reality brings together contributors from across the social and human sciences who argue that understanding nature as "socially constructed" provides serious analytical and political hope in a world faced with the urgent task of building survivable futures. It seeks to put the various existing perspectives on social nature to work to identify some of the major ways in which, and some of the specific sites at which, nature is being remade at the millenium. After an introductory chapter in which the main contemporary perspectives on nature's social constitution are examined, the chapters are divided into two parts. Part 1 establishes that contemporary social natures are at once "capitalized" and "enframed." The objects of the most intimate political-economic transformations, these natures are also discursively delimited in ways every bit as material and consequential. Highlighting this conjunction of capitalization and cultural enframing, the authors identify several key arenas in which future natures are being made. In Part 2 the limits of this making are addressed by contributions which consider the liveliness and hybridity of contemporary social natures. Showing that society and nature are continuously knotted together, authors demonstrate how organic and inorganic entities remain critically important as actors in nature's transformation.

Drawing on a variety of theoretical perspectives, the authors share a common belief that nature is both the instrument of social power as well as the font of a potentially liberatory politics. By complicating our notions of nature's social remaking, all the authors point towards a "politics of nature" quite different from that founded on the nature–society dualism so ingrained in everyday thinking. No longer able to be contained in a discrete, separate ontological space, they argue that today the remaking of nature has much wider consequences, for it impinges on virtually all aspects of social reality. In short, nature at the millenium raises pointed and vital questions about the worlds we inhabit and those we wish to inhabit in the future.

PREFACE

Now and then, certain events occur which, with wonderful clarity, provide a capsule illustration of much wider themes and concerns. In 1997 one of the most highly publicised environmental issues in Britain was the impending destruction of an area of woodland. It was to be cleared for the construction of a new runway at the country's second largest airport, located in south Manchester. In a series of stage-managed incidents designed to rally public support, radical environmentalists intent on "saving" the woodland dug tunnels and living spaces beneath it in order to hinder the construction that would eventually take place. The evictions which followed were well covered in the press and generated a lot of sympathy for the environmentalists and the forest. At the same time, the airport agreed – in what would have been one of the most ambitious and costly environmental preservation operations in British history – to literally transplant a portion of the forest lock, stock and barrel to a nearby location. However, this plan has been subsequently abandoned. The reason is that in the opinion of the ecologists employed by the airport, the considerable impact of the environmentalists' living, walking, eating and sleeping in the woodland has "damaged" it beyond repair and defiled its "natural" state.

The most obvious interpretation of these events would be to point to the irony of the fact that by virtue of their very own actions the environmentalists made the "first nature" they intended to defend into a "social nature": a nature inhabited and manipulated, a nature no longer pristine and pure. But it strikes us that the less obvious irony here is that the woodland was a social nature *from the very start*. Surveyed, mapped and culturally ordered by the environmentalists, the press and the airport authority, the woodland was continuously spoken for, represented and scripted from the very moment it entered the public domain as an "environmental issue."

Over the last decade or so it has been exciting to engage with a new and burgeoning literature across several disciplines which takes it as axiomatic that "nature" is indeed social through and through – and which traces the sometimes dark, sometimes liberatory consequences of this insight. As we look to a future replete with dangers and possibilities, there is a pressing need to convene an ongoing conversation on the modalities and consequentiality of the social natures we are presently making and inhabiting.

We are very grateful to the contributors to this book, not only for the chapters they have produced but also for believing in the value of an interdisciplinary inquiry into social nature at the millenium. While no single volume can possibly do justice to a topic of this size, we hope the book reflects something of the multiplicity and importance of the natures we construct, contest and must negotiate with. Many thanks too must go to Derek Gregory for insisting we turn a germ of an idea into the present volume, and to Tristan Palmer for his enthusiasm and for giving us the go ahead. Along the way several people gave indispensable assistance. David Demeritt, our friend and interlocutor, has been a constant source of ideas and good advice. Steve Reddy's facility with computers was invaluable and we thank him greatly. To Charis Cussins goes an enormous vote of thanks for translating with great skill and sensitivity Bruno Latour's essay to an incredibly tight deadline. We also acknowledge Mattamy Homes for permission to reproduce the advertisements in Chapter 5 by Roger Keil and John Graham. Finally, we must mention Valerie Rose, Sarah Lloyd and Sally Carter, our editors at Routledge, who inherited this project from Tristan Palmer. For their patience, trust and good humour we thank them warmly.

Bruce Braun and Noel Castree
May 1998

Part 1

INTRODUCTION

1

THE CONSTRUCTION OF NATURE AND THE NATURE OF CONSTRUCTION

Analytical and political tools for building survivable futures

Noel Castree and Bruce Braun

NATURE AT THE MILLENIUM

Nature, it seems, is on the agenda as never before. At a time of political-economic transformations global in reach and of unrivaled dynamism, technological changes of astonishing power and unfathomable extent, and apocalyptic pronouncements of ecological catastrophe and cultural crisis, the matter of nature has become a pressing issue, yet one bewildering in its complexity. From biotechnology to "wilderness" preservation, from the exciting medical promises and dark eugenic possibilities of the Human Genome Project to the moral imperatives and neo-imperialist rhetorics mixed together in discourses of "biodiversity," and from the complex politics of deforestation in India to the equally important struggles over models of global warming in Washington, nature is something imagined and real, external yet made, outside history but fiercely contested at every turn. It is at once everywhere and nowhere, the foundation for all "life" and the elusive subject of theoretical and political debate. As we face the twenty-first century, the matter of nature is no less difficult than Raymond Williams (1976) found it more than twenty years ago, and perhaps far more consequential.

That it should be both at once has everything to do with changes occurring at various levels in *fin-de-siècle* societies. The intense global economic restructuring of the last two decades has drastically reorganized networks of production and consumption at a variety of scales and sites, with the result that old industry–environment articulations have been intensified and new ones generated. Previously non-industrialized territories are today industrialized, while elsewhere formerly productive landscapes languish or become remade both materially and semiotically through the practices of "ecological restoration." With the fall of

3

actually existing socialisms and in a neo-liberal climate of "market triumphalism" (Peet and Watts 1996), local and global natures in both the developed and less developed worlds increasingly reflect strategies of accumulation. In turn these strategies have brought with them myriad environmental problems, problems today contested by the baffling array of environmental non-governmental organizations (NGOs), new social movements (NSMs) and political parties which are such a part of the contemporary political landscape.

In short, global nature is increasingly remade in the image of the commodity – whether it be logging in Madagascar or the new spatial and environmental logics of eco-travel – such that increasingly the nature–society distinction appears obsolete. At the turn of the twenty-first century, nature everywhere is "enterprised up" (Strathern 1992) or "operationalized" (Rabinow 1992). Where the juggernaut of capitalist development has not yet made this complete, technological innovations – increasingly in the form of "technoscience" – stand poised to finish the task. The controversial Human Genome Project may be the most recognizable, and anxiety-producing, example – promising to reconfigure our own bodily natures – but more prosaic interventions already exist, like those surrounding the genetic manipulation of food within modern agro-food complexes (Goodman *et al.* 1987). Elsewhere, technological innovations of a no less material kind – in information technology and the media – increasingly mediate and shape our access to the world. Indeed, in our emerging "network societies" (Castells 1996) there is now the promise of what Wark (1994) calls "third nature" – that is, the simulated natures of everyday TV and magazines, games like SimEarth, or the extraordinary optics of the geographical information system (GIS) all of which provide new, powerful means of manipulating nature as "information."

More than ever before, then, nature is something made. For some, this represents the "end" of nature (McKibben 1989), a response rooted firmly in a modern dualism in which nature is seen as external to society: its other. From this perspective nature must be defended against its "destruction" by humans, and battle lines are drawn to preserve its "pristine" character. For others, humanity's relationship with nature, in all its permutations, is ineluctable and inherently subversive of the nature–society dualism. From this perspective, human intervention in nature is thus neither "unnatural" nor something to fear or decry. This does not rule out limiting human actions in specific situations, but from this perspective what is at stake is not preserving the last vestiges of the pristine, or protecting the sanctity of the "natural" body, but building critical perspectives that focus attention on how social natures are transformed, by which actors, for whose benefit, and with what social and ecological consequences.

Over the past fifteen years an exciting, diverse, interdisciplinary field of critical theory has emerged which takes the second tack, and thus seeks to explain and illustrate the many ways in which nature is constructed and reconstructed within modern and late modern societies. From anthropology to geography, environmental history to cultural studies and political economy to science studies, critical theorists like Piers Blaikie, William Cronon, Klaus Eder, Evelyn Fox Keller, Donna Haraway, Richard Lewontin, Thomas Laqueur, Michael Redclift, Sharon Traweek, Alexander Wilson and Steve Woolgar, along with many of the contributors to this volume, have made significant theoretical and empirical contributions to our collective understanding of how the environments and bodies we inhabit are fabricated at different levels, through multiple relations, by various actors and as the effects of different forms of social power.

In this ferment of political-economic, technological and intellectual change, *Remaking Reality* brings together contributors from a variety of fields who critically explore, through theoretical debate and case studies, the making and remaking of nature at the millenium. Our book title deliberately speaks to two important points which our contributors seek, in different ways, to make axiomatic. The first point is that nature, in Haraway's (1992: 296) words, "cannot pre-exist its construction": it is figure, construction, artifact, displacement. It is something made – materially and semiotically, and both simultaneously. Those, like "deep greens," who would still appeal to "nature" as a source of moral and political guidance will, of course, find this argument scandalous. Yet, along with the contributors to this volume, we argue that Haraway's insight has profound implications and embodies a liberatory potential, radically opening the field of debate and action surrounding what kinds of natures we seek. But it is an insight that must be viewed with caution. Taken too far, it risks treating nature as a tabula rasa, or suggests that there are no limits – either physical or moral – to how human interventions in nature should proceed. If this is our first point – nature is always something made – then our second is that its making is always about much more than just nature. As our contributors show, it is increasingly impossible to separate nature off into its own ontological space. Thus, the remaking of nature(s) has wider implications – it becomes, quite simply, a focal point for a nexus of political-economic relations, social identities, cultural orderings, and political aspirations of all kinds.

These two general themes are woven into all the chapters collected here. But the volume also has three more specific aims. First, given the speed and intensity of contemporary political-economic and technological change, there is a need to identify and detail new and emerging sites at which social productions of nature occur or are contested, while attending to the social, ecological and political consequences of each. Many contributors do just that, conveying a sense of the complexity and variety of nature's construction in the 1990s. Second, after nearly two decades of writing about the social constitution of nature, it seems apposite to elaborate, put to work and explore the tensions between the various existing theoretical perspectives on social nature. Rather than quest after a meta-theory that can somehow explain nature's remaking across all sites, contributors draw on a variety of theoretical tools, including political economy, post-structuralism and science studies. Which approaches are enrolled depends in part on the conversations engaged, the sites studied and the constructions of nature that seem most in need of interrogation. What emerges is a sense that social natures are made according to no one single, overarching logic. Accordingly, no one theory can or should be put forward to explain their fabrication. This is not a charter for an undisciplined theoretical pluralism; rather it is a recognition that different theoretical positions must be engaged if nature and environment at the millenium are to be properly understood. This has practical implications, of course, because, as our contributors show, the social construction of nature is above all consequential – for both humans and non-humans – and thus the question of politics flows logically from analysis and critique. The third aim, therefore, is to contribute to part of a larger project of building "survivable futures" (Katz 1995) for people and other organisms, predicated on social and environmental justice. As the contributors show, if struggles over nature and environment take on a wider resonance then they necessarily help us expand and redefine what a "politics of nature" is all about.

This introductory chapter explores in some detail three theoretical traditions which provide valuable analytical tools for understanding the mechanisms of nature's construction(s) at the end of the millenium. We begin with a critical appreciation of Marx's contribution to the questions of nature and environment as represented in recent work on the production of nature by several contemporary writers in the Marxist tradition. We regard work in political economy as vital to understanding social nature in our times, but suggest that its primary focus on nature's material transformation in capitalist production underplays other aspects of nature's making. Recently, post-structuralist interventions have drawn attention to questions of representation and what we call – following Heidegger – "enframing nature", and we argue that an appreciation of the materiality of representation allows for the recognition that nature's construction involves more than just capital and commodities conventionally understood. Post-structuralist accounts, however, risk locating agency only at the level of "culture" or "discourse," erasing the role that organisms and physical systems play in nature's remaking. The emerging field of "science studies" provides a third theoretical resource and a possible corrective to this erasure. With their focus on science-as-practice, science studies scholars like Donna Haraway and Bruno Latour have shown that what counts as nature, and nature's remaking, occur within networks that include social, technical, discursive and organic elements simultaneously. As is evident in the chapters that follow, all three traditions provide rich analytical resources, as well as important political resources, for the pressing task of building the future natures that we wish to inhabit.

THE PRODUCTION OF NATURE

Once we begin to speak of people mixing their labour with the earth, we are in a whole world of new relations between people and nature and to separate natural history from social history becomes extremely problematic.

(Williams 1980: 76)

If nature at the millenium is, more than ever, a distinctively capitalist nature – one made and remade as a commodity form within the specific logics of capitalist production, and competition accumulation – then Marxian political economy offers indispensable critical resources for making sense of and contesting those logics. Despite Marx's well-known failure to offer a systematic account of nature and environment under capitalism, the last decade or so has seen a remarkable flowering of Marxist and neo-Marxist scholarship investigating what Neil Smith (1984), in a paradigmatic formulation, called the social "production of nature" within capitalist (and semi-capitalist) social formations.[1] This Marxian work, which has made an enormous contribution to our collective understanding of social nature under capitalism, has consisted of several overlapping streams, some overtly theoretical and some more empirically grounded.

Marx and nature

Marx's scattered reflections on nature and environment under capitalism meant that it was left to later commentators – notaby Alfred Schmidt – to comb his work for insights. Schmidt's (1971) *The Concept of Nature in Marx* showed that Marx's account of capitalist nature had two sides: on the one hand, a critique of representations of "nature" within bourgeois societies, what Smith (1984: 1) called "the ideology of nature"; and on the other, a fragmented theory of nature's creative destruction under capitalism. Schmidt realized that for Marx the various bourgeois meanings of nature made it resolutely external to society ("first nature") and, at the same time and in contradiction, universal – human beings, as biological entities, were seen as "natural" too. Smith (1984) has identified the cognitive and political implications of this dual representation of nature – a representation which underpins both the "save nature" rhetorics of contemporary ecocentrics and the "manage nature" discourses of technocentrics in government, business and the like. Rendering it external, Smith argued, is doubly ideological: first, it renders non-human objects and processes intractable barriers to which humans must at some point submit, and second it denies any social relation to environment, thus ruling out humanity's creative capacities to transform it. At the same time, Smith continued, the assumption of universality is also counter-revolutionary because it implies that social relations are as immutable as natural processes. Against this, Schmidt showed that Marx insisted on the relations between environment and society, thus avoiding the schism between them without collapsing the latter into the former as in the monistic doctrine of universal nature. As Smith (1984: 18) observed, "nature separate from society has no meaning . . . The relation with nature is an historical product, and even to posit nature as external . . . is absurd since the very act of positing nature requires entering a certain relation *with* nature." More specifically, Schmidt saw in Marx a dialectic between the two: "nature is mediated through society and society through nature" (Smith 1984: 19), a complex metabolic process which Marx centers on the labour process, the point at which society systematically engages with nature. More particularly still, Schmidt identified in Marx a specifically capitalist appropriation of environment as labour value is placed on environmental goods as part of a system predicated on class relations, competition and accumulation.

Schmidt's argument, while timely, has since been charged with two major limitations. First, Smith (1984: 23–4) has provocatively argued that "incredible as it sounds, Schmidt ends up providing us with one of the most elaborate accounts of the *bourgeois* concept of nature. In Schmidt too there is an external conception of nature . . . and a universal conception." Following Adorno and Horkheimer (1972), Schmidt ends his treatise with the complaint that, despite himself, Marx's vision of a socialist transformation of environment is guilty of the "domination of nature." But for Smith this concept of domination connotes, in its very language, the externality and universality that are the hallmark of the bourgeois conceptions. In short, in his attempt to re-emphasize the realm of nature in Marx's oeuvre, Schmidt ironically underplayed the role of social relations in constituting nature and society. In the second place, Schmidt's account of the Marxist conception of nature is extremely abstract, and even his comments on use values and exchange values do not approach the kind of concrete analytics that Marx sought in *Capital*.

This is why Neil Smith's *Uneven Development* was such an important state-ment, pushing debate forward with his thesis about the production of nature under capitalism. As Smith (1984: xiii–xiv) readily conceded, this thesis "sounds . . . quixotic and . . . jars our traditional acceptance of what had hitherto seemed self-evident . . . it defies the conventional, even sacrosanct separation of nature and society, and it does so with such abandon and without shame." Nonetheless, it is of the utmost importance. First, it gets beyond the external and universal conceptions of nature, registering the redundancy of conceiving of nature as a pristine entity, untouched by human hand – conventionally known as "first nature." Second, it rightly points to the internal relations between society and nature: social projects are invariably ecological projects and vice versa (Harvey 1996). Third, accordingly it alerts us to how capitalism constructs and reconstructs whole landscapes as exchange values under the profit imperative, to how it determines particular constellations of "natural" products in particular places. This is exemplified well by recent work on the production of agrarian regions (Fitzsimmons 1986; Marsden *et al.* 1986a, b; Watts 1989, 1991), but most forcefully, perhaps, in William Cronon's marvellous *Nature's Metropolis* (see *Antipode* 1994). Fourth, it powerfully historicizes human relations with nature and thus opens up the politics and possibility of the transformation of both nature and society (see also Smith 1996). On the basis of this general proposition, Smith (1984: ch. 2) then went on to present one of the most sophisticated Marxist theorizations of nature's production under capitalism, moving with increasing historical–geographical specificity from "production in general," to "production for exchange," to, finally "capitalist production," with its specific social (class) relations and specific (profit-driven) "value relation" to environment, systematically generating patterns of geographically "uneven development." The effect was to define a powerful overarching critical research programme in which "the major [political-economic] issue . . . becomes the question of how nature is (re)produced, and who controls this process of (re)production in particular times and places" (Whatmore and Boucher 1993: 167).

Neither Smith, nor the other authors cited, were suggesting that capitalism "produces" nature in the sense of, to take an extreme example, determining how trees grow. The portmanteau term "production" seems to imply that capitalism determines every aspect of the natural world as it transforms it: right down to each particle of natural stuff. However, this is not what Smith was arguing. Rather, what the thesis of the production of nature does capture is the way in which "first nature" is replaced by an entirely different produced "natural" landscape. The competitive and accumulative imperatives of capitalism bring all manner of natural environments and concrete labour processes upon them together in an abstract framework of market exchange which, literally, produces nature(s) anew.

However, its insights notwithstanding, Smith's reconstruction of Marx arguably underplays the "materiality of produced nature" (Castree 1995) at a time when capitalism is wreaking more environmental havoc than ever. This, along with Marx's supposed "Prometheanism," is why over the last few years others have returned to Marx to enrich in more ecofriendly ways our understanding of nature's production. Notable here is O'Connor's (1988) well-known work on the "second contradiction of capitalism"; Altvater's (1993) dual thesis that discounting the future routinely leads to resource over-exploitation, while the removal and abstraction of individual commodities from their ecosystemic context frequently entails a hidden, because "unvalued," cost to those wider ecosystems; Benton's (1989) argument about the "naturally mediated unintended consequences of production"; and Harvey's (1996) contention that the distinctive spatio-temporal imperatives of capitalist valuation are anti-ecological. In addition, a number of other authors – notably, many contributors to the journal *Capitalism, Nature, Socialism* – have tried to develop specific political-economic concepts to concretise nature–society relations, while still others have begun to periodize "regimes of environmental accumulation." In this regard, Escobar's (1996) recent distinction between "modern" and "postmodern" ecological capital is a useful heuristic, marking the transition from an expansionary, anti-environmental regime of accumulation bent on "capitalising nature" to a "postmodern" one obliged to sustainably manage its own ecological future in the interests of profitability and survival.

The upshot of these various contributions towards an "ecoMarxism" has been twofold. First, they have tempered the unabashed anthropocentrism of Marx's political economy, but without evacuating it altogether. This is important, for any liberatory remaking of nature must acknowledge that it is only through socially specific structured human action and interpretation that "nature" can be known and transformed at all. Second, ecoMarxist writing has widened both

Marxian notions of political action and challenged the political separatisms of "green politics." Today, traditional concepts of class struggle simply must be linked to ecological struggles of which they are a part (Harvey 1996; Peet and Watts 1996), while the subject of "environmental politics" can no longer be simply "nature" alone.

First World agrarian political economy

If some have remained at the level of abstract theorisation, others have preferred the messier terrain of the empirical where the real complexities of nature's remaking become insistently apparent. Inspired by Kautsky's (1976) *The Agrarian Question*, the last decade or so has seen the rapid rise of an agrarian political economy, largely focussed on the First World and particularly associated with rural sociology and agricultural geography (Buttel and Newby 1980; Marsden *et al.* 1986a). Contesting the notion that agriculture is an "insulated" and "exceptional" industrial sector and pointing to the enormous importance of food production and consumption in advanced capitalist societies, a number of authors has tried to make sense of the dramatic post-war rise of "productivist agriculture" (Peterson 1990). A complex and heterogenous field of research – involving authors like Margaret Fitzsimmons, David Goodman, Terry Marsden and Sarah Whatmore – this agrarian political economy has several important dimensions, each speaking to the consequential remaking of agricultural nature in the 1990s.

The first concerns the understanding of "political economy" itself. Inspired initially by Marxism, agrarian researchers showed that its twin focus on production and distribution within distinctively capitalist economies undermined modern agriculture's supposed exceptionalism and made questions of competition, restructuring and class relations as relevant to agriculture as to manufacturing. However, pointing to the "structuralist" tendencies of classical Marxism, many researchers soon turned to more supple neo- and post-Marxist political-economic concepts in order to grasp the complexity of agricultural transformation. As Marsden *et al.* (1996) suggest, this has had four valuable features – a focus on patterns of uneven development, a focus on local specificity within broader social structures, a focus on family farms within capitalism, and a focus on the state at all levels – each contributing to a more sophisticated understanding of agricultural nature. A second dimension of concern is technological transformation. The biological foundations of agriculture make it a distinctive form of nature's production, but postwar agrarian capitalism has exhibited remarkable technological dynamism (in pesticides, herbicides, genetically manipulated seeds and the like) in order actively to reduce its biological dependency (Goodman *et al.* 1987). This dependency reduction has, significantly, entailed remarkably increased powers to "refashion nature" (Goodman and Redclift 1991) according to the requirements of socially created demand. A third aspect of agrarian political-economy is a focus on "agri-food

10

complexes" (Friedmann 1982). Farming is clearly no longer a land-based activity alone, ending at the farm gates. As farms have been drawn unevenly into wider circuits of capital a significant development has been the rise of an agri-food industry, involving ramified connections between manufacturers of agricultural inputs, farmers, corporate food retailers, food processors and consumers. Farming thus becomes a "conferred" activity, increasingly reliant on distant markets, institutions and regulators within global food networks which are constitutive of and constituted by the local (Le Heron and Roche 1996). A fourth related concern is multinational corporations. While family farms are still important, the power to control agri-food complexes rests increasingly with multinational companies (MNCs), that largely control agricultural inputs, farming (e.g. through "contract farming"), food processing, and food retail. Finally, in an attempt to categorize and periodize these various forms of change, several authors have sought to theorize "international food regimes" (McMichael 1994) at the *fin de siècle*, paralleling efforts in studies of manufacturing to specify post-Fordist "regimes of accumulation."

This is only a summary account, but agrarian political-economy has pushed forward our understanding of late capitalist agricultural nature by putting abstract theory to the test of empirical specificities and, in the process, has generated more sensitive meso-level political-economic concepts. Capitalist agriculture emerges as a highly dynamic flashpoint of ecological transformation, technological innovation, global–local connections, and social relations of power. There remains much to be done – not least a further specification of local–global dialectics (Le Heron and Roche 1996), of individual agency within broader structures (Pile 1990), of nature's agency (Fitzsimmons and Goodman, this volume), of the discursive construction of food and agro-production (Arce and Marsden 1993), and of regimes of late capitalist agro-accumulation – but the enormous consequentiality of remaking agricultural natures is increasingly clear.

Third World political ecology

If agrarian political economy has dynamized the study of agricultural nature, "Third World political ecology" (Bryant 1992) lends that study greater ecological sensitivity and a sharper political edge in quite different geographical settings. Emerging as a concept in the mid-1970s, political ecology is perhaps most closely associated with the pioneering work of Blaikie (1985) and Blaikie and Brookfield (1987). For them, political ecology combines the concerns of ecology with "a broadly defined political-economy" (1987: 17). Consequently, rather than seeing Third World environmental problems as those of resource scarcity, overpopulation or ignorance, they focussed instead on the articulation of wider political and economic constraints with specific environmental uses. At the centre of this conjunction was the "land manager" whose relationship to environment must, Blaikie and Brookfield (1987: 239) insist, "be understood in a historical, political and economic context."

Since these beginnings political ecology has blossomed into a large but by no means coherent field (its leading practitioners today including diverse scholars like Nancy Peluso, Ramachandra Guha, Susanna Hecht, Michael Redclift and Michael Watts), lent integrity less by any theoretical confluences than by the sharing of common themes of inquiry (see, for instance, the essays in Peet and Watts 1996). Nonetheless, these themes encompass important and distinctive aspects of social nature in Third World settings, of which three stand out. First, given the preponderance of rural and agricultural workers in less developed countries (LDCs) the question of land use and management is particularly vital to many Third World livelihoods. Second, with economic globalization, many pre-modern agricultural production systems have been uneasily incorporated into global circuits of capital so that the question of the articulation of different social formations is especially germane. Third, it follows that when land degradation does occur it frequently becomes a highly politicized issue that, literally, can have life or death consequences.

Focussing on these and other features has generated important conceptual insights, many of which were already apparent in Blaikie and Brookfield's *Land Degradation and Society* (1987). First, by focussing on political-economic and ecological "marginality" – rather than ecological "limits" – they showed that marginality is both cause and effect of land degradation. Second, by inserting local land use into its social relational context they showed how the land manager is compelled to "transmit" excessive production pressures onto the land – pressures often issuing from distant sources and from the global level. Finally, hinting at later concerns about nature's discursive construction (see the third section below), Blaikie and Brookfield pointed to how perceptions of land degradation vary depending on the structured positionality of the viewer. Together, these insights radically undermined the idea that even in Third World settings nature was somehow an external and absolute limit to population growth by focussing instead on the political ecology of resource use. In particular they have affirmed poverty as the major cause of environmental deterioration, an argument which has lain at the heart of subsequent political-ecological research in all its forms (e.g. Hecht and Cockburn 1989; Redclift 1987; Stonich 1989; Peluso 1993). That these forms are varied is in many respects an advantage, for what is lost in theoretical integrity is made up for in developing specific areas of concern, such as an appreciation of regional ecological specificity (e.g. Bunker 1985), of different modalities of economic relations (e.g. Watts 1983), and of "development" itself as part of the land degradation problem (e.g. Collins 1987). Yet, as we approach 2000, political ecology also needs to build on these foundations to refine concepts – particularly regarding chains of causality in land degradation and the more careful theorization of production (Peet and Watts 1996: 7–8) – to look beyond the rural, to look at other environmental issues, and to draw out the political implications of its analyses if the Third World is to build socially and ecologically just "futurenatures" (Bryant and Bailey 1997).

New directions

These debates on Marx and nature, agrarian political economy and political ecology are clearly not of a piece but do amount to a significant, if loosely structured, research programme on nature's production within capitalist and semi-capitalist societies. However, as capitalist nature has continued to change during the 1990s a number of important new avenues await much-needed exploration – involving both new sites of nature's production and new mechanisms of nature's production hitherto ignored or marginalized by Marxian work.

The first is ecology. With the possible exception of political ecology, work on the production of nature – even in its ecoMarxist incarnations – has offered only relatively weak understandings of the nature and materiality of transformed environments. However, two related areas of scholarship now offer the prospect of a more thoroughgoing appreciation of nature's materiality, without reverting to modernist dualisms. One is "environmental history," a now flourishing subdiscipline prioritizing nature's agency in history (Worster 1988), the other the "new ecological science" (Zimmerer 1994) which has overturned outdated notions of ecosystem stability to focus on chaos, disequilibrium and instability (Botkin 1990).

Second, with the realization that capitalist economies are "embedded" (Granovetter 1994) has come the insight that political-economy and ecology of nature is as much a question of gender and ethnicity as it is of class (see the essays in Rocheleau *et al.* 1996). Agrarian political economy and Third World political ecology have already made significant contributions to mapping these three axes of social identity and power as they coalesce (Whatmore 1991; Carney 1996), a conjunction that demands a "messy Marxism" where "production" and "labour" are both gendered and ethnicized.

Third, there is the question of discourse and representation. Theoreticians like Smith, by focussing on representation as negative (the "ideology of nature") have underplayed the proactive nature of representation, for the production of nature is both material and discursive (Castree 1997). As Gramsci (1971) argued a long time ago, struggles over meaning are every bit as "material" and important as practical struggles. Some of the newer work in political ecology has begun to demonstrate this very well (e.g. Moore 1996), but as "environmental discourse" proliferates, new efforts must be made to map discursive constructions and their power geometries (see e.g. Buttel and Taylor 1994). In addition, efforts must also be made to understand the specific mechanisms of this construction and, as we show in the next section, post-structuralism offers valuable theoretical resources here.

Fourth, despite its concern with political economy, Marxian work on nature's production has, ironically, said relatively little about the question of "politics." Politics does not just reside either in class action or state action, but has to be redefined in light of the "nature politics" of the *fin de siècle*. As recent work on political ecology shows well, the classical Marxian notions of socialist struggle are complicated by new political forms like NSMs, NGOs, local coalitions and special interest groups within civil society (Laclau and Mouffe 1985), and by quotidian "hidden" political activites like those "weapons of the weak" charted by Scott (1985) and others.

Finally, the literatures reviewed above tend to focus on human–environment relations. But, important as this focus is, it risks overlooking two of the most distinctive and dramatic arenas in which late capitalist nature is produced: "big science" and hi-technology and the media. Both have distinctive political economies and both, as we noted earlier, are increasingly important aspects of the social production of nature, semiotically and materially. Yet the modalities of these hi-tech natures, along with their consequences for specific ecologies and bodies, have not yet received the sustained attention they deserve from Marxist theorists. Perhaps the site where the need for such inquiry is most pressing is one that, on the face of it, seems too "natural" to be artifactual: the body. Yet today, from biomedicine, to fashion, to media "bodyscapes," the body is, of course, firmly situated within the complex technical, discursive and productive landscapes of late capitalism. Harvey (1998) has suggested one direction that such inquiry might take. For him the body must now be seen as an "accumulation strategy": the site, that is, of an astonishing array of commodification practices which connect specific bodies to wider relations of capitalism. As an "unfinished" project, the body is seen here as remade in historically and geographically specific modalities. And these are modalities every bit as semiotic as they are material: for changing images of the body are inextricably entwined with interventions in the body in corporeal accumulation regimes.

The possibilities here are legion. The former topic has, for example, been explored by Emily Martin (1987, 1994, this volume) who has mapped changing perceptions of the body in medical discourse and popular culture, from bodies "suited for and conceived in terms of the era of Fordist mass production to bodies suited for and conceived in the terms of the era of flexible accumulation" (1992: 121). The latter topic has been addressed in recent work on reproductive processes by Adele Clarke (1995). She distinguishes "modern" (or Fordist) means of regulating and capitalizing bodies and bodily practices which focussed on the "industrialization" of reproductive processes (as well as the mass production and consumption of contraceptives, commercial menstrual products and so on), from "postmodern" (or post-Fordist) means that are characterized by individualization and deep intervention along the lines of fetal surveillance, surgical interventions in utero, genetic testing, and so on, signaling a shift from Fordist mass production and distribution to "niche" body markets. In each the body becomes an accumulation regime, but the differences are important. In one, the regulation of the body and bodily practices becomes commodified; in the other, the body itself is capitalized and remade.

ENFRAMING NATURE: CULTURAL INTELLIGIBILITY AND ECONOMIC AND POLITICAL CALCULATION

> It is necessary to reiterate the connections between the making and evolution of nature and the making and evolution of the discourses and practices through which nature is historically produced and known.
>
> (Escobar 1996: 46)

If today, then, the human body is an "accumulation strategy" it has become so not only because of important shifts at the level of production and consumption, but also in part because the body is now being made culturally intelligible – and thus available to forms of economic and political calculation – in new and often befuddling ways. Whether it be ultrasound imaging or mapping the human genome, new reproductive technologies or cybernetic systems which reconfigure the body as "information," the body is increasingly a "material-semiotic" object known in such ways that it can be changed (Haraway 1991, 1997; Rabinow 1992; Clarke 1995).

Perhaps more than at any other site, the artifactual body demonstrates the intertwined relations between capital, discourse and technology in the reinvention of nature at the beginning of the twenty-first century. This represents a significant challenge to understanding nature's social production and, accordingly, over the past fifteen years other critical approaches have emerged that owe less to Marx and political economy, and more to post-structuralism and to the sociology of scientific knowledge (SSK) and science studies. While they overlap, we choose to treat them as separate strands that provide their own unique insights into nature at the millenium.

Historicizing appearances: nature and representation in late modernity

> Nature is itself an entity which shows up within the world and which can be discovered in various ways and at various stages.
>
> (Heidegger 1962: 92)

The relatively slight attention paid to the constitutive role of discursive practices in Marxian theories of nature's social production should come as little surprise. Marxist theory – dialectics notwithstanding – has for the most part relied upon realist epistemologies where "nature" and "culture," and "scientific knowledge" and "politics," are kept separate or "regionalized" (Latour 1993; Castree 1995). Nature is assumed to be something that is unproblematically "ready-at-hand" to human actors; while its social transformation may be seen as historical, its "materiality" is not. Equally as important, Marx grounded human interaction with "nature" in biology. As Marx explained, humanity's relation with nature

15

was, in the first instance, practical rather than theoretical; it was related to, and determined by, quasi-transcendental "needs" rather than constituted and mediated through language:

> Men do not in any way begin by "finding themselves in a theoretical relationship to the things of the external world." Like every animal, they begin by eating, drinking, etc. That is, not by "finding themselves" in a relationship but by behaving actively, gaining possession of certain things in the external world by their actions, thus satisfying their needs. (They thus begin by production.)
>
> (Marx 1975: 190)

At one level the argument appears self-evident. The body has certain conditions for survival; we know that people risk death without sufficient food, a supply of oxygen, or protection from exposure. Thus, nature is continuously appropriated – and remade – as part of the "species existence" of humans. But this tells us little about how "needs" become defined, or, for that matter, what objects are taken up to meet "needs," or even how this is organized socially. As Baudrillard (1975) has shown, such questions are not so easily resolved, since both "needs" and the "use value" of things (or nature) are in part constituted culturally. For Baudrillard, it is not only capitalist modernity – with its logic of exchange value – that infuses a level of culture (or ideology) into what was previously an unmediated biological relation with nature.[2] Our relation with things is always already a sign relation; discursive relations and representational practices are constitutive of the very ways that nature is made available to forms of economic and political calculation and the ways in which our interventions in nature are socially organized.

The insights of post-structuralism have in recent years allowed us better to attend to these discursive mediations. They occur at a variety of levels, some more apparent than others. Arturo Escobar (1996), for instance, has shown how forms of capitalist development are always inflected in and through discursive formations, evident today in such discourses as "sustainable development" and "biological conservation" which authorize new processes of capitalizing nature. Far from being merely ideology, these discourses contribute to the very "production of production conditions." In other words, they mediate and organize human interventions in nature, including along which lines capital flows. This has important consequences, not only for which ecologies are transformed and how, but also, as Escobar intimates in his discussion of biodiversity conservation in Columbia, for whose relations to nature become foregrounded or displaced.

Interventions like Escobar's have challenged political ecologists to attend to how discursive relations – and not just market relations – organize social and ecological change and represent promising avenues of inquiry that bridge political economy and post-structuralism. Yet, we wish to extend Escobar's point further. The social production of nature is not just "mediated" by moral and political discourses like "sustainability"; rather, discursive relations infuse our relation with nature at every turn, including even at the micro level of knowledge and practice. This is different from saying that discourse is all there is, or that the world has no objective existence (cf. Haraway 1997; Demeritt, this volume). Rather, it is to insist that what counts as "nature," and our experience of nature (including our bodies), is always historical, related to a configuration of historically specific social and representational practices which form the nuts and bolts of our interactions with, and investments in, the world. Discourses like "sustainability" are important to the extent that they organize our attitudes towards, and actions on, nature. But, as a number of writers now insist, nature's materialization – its very visibility and availability – also requires careful attention, and is equally as important for how nature is known and remade with particular social and ecological consequences.

For many this "deep discursivity" is a challenging insight and it therefore merits further attention. Arguments which stress the constitutive role of discursive practices in the social production of nature can be traced to a crucial distinction between language as "instrumental" and as "expressive constitutive" (Taylor 1992; see also White 1991). The former – taking its initial form with Descartes – understands language as arising subsequent to, and as a means of organizing, our experience of an external reality. By this view, language serves to coordinate our actions in a world whose framework precedes or is independent of language, but to which language can be made to correspond. The latter – traced through Herder, Nietzsche, and Heidegger – understands language as disclosing a world of objects and involvements, thereby constructing a discursive field shared by both subjects and objects. Here, language is seen as the level at which social identities, and forms of subjectivity and intentionality are constituted. So, while in the former view language comes to reflect a pre-given reality; in the latter, language renders visible a "world" into which humans are "thrown."

Martin Heidegger is perhaps most closely associated with the view that in modernity we have forgotten the second – "enframing" – function of language,[3] and he is also largely responsible for drawing attention to the consequences of modern or subject-centered representational epistemologies. By placing a transcendental human consciousness at the centre, as does the instrumental view, human temporality becomes erased. In short – Heidegger warned (1962) – human knowing becomes dehistoricized, and the "order" of the world is seen as something "discovered" rather than something that itself requires explanation. This carries with it significant costs; indeed, Heidegger (1977) considered the lack of attention paid to the temporality of "being" (or the throwness of the subject) to be one of the most dangerous features of modernity, since it allowed moderns to mistake their "ordering" of the world with the world itself, and therefore rendered them unable to think the limits of representation – its closures and absences.[4] This is not merely an arcane philosophical debate but one with important moral and political dimensions. By stressing the temporal dimensions

17

of enframing, Heidegger showed knowledge about the world to be situated in specific historical contexts, and perhaps more important, to be always partial. In Heidegger's words, representation involved both "unconcealment" (bringing things into presence) and "concealment" (excluding other possible appearances). Or, as Foucault would later show, what was "visible" at any specific conjuncture was related to the production of "spaces of visibility," which necessarily also involved producing "spaces of *in*visibility" (see Foucault 1977; Deleuze 1988; Rajchmann 1988).[5]

In short, nature – like all objects – is an entity which "shows up" within the world, but only in certain ways and not others. Forgetting this, Heidegger argued, gave to modern representation the character of *gestell* – a "stamping" – whereby representation took on a normalizing function, fixing identities as immutable, something captured especially well in Gayatri Spivak's (1985) notion of "epistemic violence." For this reason Heidegger (like Nietzsche and later Foucault, Derrida and Spivak, among others) sought to disrupt the familiar story of the Enlightenment. Far from a progressive movement from tradition and metaphysics into the "clear light" of Reason (unencumbered by history) Heidegger characterized modernity as carrying with it its own unexamined theology (cf. Mitchell 1988) which obscured the ways in which things were given to subjects and the relations of power that necessarily infused all identities.

Several consequences follow from historicizing nature's appearances. First, it allows us to recognize the irreducible presence of the discursive at all levels of human relations with nature – nature is constructed "all the way down." This does not mean that only through language and ideas do things have objective existence, only that, in Derrida's later – and oft-misunderstood – formulation, there is no "outside" a general textuality, no "getting beyond" the epistemo- logical clearings in which we stand from which to obtain certain knowledge. Here we find a crucial difference between Marx and Heidegger that remains a point of contention in debates over nature's social production. Whereas Marx argued that humans did not first encounter the external world through a "theoretical relationship" but a "practical" one, Heidegger suggested otherwise:

> That which is does not come into being at all through the fact that man first looks upon it, in the sense of a representing that has the character of subjective perception. Rather, man is the one who is looked upon by that which is; he is the one who is – in company with itself – gathered together toward presencing, by that which opens itself.
>
> (Heidegger 1977: 131)

Humans may have a practical interest in nature, but, following Heidegger, our interventions presuppose nature's intelligiblity within specific discursive regimes and orders of knowledge.

Second, just as there is no "getting beyond" discourse, so also, discursive relations are everywhere intricately involved in the material transformation of nature. Indeed, it is precisely because nature exists as *trópos* (Haraway 1992), that it can be remade as commodity, just as the tropes of Orientalism underwrote European imperialism and channeled its development along various lines. The social production of nature occurs within wider discursive fields in and through which "things" are rendered visible and available to forms of calculation. In other words – to follow Haraway's (1997) reworking of Ian Hacking's (1983) phrase – representation is intervening.

Third, post-structuralist accounts of nature's construction place attention firmly on the operation of power and widen what is taken to be the domain of politics. Power – as Foucault (1977) so brilliantly showed – is not only, or even primarily, something "held," as in models of sovereign power. Rather, power is diffuse – it operates unannounced in myriad social practices, including those we take as "merely" discursive. Indeed, it is precisely because we mistake our ordering of appearances for the world itself, unaware of how our knowledges reflect their social context, that power relations become naturalized in our representations of nature. The significance of representational practices there-fore lies not only in that they disclose a "world," but that representation is a worldly practice. Representational practices are material at the same time as they materialize; they are deeply embedded in social – and ecological – relations at the same time as they render "society" and "nature" intelligible. Thus, when Heidegger defines "thinking" as attending to the clearing in which things come into presence, he gestures towards an analytical space that is also a space of politics – although not a politics recognizable by familiar road signs. In this light, Donna Haraway's (1992) contentious claim that "nature" cannot pre-exist its construction is not an idealist position that denies the reality of a physical world, but rather an epistemological intervention that is deeply infused with political intent. For feminists like Haraway and Judith Butler – in dialogue with anti-racist and queer activism – politics involves disrupting the "self-evidence" of identities like "nature" and the "body." In Butler's words, "to call a presupposition into question . . . is to free it from its metaphysical lodging in order to understand what political interests were secured in and by [its] metaphysical placing, and thereby to permit the term to occupy and to serve very different political aims" (1993: 30).

Finally, prefiguring what has become a common theme in post-structuralist writing, Heidegger's anti-foundationalism provides an alternative to those modern subject-centered epistemologies in which the truthfulness and certainty of knowledge claims involved setting aside presuppositions (getting clear of our historical "situatedness"). Knowledge – including our knowledge of nature – is always social and historical and this has implications for the claims we make about nature. Indeed, configurations of specific social, institutional and technological practices are precisely what enable us to build our accounts of the world. As Haraway (1991) has persuasively argued, responsibility thus lies in taking seriously these enabling conditions, attending to how the ways in which our constructions of nature occur within, and are suffused with, relations of class, gender, race and sexuality.

Effects of power and the domain of politics: constructing and contesting nature's materialization

Queering what counts as nature is my categorical imperative.

(Haraway 1994: 60)

Post-structuralist accounts of nature's construction are important, but they are not without their own problems. If we turn to a prominent example their strengths and weaknesses can be made apparent. Judith Butler's *Bodies that Matter* (1993) is arguably one of the most controversial post-structuralist accounts of how "nature" – in this case the "body" – comes to "matter" (a word whose double meaning neatly captures Butler's constructivist approach). It is worth exploring at some length.

In the book, Butler elaborates a theory of the "performativity" of sexual identity, part of a wider debate within feminist and queer theory over how gendered and sexed identities are constituted and lived. Performativity here should not be confused with free choice, as if one simply decided on sexual identity as one decides on what clothes to wear in the morning. Rather, performativity refers to how sexual identities are constituted (or "stabilized") in and through regulatory apparatuses, what Butler calls "citationality" (ritualized repetition or the iteration of cultural norms). As we shall see, it is precisely in performativity (seen as incomplete or imperfect iteration) that Butler also locates the disruption of what qualifies as a "viable" body within heterosexual cultural formations. For our purposes, the value of Butler's account lies in the crucial questions she raises about the body and its materiality or, more precisely, how the body is "materialized" along culturally and historically specific lines.

Like many other feminist scholars, Butler begins by placing in question the assumption that gender stands to sex, as culture does to nature or, in somewhat different terms, that gender is a discursive construct (culture) imposed upon pre-discursive sexual difference (biology). Rather, in what is by now the signal maneuver of post-structuralism, Butler places in question the self-evidence of the body-as-received and explores instead the conditions of possibility for speaking about and experiencing the sexed body, resulting in an account that turns on how a domain of "intelligible bodies" is constructed as well as "unthinkable, abject, unliveable bodies" which are excluded from, but haunt the former (see also Grosz 1994). It is important to highlight the political stakes involved. As Cheah (1996: 110) explains, writing the body as "constructed" disrupts post-Cartesian mechanistic accounts which understand the body as an immutable "natural entity," governed by natural laws of causality, or, in other cases, disrupts teleological accounts where intelligibility and matter are united in a body which strives toward an internally prescribed final goal. Such accounts are harmful, especially for women. As Cheah notes, they reduce the body to a passive object rather than a locus of power and resistance. But equally as important, they mistake biological discourse for the world itself (Haraway 1992), which, in the midst of a deeply phallocentric and heterosexist culture, is to remain inattentive to the deeply problematic closures and relations of power enacted in the body's intelligibility.

For Butler, writing that the body – or sexual difference – is "indissociable from discursive demarcations" (p. 2) is not the same as claiming that discourse causes sexual difference, or that bodies are not material. Rather it is to claim that how this materiality is understood, and how it becomes part of the ways that bodies are regulated and experienced, pivots on the sedimentation of historically specific modalities of how the body is "framed":

> The category of "sex" is, from the start, normative; it is what Foucault has called a "regulatory ideal". In this sense, then, "sex" not only functions as a norm, but is part of a regulatory practice that produces the bodies it governs, that is, whose regulatory force is made clear as a kind of productive power, the power to produce – demarcate, circulate, differentiate – the bodies it controls. Thus, "sex" is a regula-tory ideal whose materialization is compelled, and this materialization takes place (or fails to take place) through certain highly regulated practices . . . In this sense, what constitutes the fixity of the body, its contours, its movements, will be fully material, but materiality will be rethought as the effect of power, as power's most productive effect . . . [Bodies emerge from] a process of materialization that stabilizes over time to produce the effect of boundary, fixity, and surface we call matter.
>
> (Butler 1993: 1–2, 9)

The "matter" of the body is never pre-given, but instead takes form in and through a variety of discursive practices that regulate normative understandings of the sexed body, and are forcibly reiterated through time. It is important to follow Butler's Foucaultian emphasis on the productivity of power in order to grasp the full significance of her argument. Butler is not simply locating the misrepresentation of the "real" body in ideology or its repression by forms of sovereign power. For Butler, power does not repress, it produces; and what it produces are particular effects of truth, forms of subjectivity, and modes of materializing and experiencing bodies. In short, the very possibility of the "lived" body – or of any "lived" body – rests in the iteration of certain ways in which the intelligibility of the body has been circumscribed and "given to" subjects, simply on account that without these "we would not be able to think, to live, to make sense at all" (Butler 1993: xi). In a sense, the sexed body is one of those things that we cannot do without.

For Butler, understanding the "body" as performative suggests the possibility of its resignification, a topic to which we will turn later. But for Butler's critics – and there are many – her argument is problematic because she appears to claim that the "body" is merely an effect rather than the cause of its representation. Butler insists that hers is not an idealist position. "Construction" in her account points not to the willful "inventing" of bodies, but to deeply sedimented regulatory modes of conferring materiality upon bodies, an approach that highlights relations of power and one that, according to Butler, retains an essential link between the body's "appearance" (or form) and its "matter." Butler's preferred description is thus not "construction," but "constitutive constraint":

> If certain constructions appear constitutive, that is, have the character of being that "without which" we could not think at all, we might suggest that bodies only appear, only endure, only live within the productive constraints of certain highly gendered regulatory schemas . . . The discourse of "construction" . . . is perhaps not quite adequate to the task at hand. It is not enough to argue that there is no prediscursive "sex" that acts as the stable point of reference on which, or in relation to which, the cultural construction of gender proceeds. To claim that sex is already gendered, already constructed, is not yet to explain in which way the "materiality" of sex is forcibly produced. What are the constraints by which bodies are materialized as "sexed," and how are we to understand the "matter" of sex, and of bodies more generally, as the repeated and violent circumscription of cultural intelligibility? Which bodies come to matter – and why?
>
> (Butler 1993: xi–xii)

Butler's approach has attractive analytical and political advantages. By recasting the body as an effect of power, her account highlights the ways that power, knowledge and social practice are intricately interwoven in how bodies come to "matter." Further, it retains, or at least appears to retain, the centrality of the physical body, even as it historicizes its appearances. "Matter" remains central to "materialization" to the extent that Butler's account takes the cultural intelligibility of the "sexed" body to be a delimiting and constraining that occurs in and through discourses of what counts as the body and sex. In other words, the "body" is produced by repeated and violent circumscription: by exclusion, erasure, foreclosure, or abjection. Butler argues that there is therefore always a material "outside" to the body's social intelligibility, an "outside" that defines negatively what counts as the body. If this sounds remarkably similar to Heidegger's account of "enframing" it is not only coincidental, for Butler operates with many of the same distinctions between presence/absence, visiblity/invisibility, unconcealment/concealment, light/shadow, that organize Heidegger's account of the intelligibility and availability of the "object."

Butler continuously draws attention to foreclosure, to how the "natural" body is constituted by excluding other "natures," and the way that the ontologizing of what is rendered "present" simultaneously hides the relations of power that draw these limits. Indeed, it is precisely when the body is taken for granted – when it appears outside discourse and power (or history) – that relations of power are most "insidiously effective" (1993: 35). Disrupting this ontologizing move thus becomes central to any refiguring of the body.

Yet, accounts like Butler's also have serious limitations. At first glance, it appears to avoid a strict linguistic constructivism; after all, "bodies that matter" are still bodies; it is only how they matter that is an "effect" of closure (power) and thus the "constructed" body is not, as some critics charge, an imaginary construct unrelated to the physical world.[6] The "real" continues to exist, only it lies "beyond" the imaginary/symbolic nexus, never immediately accessible as a "thing-in-itself." But one might question whether such accounts provide as adequate – and politically useful – an approach to the physicality of bodies as writers like Butler contend. If the "matter" of bodies is something that is made culturally intelligible only through exclusion, for instance, how are we to imagine a more complete, less constrained, physicality except as an inaccessible "shadow" that lies "outside" discourse? How are alternate "materializations" enabled?

This remains one of the most trenchant problems haunting post-structuralist approaches. Butler responds to this problem in two ways. First, she explains that the "materialization" of the body occurs through a "forcible reiteration" of regulatory norms, but that this materialization is never complete since bodies "never quite comply with the norms by which their materialization is impelled" (1993: 2). But again this begs the question, if there is no "outside" to citationality how does this "never quite" compliance become manifest? Does this not require acknowledging, at least at some level, that matter precedes form? Butler answers by locating counter-hegemonic articulations in the gaps and fissures within discursive formations. In short, by invoking notions of multiplicity and unfixity, Butler claims that it is the "instabilities" within materializations of the body that "spawn rearticulations that call into question the hegemonic force of that very regulatory law" (p. 2).[7] Yet, this too can be seen simply to remain fully internal to the "cultural," a point Butler acknowledges without apology: "To posit by way of language a materiality outside of language is still to *posit* that materiality, and the materiality so posited will retain that positing as its constitutive condition" (p. 30).

It is this "internalist" account that Cheah (1996) suggests results in a formulation that is essentially neo-Kantian, where cultural intelligibility occurs on a plane distinct from things. Drawing on Foucault, for instance, Butler writes that "ontological weight is not presumed but always conferred" (1993: 34), a statement that accords dynamism only to form and not matter. This differs from Kantian form/matter distinctions, Cheah suggests, only in that form is now seen as an instrument of power.[8] Indeed, Butler equivocates on this point. She writes, for instance, that "discursive possibilities opened up by the *constitutive outside of hegemonic positions* . . . constitute the disruptive return of the excluded [nature] within . . . the symbolic" (p. 12, emphasis added) and states that her purpose is "to understand how what has been foreclosed or banished from the proper domain of "sex" [nature] . . . might at once be *produced as a troubling return*, not only as an imaginary contestation that effects a failure in the workings of the inevitable law, but as an enabling disruption, the occasion for a radical rearticulation of the symbolic horizon in which bodies come to matter at all" (p. 23, emphasis added). And yet, in the final analysis, in Butler's account the "constitutive outside" disrupts only through the instabilities of reiteration (only through more culture).[9]

Like other post-structuralist writers, Butler draws attention to the relations of power involved in the appearance of order. There can be no "nature" apart from its presencing. The implications for a "cultural politics" of nature should be readily apparent. "This unsettling of 'matter'," Butler writes (1993: 30) "can be understood as initiating new possibilities, new ways for bodies to matter." Yet, with post-structuralist accounts arise other difficult questions. As seductive as Butler's call for rearticulating the body may be, it is difficult to understand how she imagines this project proceeding. Caught between a political desire to open the body to counter-hegemonic materializations and a theoretical account that refuses to privilege a prediscursive realm (thereby figuring the materialized body as something defined only negatively through exclusion), Butler's account falters precisely at the moment of trying to imagine a site from which the body can be made to matter differently.[10] What is lacking, as Cheah suggests, is a recognition of a causal and dynamic relation between intelligibility and materiality.[11]

The limitations of post-structuralist accounts of "nature's" social construction notwithstanding, such approaches have placed considerable attention on the relation between discourse and materiality. Writing nature's "positivity" as something achieved through modes of delimiting and normalization rather than something found in journeys of discovery, has become increasingly central to critical studies of "nature," evident in the work of Haraway (1989, 1991, 1997), Martin (1994), Ross (1994), Cronon (1995) and Keller (1995) among others. This has occurred across numerous sites. A number of writers have shown the ways in which "forests" – especially "tropical rain forests" – are imbued with cultural meanings: as sites of fantasy and terror; as seen through edenic narratives of origins and purity; as layered with the rhetorics of nationalism; or even as understood through contesting paradigms in ecosystem ecology (see Botkin 1990; Harrison 1992; Schama 1995; Slater 1995; Willems-Braun 1997). Each is a way of valuing; none leave the forest untouched, and often these narratives contain social as well as ecological consequences (see, for instance, Hecht and Cockburn 1989). Cronon's (1995) eloquent critique of one of North American's most deeply held and emotional investments – the idea of "wilderness" – provides a striking example of the analytical and political stakes involved in disrupting the "self-evidence" of nature. Cronon shows "wilderness" – an identity iterated often in American culture – to be both the projection and inscription onto nature of

24

forms of Romanticism together with frontier ideology, with the result that today "nature" – especially in the western half of the continent – is being remade in the image of these narratives. As Cronon also shows, this fascination with "wilderness" is not without consequences elsewhere, and these are both social and ecological in character. With such attention paid to "defending" the "last vestiges" of wilderness, already modified landscapes – and especially urban natures – receive far less attention, even though these now comprise the vast majority of the earth's lands, and are most closely related to issues of human health. Indeed, this quickly draws in questions of class and race, a point made by members of the far less exotic "environmental justice" movement (Alston 1990; Bullard 1990; Di Chiro 1995) who refuse to separate "nature" and "culture," but seek to build survivable futures characterized by environmental and social equity.

Finally – and perhaps most prolifically – feminist scholarship has drawn attention to the many ways that patriarchal culture is inscribed on the female body. Nellie Oudshoorn (1994), for example, has discussed the cultural assumptions that have underwritten the "medicalization" of women's bodies, and charted some of the consequences. The ways in which women's bodies – rather than men's – were mapped in medical and scientific discourse as "hormonal" bodies allowed for subsequent forms of social and sexual regulation (through contraceptive technologies), and also made the female body an "accumulation strategy" (through marketing products regulating the "hormonal" body). Likewise, Martin (1987) has shown the implicit (and often explicit) gendering of reproductive narratives in biology (see also Hartouni 1994). More recently, Martin has mapped parallels between economic and medical discourses on "flexible" bodies (Martin 1994), and has suggested (Martin, this volume) that different ways of "framing" the body (as machine, or as flexible) may authorize different normalizations and regulations of the body. Indeed, in these interventions Oudshoorn's phrase "beyond the natural body" takes on two distinct but related meanings: the body that we equate with nature is already deeply cultural (it is given form and meaning through particular discursive practices); and, the body is increasingly artifactual (it is remade materially in new and unique ways through an array of social practices that seize upon, and are enabled by, the body's cultural intelligibility).

Post-structuralist arguments about the "construction" of nature have drawn much-needed attention to the "cultural politics" of nature. They have also elicited angry charges that "reality" is taken to be merely a matter of desire, opinion, speculation or fantasy (see Gross and Levitt 1994). Yet, with Haraway, we argue that stressing the contingency of what counts as "reality" is not a denial of nature's materiality, but rather, a recognition that nature "is collectively, materially, and semiotically constructed – that is, put together, made to cohere, worked up for and by us in some ways and not others" (Haraway 1997: 301 n.12). On how this occurs, Haraway suggests, much turns – even the earth itself.

25

NETWORKS, ACTANTS AND
TECHNOSCIENCE: BUILDING HYBRIDS

Post-structuralist accounts of nature's construction provide provocative arguments for the centrality of discursive practices for both nature's intelligibility and its availability to forms of instrumental power. Yet – as evident in the work of Butler – such accounts risk privileging active form over passive matter. A potential path beyond post-structuralism's "Leibnizian conceit" (Harvey 1996) may be found in the field of the sociology of scientific knowledge (SSK), or simply, "science studies." Already this is a vast terrain that draws on a variety of theoretical and philosophical traditions. Most practitioners share with post-structuralism the notion of nature's "constructedness," but they cast their net far wider than post-structuralism's narrow focus on discourse, or the "rules" that govern which statements can be made about things at any given moment. Instead, attention is paid to the intersection and simultaneity of multiple material and discursive practices: from the social, economic and institutional relations organizing scientific inquiry to the inscription devices and visual technologies that increasingly generate new possibilities for envisioning and intervening in the world; from the metaphors and narratives that help determine what are interesting questions to the "objects" of inquiry themselves, which – as Callon (1986) and Haraway (1992) argue – do not hold still, but provide surprises of their own. Thus, as Haraway notes, what counts as "nature" always has multiple dimensions – mythic, textual, technical, political, organic, and economic – which "collapse into each other in a knot of extraordinary density" (Haraway 1994: 63).

Such knots are difficult to disentangle. Yet, it is precisely exploring these threads and interconnections that makes SSK such a rich resource. As is evident in many of the chapters in this volume, SSK has begun to reshape how we talk about the social constitution of nature. There are many advantages to SSK approaches, but here we want to highlight four. First, with its emphasis on knowledge production as "worldly," SSK scholars trace nature's "emergence" in specific, historical practices (fieldwork, the laboratory, writing), not in order to dismiss or minimize science, but in order show how these world-changing knowledges are made in social and institutional contexts saturated with relations of power. Second, although the practice of science is fully social (and deeply technical), it still revolves around interactions with a physical world, and thus the knowledges produced cannot be reduced fully to the social. While science studies shares with post-structuralism a concern over how nature is rendered intelligible in certain ways and not others, it has the potential to provide more subtle and complex empirical accounts of how specific knowledges come to be held without compromising the sense that the materiality of nature is itself central to our knowledge of it (Callon 1986; see also Hayles 1991). Third, many SSK scholars attend not only to the myriad ways in which technoscience renders nature intelligible, but also to how our world – and nature – is increasingly an artifact of technoscience. Nature is today becoming a hybrid of machines and organisms, a hybridity entailing new social and political relations. Indeed, to rephrase Baudrillard (1983), today science precedes reality; things appear in technoscientific networks – as models or as information – before they appear as objects in the world. Although it is important not to conflate epistemology and ontology (the notion that our ideas generate the world), in technoscientific networks epistemology and ontology increasingly implode – the world "outside" the laboratory comes to mirror the world "inside" (Latour 1988). Finally, science studies scholars like Bruno Latour have also placed in question the autonomy of the "social," which is progressively stitched together through things (including not only technical objects, but also our productions of nature). This has received less attention. Where earlier sociologists talked about social relations as those between people (or classes), Latour speaks of collectivities that encompass people, things and machines. It follows, then, that politics is not solely something that occurs between people, but that the "quasi-objects" produced in modernity carry with them their own political rationality (see also Mumford 1964; Winner 1986) and thus that both analysis and politics must take seriously the complex mediations and networks that weave natural and social worlds into a "seamless fabric" (Latour 1993). In what follows we explore some of these themes in greater detail.

The social construction of scientific knowledge

Science is shadowed, at a constant distance, by its own anthropology.
(Serres 1987: 41, quoted in Latour 1990: 145)

If any one position characterizes science studies, it is that scientific knowledges are made in historically specific, socially situated practices, rather than "found." According to this view, science can therefore be studied using the tools of anthropology, sociology and so on. Indeed, one of Bruno Latour's most important interventions has been to show the relevance of ethnographic studies of "science in action," in a sense bringing anthropology home from the tropics. As Latour (1993) explains, in the past anthropology and Western science shared a central assumption: that a "great divide" separated "modern" from "primitive" cultures, and that this divide was essentially that between "reason" and "myth." Systems of belief about the external world found in "pre-modern" cultures ("ethnoscience") could be explained by reference to culture, religion or politics. Our beliefs ("Science"), on the other hand, were assumed to exist independent from culture, religion or politics, and thus could be explained by reference to Nature alone.

In a consequential early study of neuroendocrinologists at the Salk Institute in California, Latour and Woolgar (1979) showed that the air-conditioned, sterile laboratories of modern science were equally susceptible to anthropological inquiry, since necessarily they embodied social, technical, economic and institutional relations. There are now numerous ethnographies of science (for surveys see Rouse 1992; Haraway 1994) as well as historical accounts of the social construction of scientific knowledges (e.g. Shapin and Schaffer, 1985; Haraway 1989) in which historians and social scientists have sought to produce finely textured accounts of the generation of "truths" about nature. Traweek's *Beamtimes and Lifetimes* (1988) provides one example: In a study of high energy physics laboratories in the USA and Japan, she showed that cultural and political considerations made a great difference in what knowledges were produced, in part because the detectors used by scientists took different forms depending on the funding and research contexts in each country. Likewise, Haraway (1991) has shown – as has Treichler (1987) – the various ways in which culture has infused HIV – research. In each instance, attention has been paid to the practices through which certain knowledges, rather than others, attain the status of "facts," following the principle set forward by members of the "Edinburgh School" who made symmetry in explanation a central principle in science studies. According to this principle, both "valid" and "false" knowledges must be explained in the same terms: one cannot be explained through reference to nature and the other through reference to culture (or ideology). Rather, both must be seen as a result of historically specific social and scientific practices.

Essentially, science studies remains agnostic on the question of "truth," although as Haraway (1997: 137, 301 n.12) explains, this should not be confused with a relativism that views all knowledges as somehow "equal." Reality, Haraway argues, is not a "subjective" construction, but rather a "congealing of ways of interacting" and these require a dense array of bodies, artifacts, minds, collectives, etc. Thus, "reality is eminently material and solid, but the effects sedimented out of technologies of observation/representation are radically contingent in the sense that other semiotic-material-technical processes of observation would (and do) produce quite different lived worlds" (302 n.12). Far from devaluing scientific inquiry, Haraway and other SSK scholars seek to rework the stories we tell about science – often for explicitly political reasons. By holding to a philosophy of science that maintains that science occurs in a sphere hermetically sealed from the "polluting" influence of society, we fail to see how science necessarily and always builds its knowledges through social means. The issue is to see how this matters.

Part of the project of science studies thus involves reworking our under-standings of "objectivity." Objectivity figured as disembodiment, Haraway (1991) argues, is an escape from the materiality of knowledge–production (as if our observations could somehow be independent of the observer!), and thus erases history from representation (precisely the move Heidegger describes as our forgetfulness of Being). In contrast, "strong objectivity" – borrowing from Harding (1986) – has everything to do with taking responsibility for the ways that our knowledge–production practices are "situated" in particular historical moments and social–spatial contexts, acknowledging rather than abdicating our responsibility for the ways that knowledge reflects its enabling conditions.

Nature, science and the construction of "society"

Issues surrounding the social construction of scientific knowledges are of significance not only in terms of epistemology. They relate also to how nature and society are today reconfigured. This can be shown in two ways. First, if as Rabinow (1992: 236) suggests, "representing and intervening, knowledge and power, understanding and reform, are built in, from the start, as simultaneous goals and means," then how future natures are produced will respond in part to what Bachelard referred to as the "temporality" of science (which, he claimed was not reducible – but also not unrelated – to the temporal rhythms of other "levels" of social life). Indeed, the "world-changing" character of technoscience is progressively evident as we face the twenty-first century (although, contra Bachelard, the relays between science, economy, culture and state are arguably more intense and more frequent since World War II: see Heims 1980; Haraway 1982). Second, how technoscience makes possible our interventions in nature increasingly has social implications. This is plainly evident in such fields as genetics, but the point can be applied more broadly. As Latour (1993) notes, while sociologists of scientific knowledge (SSK) scholars have insisted on symmetry in explanation for "truth" and "falsity," they have remained largely asymmetrical at another level. In their accounts science is shown to be explained by reference to society, but society itself remains "solid," impervious to all but "internal" explanation. Yet, it is abundantly clear that technoscience and its artifacts are central to remaking society and nature simultaneously. Indeed, arguments surrounding the embeddedness of science and technology would matter little if it were not also that society is built, in part, through science and technology. But it is precisely the ways in which society is constructed through, or in relation to, things (microbes, door closers, machines, and so on) along with the various ways that science is the cause rather than medium of nature's representation that, as moderns, we are unable to see, since we live within a modern "constitution" that assigns "nature" and "culture" to two distinct realms, and similarly situates "knowledge" in one (nature) and "politics" in another (culture). This modern constitution, Latour argues, allows technoscience to build both nature and society simultaneously, but in ways that remain relatively unexamined.

For this argument Latour relies extensively on Shapin and Schaffer's (1985) *Leviathan and the Air Pump*. In this important study, Shapin and Schaffer located the separation of science and politics in the seventeenth-century dispute between Hobbes and Boyle over where political and epistemological authority was located. As they show, Boyle sought to locate authority in nature which could be known through scientific observation. In his experiments with the air pump, Boyle constructed the "laboratory" as a separate, regulated arena (with its own rules for witnessing and deciding on matters of fact). What was problematic, as Shapin and Schaffer explain, was that this rendered invisible the social and political relations involved in producing representations of nature. Hobbes, on the other hand, rejected Boyle's theatre of proof, which he thought simply led to the political fragmentation that occurred when people could petition entities other than civil authority – like "Nature" or "God." For Hobbes, there could exist only one Knowledge and one State, both residing in the Sovereign designated by consent through the social contract. As Latour explains, SSK scholars have shown the "political" erasures that occurred in Boyle's laboratories, but have failed to recognize that Hobbes also constructed his own hermetically sealed sphere: "society." By constructing society as consisting only of humans (with its own science of representation – politics), Hobbes erased the multiple mediations of facts, things and machines! As Latour summarizes:

> It is not that Boyle invents scientific discourse and Hobbes political discourse, it is that Boyle invents a political discourse where politics should not count and that Hobbes devises a scientific politics where experimental science should not count. In other words, they are inventing our modern world, a world in which the representation of things through the medium of the laboratory is forever severed from the representation of citizens through the medium of social contract ... From now on every one should "see double" and make no direct connection between the representation of non-humans and the representation of humans, between the artificiality of the facts and the artificiality of the Body Politic.
>
> (Latour 1990: 155–6)

The first half of this formulation – that concerning Boyle – is now commonplace. The second, we wish to argue, has received less attention, but is in part what renders the first crucial. Society is not simply the association of "naked individuals" – as Hobbes implied; rather, social actions always mobilize things (like nature) which are "actants" in their own right.

Latour makes this argument explicitly in his *Pasteurization of France* (1988). In late nineteenth-century France, Latour explains, a world "with" microbes resulted in very different forms of governmentality and social regulation, and different orders of behaviour, than did a society "without" microbes. In a social context where discourses on hygiene had already become dominant, Pasteur's "discovery" of the microbe was not socially innocent, but reconfigured a wider network of forces:

> The Pasteurians provided neither the level nor the weight nor even the worker who did the work, but they provided the hygienist with a fulcrum. To use another metaphor, they were like the first observation balloons. They made the enemy visible.
>
> (Latour 1988: 34)

In short, after Pasteur, microbes became social actors in their own right. On the one hand they strengthened the social and rhetorical positions of the hygienists for whom microbes provided an ally, and they also elevated the position of Pasteur, who became a central figure in French society. But Latour's point goes beyond this: by seeing the health of populations through the lens of microbes, many other elements of social (and material) life became rearranged. As one commentator from the period noted, "Ignoring the danger of the microbe awaiting us, *we have hitherto arranged our way of life without taking any account of this unknown enemy*" (Leduc: 1892: 234, quoted in Latour 1988: 35, emphasis added). Representation begets intervention; microbes become actants, they reconfigure both a semiotic terrain (environments suddenly were framed within different economies of signification) and, ultimately, a physical terrain. Put baldly, in nineteenth-century France, nature and society were both remade in the "image" of microbal theories of disease.

Importantly, Latour shows not only how actants like microbes become entangled in social life, but that the very possibility of social life proceeds through and is enabled by such actants:

> There are not only "social" relations, relations between man and man . . . for everywhere microbes intervene and act . . . We cannot form society with the social alone. We have to add the action of microbes. We cannot understand anything about Pasteurism if we do not realize that it has reorganized society in a different way. It is not that there is a science done in the laboratory, on the one hand, and a society made up of groups, classes, interests, and laws, on the other. The issue is at once much more simple and much more difficult.
>
> (Latour 1988: 35)

Elsewhere, Latour has drawn out the consequences more explicitly, arguing that without the many objects that gave to "society" its durability as well as its solidity, what we have commonly taken to be the traditional domain of social theory – empire, classes, professions, organizations, States – become "so many mysteries" (Latour 1993: 120). In short, the solidity of "society," even the constitution of the "subject" cannot be thought apart from things, whether these be "nature" or "machines."

Amodernity and the analytics and politics of quasi-objects

One of the distinct advantages of Latour's "symmetrical" approach to science and society is that it highlights the analytical and political stakes involved in how boundaries are drawn between science and society, nature and culture. Blurring how these boundaries are drawn serves a wider project of showing the relations of power involved in how social natures are built at the close of the twentieth century. Until now, these relations have remained invisible as a result of our adherence to what Latour (1993) calls the three "guarantees" of modernity. As

31

we noted earlier, Boyle assumed that scientists were merely modest witnesses to Nature (cf. Haraway 1997), and this provides the first guarantee: "*it is not men who make Nature; Nature has always existed and has always already been there; we are only discovering its secrets*" (Latour 1993: 30). Hobbes assumed that citizens alone spoke with one voice through their representative (or sovereign). This provides the second: "*human beings, and only human beings, are the ones who construct society and freely determine their own destiny*" (Latour 1993: 30). These are underwritten by a third guarantee: the absolute separation of nature and society. Here Latour makes the analytical and political stakes clear:

> The overall structure is now easy to grasp: the . . . guarantees taken together will allow the moderns a change in scale. *They are going to be able to make Nature intervene at every point in the fabrication of their societies while they go right on attributing to Nature its radical transcendence; they are going to be able to become the only actors in their own political destiny, while they go right on making their society hold together by mobilizing Nature* . . . The essential point about this modern Constitution is that it renders the work of mediation that assembles hybrids invisible, unthinkable, unrepresentable . . . Everything happens in the middle, everything passes between the two, everything happens by way of mediation, translation and networks, but this space does not exist, it has no place. It is the unthinkable, the unconscious of the moderns.
>
> (Latour 1993: 32, 34, 37, emphasis added)

Latour argues that we need to be "amodern," or, in other words, we need to retie rather than endlessly attempt to untangle, the Gordian knot between nature and society so as to recognize this "middle kingdom" of quasi-objects. As Goodman and Fitzsimmons (this volume) show in their discussion of agro-food systems, this allows entirely new actors and relations to be brought into view and new political interventions imagined. Although until now Latour has been wary of programmatic statements about what a politics of mediation, translation and networks entails (although see Latour, this volume), his intervention in our self-understanding as "moderns" (as believing in the absolute separation of nature and culture, science and society), is rich with analytical and political possibilities, and many contributors to this volume take these up explicitly (see the chapters by Demeritt, Watts, and Goodman and Fitzsimmons).

According to Latour, Haraway and other SSK scholars, tracing networks is where political hope lies. Recognizing the complex intertwinings of nature, culture, science and technology allows us to see the various ways that it is impossible to change the social order without at the same time modifying the natural order, and vice versa. Haraway and Latour refer to these hybrid nature-cultures as "illegitimate couplings," "monsters" or "quasi-objects" – entities that until now have had no standing in modernity's accounts of itself, yet all the while have silently worked to organize social and ecological life. By rendering these mixtures "unthinkable," Latour (1993: 42) explains, the moderns allowed for their proliferation but were unable to trace their consequences. Being amodern thus requires re-skilling, fashioning new political lenses with the hope of "seeing" differently. For amoderns, no longer held in thrall by modern stories of the absolute separation of nature and society, monsters become "visible and thinkable" and explicitly pose serious dilemmas for the social order. This all makes for a politics with which we are unfamiliar: neither technophilic nor technophobic, interested neither in preserving Eden nor rendering everything as resource, but attentive simply to the social and ecological consequences that everywhere are intertwined in everyday practices. For SSK scholars, tracing the Ariadne threads that lead through the linked worlds of science, culture, nature and politics allows us to see where responsibility lies, what stories need to be told, and what differences these make.

CONCLUSION: TOWARD A POLITICAL THEORY OF SOCIAL NATURE

> The intertwinings of social and ecological projects in daily practices as well as in the realms of ideology, representations, esthetics, and the like are such as to make every social (including literary or artistic) project a project about nature, environment, and ecosystem, and vice versa.
>
> (Harvey 1996: 189)

In the preceding pages we have outlined some analytical and political tools available for interrogating nature at the millenium. Yet, if nature at the dawn of the twenty-first century is resolutely social this does not mean that the modern dualism between "nature" and "society" no longer retains a hold on our imagination. Indeed, the opposite may be the case: today we hear regularly of the "death of nature" or the "end of nature," and now as often as ever before "nature" is seen as a refuge – a "pure" place to which one travels in order to escape from society. Along similar lines, deep green environmentalism shuttles between apocalyptics and melancholy, mourning the loss, or desperately seeking to preserve (or at least witness!), the last remnants of a "pristine" nature. And yet, as Neil Smith (1996: 41) has recently reiterated, this desire to "save nature" is deeply problematic, since it reaffirms the "externality" of a nature "with and within which human societies are inextricably intermeshed." There are, to be sure, reasons to limit or regulate human interventions in specific environments which can be justified on both ecological and social grounds. But to focus on preserving a nature that "excludes" humans is today a self-defeating strategy – it is, as Smith argues, to save something that is no longer recognizable, if it ever was, while at the same time shifting attention from some of the most pressing and interlinked social and ecological problems that face late capitalist and technoscientific cultures. Indeed, by rendering nature as something "external" to be saved "from" humans, we erase its social and discursive constitution, with the result that the nature to be preserved simply reflects our own social values and anxieties – it becomes a "fun-house reflection of ourselves" (Haraway 1992: 296).

The crucial issue therefore, is not that of policing boundaries between "nature" and "culture" but rather, of taking responsibility for how our inevitable interventions in nature proceed – along what lines, with what consequences and to whose benefit. As Smith (1996: 49) explains, we need a "political theory of nature": one which expresses the inevitability and creativity of our relationships with nature; which recognizes the destructive dynamics embodied in capitalist modes of production; which accounts for how relationships with nature are differentiated according to gender, class, race, and sexual preference; which accepts the implausibility of a nature "autonomous" from culture; and which, finally, helps us unlearn the "instinctive romanticism" which pervades treatments of nature in bourgeois and patriarchal society.

Our intention in this volume is to begin the task of developing an analytics and politics directed toward constructing survivable futures at the dawn of the twenty-first century. Contributors to this volume care deeply about the natures that are being built today and tomorrow in the midst of social, cultural and technological changes of unprecedented scope. Indeed, precisely for this reason, many find hope in social and political theories that eschew a nature–society split and instead insist on seeing the two as continuously constituted through the other – nature made artifactual, just as society is made natural. The costs of retaining the dualism have become too high; as Latour explains, too much is left unseen. Indeed, to claim that nature is everywhere becoming artifactual is not, as some suggest, to argue against nature or for its "destruction," for social natures are no less ecological than so-called "natural" systems from which humans have, at least in theory, been excluded. Indeed, as Harvey (1996: 186) provocatively notes, New York City is an "ecosystem"; and, if we truly are concerned about the natures that we are building, we need to get out of the habit of excluding urban historical geographies from our environmental histories. In short, we need to open the social fabrication of nature to interrogation at every level.

This is neither a simple, nor a singular, task. Nature is multiple; its social production proceeds according to no single temporality, occurs with no one underlying logic, follows no unified plan. Accordingly, struggles over the social production of nature are multifaceted; they occur at various levels, involve a large cast of actors (not all of which are human), and follow a plurality of social and ecological logics that cannot be reduced to a single story. This calls forth analytical and political responses that are multiple and often discontinuous: tracing circuits of capital, contesting the social and ecological logics of particular production processes and articulating relations between class, race, gender and ecology. Or it may take the form of showing how "nature" is mediated or constituted discursively, locating the consequences of this in terms of gender, race, or sexuality, and where necessary disrupting the "self-evidence" of nature-as-received in order to open space for different constructions less implicated in relations of domination. And elsewhere it may require tracing networks, attending to the "quasi-objects" so long excluded from political discourse but through which society and nature are churned up daily, and taking seriously the simultaneity of social, technical, discursive, political and organic elements in any and all productions of nature. These themes surface in various forms, and often conjoined, throughout the volume, and in large measure they cross-cut the two-part organization of the book. Broadly, however, chapters have been grouped according to one of two themes. Those in Part 2 – Capitalizing and Enframing Nature – explore the social, discursive and ideological practices that inform and organize nature's remaking in the wider context of late capitalism. Those in Part 3 – Actors, Networks and the Politics of Hybridity – trace the various networks (social, institutional, technical as well as discursive) and the many actors/actants (humans, machines, nature, etc.) that are mixed together in specific social productions of nature, with an eye to their social, economic and ecological consequences. This schematic division is, of course, somewhat arbitrary. What all contributors share is a commitment to rethinking environmental politics along lines aligned with social justice and in ways appropriate for the social natures of the twenty-first century. Revolutionary environmentalism, after all, cannot be about reinstating "nature." That time has passed, if indeed it was ever here. Rather, it must be about a continuous vigilance to the sorts of natures we are producing – to how reality is being remade. For this, we need to fashion new – or refashion old – analytical and political tools, tools for making the future natures that we wish to inhabit.

NOTES

1 It is no coincidence that the first stirrings of Marxist interest in social nature came in the wake of the neo-Malthusian debates of the early 1970s. Focussed particularly on the Third World, the frequently outspoken statements by Commoner, Ehrlich, Hardin and Meadows *et al.* pointed to "natural limits" to population growth and blamed peasant ignorance for Third World famine, soil erosion and the like. That such blame-the-victim thinking – with its baggage of population policies and "positive checks" – was so much a part of everyday "common sense" in the early years of environmental concern in the capitalist world made David Harvey's (1974) provocative inversion of Malthusian thinking that much more significant. Arguing for a paradigm shift in our conception of human–environment relations, Harvey unravelled the absolutisms of "population problem" thinking to argue that "resources" are always socially defined and that "scarcity" is always socially produced. By thus drawing attention to the political economy of resource use, Harvey rephrased the scarcity-of-resources view to argue instead that "There are too many people in the world because the particular ends we have in mind . . . and the materials available in nature, that we have the will and the way to use, are not sufficient to provide us with those things to which we are accustomed" (Harvey 1974: 274).
2 Production-for-exchange in modern, industrial capitalism is clearly organized in a

historically specific way, involving complex relations of production and consumption, and responding to very different spatial and temporal rhythms. This results in very *different* productions of nature than other historical modes, but such productions of nature, while different in form, are not different in type (i.e. are not *more* mediated) than what Neil Smith (1984) offers as an abstraction called "production-in-general." The result is that the critique of the production of nature in advanced capitalism cannot take as its starting point a "natural" relation to the environment from which capitalist modes of production depart, but rather, can only proceed relationally, as a comparison of the different socio-economic and cultural logics organizing nature's production, and with what social and ecological effects.

3 It is important to keep in mind that while for Heidegger language was the "house of Being," this does not mean that he understood the "appearance" of things through a simple linguistic constructivism. Rather, as Joseph Rouse (1987) explains, nature "shows up" or "emerges" in and through a configuration of practices, including Science (see also Demeritt, this volume).

4 In modernity, Heidegger argued, the historical specificity of how things are "given" to human subjects recedes and becomes the "invisible shadow that is cast around all things everywhere. By means of this shadow the modern world extends itself out into a space withdrawn from representation, and so lends to the incalculable the determinateness peculiar to it, as well as a historical uniqueness . . . In truth, however, the shadow is a manifest, though impenetrable, testimony to the concealed emitting of light. In keeping with this concept of shadow, we experience the incalculable as that which, withdrawn from representation, is nevertheless manifest in whatever is, pointing to Being, which remains concealed" (1977: 135, 154).

5 Rajchman (1988) described Foucault's writings as the "art of seeing." Foucault, Rajchman claims, was not interested simply in what things looked like, but "how things were *made* visible, how things were *given* to be seen, how things were '*shown*' to knowledge or power . . . [how] things became *seeable*" (p. 91, original emphasis).

6 Butler (1993: 5) is herself highly critical of strong constructivist arguments:

> When the sex/gender distinction is joined with a notion of radical linguistic constructivism, the problem becomes even worse, for the "sex" which is referred to as prior to gender will itself be a postulation, a construction, offered within language, as that which is prior to language, prior to construction. But this sex posited as prior to construction will, by virtue of being posited, become the effect of that very positing, the construction of construction. If gender is the social construction of sex, and if there is no access to this "sex" except by means of its construction, then it appears not only that sex is absorbed by gender, but that "sex" becomes something like a fiction, perhaps a fantasy, retroactively installed at a prelinguistic site to which there is no direct access.

7 In many ways this is the politics of the "catachresis" – that which disrupts by not speaking the proper (hegemonic) name.

8 Moreover, as Grosz (1994: 21) notes, this assumes a certain immutability to culture: "Culture [discourse] itself can only have meaning and value in terms of its own other(s): when its others are obliterated – as tends to occur within the problematic of social constructivism – culture in effect takes on all the immutable, fixed characteristics attributed to the natural order."

9 Cheah (1996: 120) reveals some of the ontological contradictions involved:

> By defining "constitution" as repeated identification, Butler confines the term within an ideational scenario. Consequently, she hesitates before the question of what sustains the causal power of intelligible form over matter,

the question of what allows intelligible form to materialize as matter in general. Yet, this question seems unavoidable. For if, "'to be constituted' means 'to be compelled to cite, to repeat or to mime' the signifier itself," it must be asked: what are the ontological conditions under which the compulsion to identify can take place? . . . Does not the concept of morphology as a mediating term between psyche and matter presuppose this question of the causal power of ideas over matter and vice versa but cannot ask it? . . . if *nomos* or *tekne* can become *physis*, then must there not be another non-anthropologistic level of causality and naturalist teleology, of which the performativity of language would only be a case?

Cheah draws the following implications from his argument: "This would imply that political change can no longer be understood as a function of sociohistorical form qua the sole principle of dynamism. Instead, the category of the political itself needs to be rethought outside of the terms of history and culture, which are its time-honored cognates" (1996: 120).

10 Indeed, Butler (1993: 206) in many ways is caught in the same "prepolitical pathos" that she feels marks so many psychoanalytic accounts of the "subject":

As resistance to symbolization, the "real" functions in an exterior relation to language, as the inverse of mimetic representationalism, that is, as the site where all efforts to represent must founder. The problem here is that there is no way within this framework to politicize the relation between language and the real. What counts as the "real," in a sense of the unsymbolizable, is always relative to a linguistic domain that authorizes and produces that foreclosure, and achieves that effect through producing and policing a set of constitutive exclusions. Even if every discursive formation is produced through exclusion, that is not to claim that all exclusions are equivalent: what is needed is a way to assess politically how the production of cultural unintelligibility is mobilized variably to regulate the political field . . . To freeze the real as the impossible "outside" to discourse is to institute a permanently unsatisfiable desire for an ever elusive referent: the sublime object of ideology. The fixity and universality of this relation between language and the real produces, however, a prepolitical pathos that precludes the kind of analysis that would take the real/reality distinction as the instrument and effect of contingent relations of power.

11 Cheah suggests that Derrida presents a way out of this dilemma by undoing the founding oppositions between nature and it others. Thus, he quotes affirmatively Derrida's statement that: "Culture [is to be thought] as nature different and deferred, differing-deferring; all the others of *physis* – *techne, nomos, thesis,* society, freedom, history, mind, etc. as *physis* different and deferred, or as *physis* differing and deferring. *Physis* in *différance*" (Derrida 1982: 17). Yet, it is not clear where this situates dynamism. In other words, even if one refuses to situate culture "outside" *physis*, culture becomes that part of *physis* that is dynamic, to the exclusion of "other" non-cultural elements.

REFERENCES

Adorno, T. and Horkheimer, M. (1972) *The Dialectic of Enlightenment*, trans. J. Cumming, New York: Herder and Herder.

Alston, D. (ed.) (1990) *We Speak for Ourselves: Social Justice, Race and Environment*, Washington: Panos Institute.

Altvater, E. (1993) *The Future of the Market*, London: Verso.

Antipode (1994) "Special Issue on *Nature's Metropolis*," 26.

Arce, A. and Marsden, T. (1993) "The social construction of international food," *Economic Geography* 69: 293–311.

Baudrillard, J. (1975) *The Mirror of Production*, trans. M. Poster, St. Louis: Telos Press.

—— (1983) *Simulacra and Simulations*, trans. P. Foss, P. Patton and P. Beitchman, New York: Sémiotexte.

Benton, T. (1989) "Marxism and natural limits," *New Left Review* 178: 51–86.

Blaikie, P. (1985) *The Political-Economy of Soil Erosion*, London: Methuen.

Blaikie, P. and Brookfield, H. (1987) *Land Degradation and Society*, London: Methuen.

Botkin, D. (1990) *Discordant Harmonies*, New York: Oxford University Press.

Bryant, R. (1992) "Political ecology," *Political Geography* 11: 12–36.

Bryant, R. and Bailey, S. (1997) *Third World Political Ecology*, London: Routledge.

Bullard, R. (1990) *Dumping in Dixie: Race, Class, and Environmental Quality*, Boulder: Westview Press.

Bunker, S. (1985) *Underdeveloping the Amazon*, Chicago: University of Illinois Press.

Butler, J. (1993) *Bodies that Matter: On the Discursive Limits of Sex*, New York: Routledge.

Buttel, F. and Newby, H. (1980) *The Rural Sociology of Advanced Societies*, Montclair: Allenheld Osmun.

Buttel, F. and Taylor, P. (1994) "Environmental sociology and global environmental change: a critical assessment," in M. Redclift and T. Benton (eds) *Social Theory and the Global Environment*, London: Routledge.

Callon, M. (1986) "Some elements of a sociology of translation: domestication of the scallops and the fishermen of St. Brieuc Bay," in J. Law (ed.) *Power, Action and Belief: A New Sociology of Knowledge?*, London: Routledge.

Carney, J. (1996) "Converting the wetlands, engendering the environment," in R. Peet and M. Watts (eds) *Liberation Ecology*, London: Routledge.

Castells, M. (1996) *The Rise of the Network Society*, Cambridge, MA: Harvard University Press.

Castree, N. (1995) "The nature of produced nature," *Antipode* 27: 12–48.

—— (1997) "Nature, economy and the cultural politics of theory: 'the war against the seals' in the Bering Sea, 1870–1911," *Geoforum* 28: 1–20.

Cheah, P. (1996) "Mattering," *Diacritics* 26: 108–39.

Clarke, A. (1995) "Modernity, postmodernity and reproductive processes, *c*.1890–1990," in C. Gray, H. Figueroa-Sarriera and S. Mentor (eds) *The Cyborg Handbook*, New York: Routledge.

Collins, J. (1987) *Unseasonal Migrations*, Princeton: Princeton University Press.

Cronon, W. (1995) "The trouble with wilderness; or, getting back to the wrong nature," in W. Cronon (ed.) *Uncommon Ground: Toward Reinventing Nature*, New York: W. W. Norton.

Deleuze, G. (1988) *Foucault*, Minneapolis: University of Minnesota Press.

Derrida, J. (1982) *Margins of Philosophy*, trans. Alan Bass, Chicago: University of Chicago Press.

Di Chiro, G. (1995) "Nature as commodity: the convergence of environment and social justice," in W. Cronon (ed.) *Uncommon Ground: Toward Reinventing Nature*, New York: W. W. Norton.

Escobar, A. (1996) "Constructing nature: elements for a post-structural political ecology," in R. Peet and M. Watts (eds) *Liberation Ecology*, London: Routledge.

Fitzsimmons, M. (1986) "The new industrial agriculture," *Economic Geography* 62: 334–53.

Foucault, M. (1977) *Discipline and Punish: Birth of the Prison*, London: Penguin.

Friedmann, H. (1982) "The political-economy of food: the rise and fall of the post-war international food order," *American Journal of Sociology* 88 (suppl.): 248–86.

Goodman, D. and Redclift, M. (1991) *Refashioning Nature*, London: Routledge.

Goodman, D., Sorj, B. and Wilkinson, J. (1987) *From Farming to Biotechnology*, Oxford: Blackwell.

Gramsci, A. (1971) *Selections from the Prison Notebooks*, New York: International Publishers.

Granovetter, M. (1994) "Economic action and social structure: the problem of embeddedness," *American Journal of Sociology* 91: 481–510.

Gross, P. and Levitt, N. (1994) *Higher Superstition: The Academic Left and its Quarrels with Science*, Baltimore: Johns Hopkins University Press.

Grosz, E. (1994) *Volatile Bodies: Toward a Corporeal Feminism*, Bloomington: Indiana University Press.

Hacking, I. (1983) *Representing and Intervening: Introductory Topics in the Philosophy of Natural Science*, Cambridge: Cambridge University Press.

Haraway, D. (1982) "The high cost of information in post-world war II evolutionary biology," *Philosophical Forum* 13: 244–78.

—— (1989) *Primate Visions: Gender, Race and Nature in the World of Modern Science*, New York: Routledge.

—— (1991) *Simians, Cyborgs and Women: The Reinvention of Nature*, New York: Routledge.

—— (1992) "The promises of monsters: a regenerative politics for inappropriate/d others," in L. Grossberg, C. Nelson and P. Treichler (eds) *Cultural Studies*, New York: Routledge.

—— (1994) "A game of cat's cradle: science studies, feminist theory, cultural studies," *Configurations* 1: 59–71.

—— (1997) *Modest Witness@Second Millennium*, New York: Routledge.

Harding, S. (1986) *The Science Question in Feminism*, Ithaca: Cornell University Press.

Harrison, R. (1992) *Forests: The Shadow of Civilization*, Chicago: University of Chicago Press.

Hartouni, V. (1994) "Breached birth: reflections on race, gender, and reproductive discourse in the 1980s," *Configurations* 2: 73–88.

Harvey, D. (1974) "Population, resources and the ideology of science," *Economic Geography* 50: 256–77.

—— (1996) *Justice, Nature and the Geography of Difference*, Oxford: Blackwell.

—— (1998) "The body as an accumulation strategy," *Society and Space*, forthcoming.

Hayles, N. K. (1991) "Constrained constructivism: locating scientific inquiry in the theatre of representation," *New Orleans Review* 18: 76–85.

—— (1995) "Searching for common ground," in M. Soulé and G. Lease (eds) *Reinventing Nature? Responses to Postmodern Deconstruction*, Washington: Island Press.

Hecht, S. and Cockburn, A. (1989) *The Fate of the Forest: Developers, Destroyers, and the Defenders of the Amazon*, London: Verso.

Heidegger, M. (1962) *Being and Time*, New York: Harper & Row.

—— (1977) *The Question Concerning Technology and Other Essays*, trans. W. Lovitt, New York: Harper and Row.

Heims, S. (1980) *John von Neumann and Norbert Wiener: From Mathematics to the Technologies of Life and Death*, Cambridge: MIT Press.

Katz, C. (1995) "Under the falling sky: apocalyptic environmentalism and the production of nature," in A. Callari, S. Cullenberg and C. Biewener (eds) *Marxism in the Postmodern Age*, New York: Guilford.

Kautsky, K. (1976) *The Agrarian Question*, selected parts, trans. J. Banaji, *Economy and Society* 5: 1–49.

Keller, E. (1995) *Refiguring Life: Metaphors of Twentieth-Century Biology*, New York: Columbia University Press.

Laclau, E. and Mouffe, C. (1985) *Hegemony and Socialist Strategy*, London: Verso.

Latour, B. (1988) *The Pasteurization of France*, Cambridge, MA: Harvard University Press.

—— (1990) "Postmodern? No, simply amodern! Steps towards an anthropology of science," *Studies in the History and Philosophy of Science* 21: 145–71.

—— (1993) *We Have Never Been Modern*, Cambridge, MA: Harvard University Press.

Latour, B. and Woolgar, S. (1979) *Laboratory Life: The Social Construction of Scientific Facts*, Beverley Hills: Sage Publications.

Le Heron, R. and Roche, M. (1996) "Globalisation, sustainability, and apple orcharding, Hawke's Bay, New Zealand," *Economic Geography* 72: 417–32.

Lowe, D. (1995) *The Body in Late-Capitalism*, Durham: Duke University Press.

McKibben, B. (1989) *The End of Nature*, New York: Random House.

McMichael, P. (1994) *The Global Restructuring of Agro-Food Systems*, Ithaca: Cornell University Press.

Marsden, T. *et al.* (1986a) "Towards a political-economy of capitalist agriculture: a British perspective," *International Journal of Urban and Regional Research* 10: 498–521.

—— (1986b) "The restructuring process and economic centrality in capitalist agriculture," *Journal of Rural Studies* 2: 271–80.

—— (1996) "Agricultural geography and the political-economy approach: a review," *Economic Geography* 72: 361–75.

Martin, E. (1987) *The Woman in the Body: A Cultural Analysis of Reproduction*, Boston: Beacon Press.

—— (1992) "The end of the body?" *American Ethnologist* 19: 121–138.

—— (1994) *Flexible Bodies: Tracking Immunity in American Culture from the Days of Polio to the Age of AIDS*, Boston: Beacon Press.

Marx, K. (1975) *Texts on Method*, T. Carver (ed.) Oxford: Blackwell.

Mitchell, T. (1988) *Colonizing Egypt*, Cambridge: Cambridge University Press.

Moore, D. (1996) "Marxism, culture and political-ecology," in R. Peet and M. Watts (eds) *Liberation Ecologies: Environment, Development, Social Movements*, New York: Routledge.

Mumford, L. (1964) "Authoritarian and democratic technics," *Technology and Culture* 5: 1–8.

O'Connor, J. (1988) "Capitalism, nature, socialism: a theoretical introduction," *Capitalism, Nature, Socialism* 1: 11–38.

Oudshoorn, N. (1994) *Beyond the Natural Body*, London: Routledge.

Peet, R. and Watts, M. (1996) "Liberation ecology: development, sustainability, and environment in an age of market triumphalism," in R. Peet and M. Watts (eds) *Liberation Ecologies: Environment, Development, Social Movements*, New York: Routledge.

Peluso, N. (1993) *Rich Forests, Poor People*, Berkeley: University of California Press.

Peterson, M. (1990) "Paradigmatic shift in agriculture: global effects and the Swedish response," in T. Marsden *et al.* (eds) *Rural Enterprise*, London: David Fulton.

Pile, S. (1990) *The Private Farmer*, Aldershot: Avebury.

Rabinow, P. (1992) "Artificiality and enlightenment: from sociobiology to biosociality," in J. Crary and S. Kwinter (eds) *Incorporations*, New York: Zone Books.

Rajchman, J. (1988) "Foucault's art of seeing," *October* 44: 88–117.

Redclift, M. (1987) *Sustainable Development: Exploring the Contradictions*, London: Methuen.

Ross, A. (1994) *The Chicago Gangster Theory of Life: Nature's Debt to Society*, London: Verso.

Roucheleau, D., Thomas-Slayter, B. and Wangari, E. (eds) (1996) *Feminist Political Ecology*, London: Routledge.

Rouse, J. (1987) *Knowledge and Power: Toward a Political Philosophy of Science*, Ithaca: Cornell University Press.

Rouse, J. (1992) "What are the cultural studies of scientific knowledge?," *Configurations* 1: 1–22.

Schama, S. (1995) *Landscape and Memory*, New York: Alfred A. Knopf.

Schmidt, A. (1971) *The Concept of Nature in Marx*, London: New Left Books.

Scott, J. C. (1985) *Weapons of the Weak*, New Haven, CT: Yale University Press.

Serres, M. (1987) *Statues*, Paris: Francois Bourin.

Shapin, S. and Schaffer, S. (1985) *Leviathan and the Air-Pump: Hobbes, Boyle and the Experimental Life*, Princeton: Princeton University Press.

Slater, C. (1995) "Amazonia as edenic narrative," in W. Cronon (ed.) *Uncommon Ground: Toward Reinventing Nature*, New York: W. W. Norton.

Smith, N. (1984) *Uneven Development: Nature, Capital and the Production of Space*, Oxford: Blackwell.

—— (1996) "The production of nature," in G. Robertson, M. Mash, L. Tichner, J. Bird, B. Curtis and T. Putnam (eds) *Future Natural*, London: Routledge.

Spivak, G. (1985) "The Rani of Samur," in F. Barker *et al.* (eds) *Europe and Its Others*, Colchester: University of Sussex.

Stonich, S. (1989) "The dynamics of social processes and environmental destruction," *Population and Development Review* 15: 269–96.

Strathern, M. (1992) *Reproducing the Future: Anthropology, Kinship and the New Reproductive Technologies*, New York: Routledge.

Taylor, C. (1992) "Heidegger, language, and ecology," in H. Dreyfus and H. Hall (eds) *Heidegger: A Critical Reader*, Oxford: Blackwell.

Traweek, S. (1988) *Beamtimes and Lifetimes*, Cambridge, MA: Harvard University Press.

Treichler, P. (1987) "AIDS, homophobia and biomedical discourse: an epidemic of signification," *October* 43: 31–70.

Wark, M. (1994) "Third nature," *Cultural Studies* 8: 115–32.

Watts, M. (1983) *Silent Violence*, Berkeley: University of California Press.

—— (1989) "The agrarian question in Africa," *Progress in Human Geography*, 13: 1–41.

Watts, M. (1991) "Geography and struggles over nature," in F. Buttel and L-A. Thrupp (eds) *The Food Question*, London: Earthscan.

Whatmore, S. (1991) *Farming Women*, London: Macmillan.

Whatmore, S. and Boucher, S. (1993) "Bargaining with nature: the discourse and practice of "environmental planning gain," *Transactions of the Institute of British Geographers* 18: 166–78.

White, S. (1991) *Political Theory and Postmodernism*, Cambridge: Cambridge University Press.

Willems-Braun, B. (1997) "Buried epistemologies: the politics of nature in (post)colonial British Columbia," *Annals of the Association of American Geographers* 87: 3–31.

Williams, R. (1976) *Keywords: A Vocabulary of Culture and Society*, London: Fontana.

—— (1980) *Problems in Materialism and Culture*, London: Verso.

Winner, L. (1986) *The Whale and the Reactor: A Search for Limits in an Age of High Technology*, Chicago: University of Chicago Press.

Worster, D. (ed.) (1988) *The Ends of the Earth: Perspectives on Modern Environmental History*, New York: Cambridge University Press.

Zimmerer, K. (1994) "Human geography and the new ecology," *Annals of the Association of American Geographers* 84: 108–25.

Part 2

CAPITALISING AND ENFRAMING NATURE

INTRODUCTION

We have named Part 2 "capitalising and enframing nature" for several reasons. As we approach the millenium nature has become commodified "all the way down". In fact, in many ways it is, to use Ross's (1994: 237) felicitous phrase, the "new box-office star" of corporate capital, the subject of new sites and modalities of accumulation. Where in the past, capital expanded outwards, as it were, to push back the frontiers of non-commodified nature (be it through commercial agriculture or logging say), today it has turned inwards to remake these social natures afresh and to commodify new ones, like the human body. For us, the term "capitalisation", with its intimation of the "deep" remaking and repackaging of social nature – in "wilderness preservation" as much as bio-technology, green consumerism, medical implants and the like – captures crisply this sense of involuted hyper-commodification. And yet if social natures are everywhere commodified, this is not simply or only a process of material transformation. As Baudrillard and others have argued, exchange values always already have sign values and their capitalisation depends on and deploys these complex discursive enframings. As Eder (1996) recently shows in his *The Social Construction of Nature*, the "world-disclosing" capacities of linguistic and cultural frames are every bit as material and important as the things they disclose. Most importantly, because those frames are both ordered and culturally specific, they can act as normalising forces in which social natures become refractions of power and inequality in society. The challenge, therefore, is to identify practices, both material and semiotic, which can contest hegemonic capitalisations and enframings of nature.

In Part 2, authors take it as axiomatic that the capitalisation and enframing of nature is simultaneous and ineffable. Using a variety of theoretical tools and case studies they demonstrate the intimate intertwinings of practices of production and consumption with those of representation and discursivity. **Cindi**

43

Katz (Chapter 2), in a wide-ranging essay, considers corporate capitalism's shift from opposing environmentalism during the 1970s to embracing it during the 1990s by making nature an "accumulation strategy". With the loss of what she calls an "extensive nature", readily at hand as a resource to be exploited in, for example, the developing countries, she notes that Western capital is now seeking to plumb an ever more "intensive nature". Suddenly "corporate environmentalism" is everywhere concerned to delimit and privatise a nature in need of "preservation" or "restoration". Considering, among other examples, the transnational Nature Conservancy and the local Central Park Conservancy in New York, she traces the classed, raced and gendered subtext of this new enclosures movement. Showing how these nature spaces are constructed in both reality and the imagination, Katz contests their ideological character and argues for a reflexive political ecology which can hold environmental, social, cultural and political concerns in common tension.

Emily Martin (Chapter 3), in a provocative contribution, points to the affinites between a shift from Fordist to post-Fordist regimes of accumulation in the Atlantic economies and from machinic conceptions of the body as a domain to be defended from external invaders (germs, bacteria, etc.) to conceptions of the body as flexibe, fluid and porous. Drawing on her anthropological work on expert and everyday notions of the "healthy body", she argues that the move away from earlier notions of bodies separate from an external world to ones which muddy the natural–human divide are only apparently liberatory. The new "flexible body", she worries, may simply reinstate earlier forms of biopolitical control in the form of new corporealities suitable for management by equally new biomedical accumulation practices.

If Emily Martin shows that today the human body is one of the new frontiers of social nature, then **Hilary Rose** (Chapter 4) tries to identify non-hegemonic, insurgent ways of defining the body. Focussing on the Human Genome Project, Rose argues that the "new genetics" is increasingly part of a "consumer culture without limits". Dissenting from the idea that modern science is a citadel, she argues that it is increasingly embroiled in wider political-economic relations in which genetic research and "therapy" are more and more marketised. By drawing historical parallels, Rose suggests that, their differences notwithstanding, the old "state eugenics" of the interwar years – in Germany and other countries – may have a more insidious contemporary in what she calls "consumer eugenics". Through a comparative look at the existing scientific and state controls on human genetic research, she then tries to assess the critical resources available in civil society for a responsible contestation of the unspoken "body norms" which the new genetics installs.

Just as the body once seemed to be the antithesis of society and social intervention, modern cities have long seemed to be the antithesis of nature. In Chapter 5 on post-Fordist urban environments, **Roger Keil and John Graham** consider contemporary attempts to "reassert nature" in major Western cities. Periodising modern urbanism, they show that these new metropolitan ecologies are intimately bound to wider shifts in urban political economy and do not represent a simple reintroduction of "first nature" into the cities from which they were historically expelled. Taking the case of Toronto, they show instead that "eco-modernisation" is consciously deployed as a strategy of socio-spatial regulation.

Moving from the city to the country, **James McCarthy** (Chapter 6) takes up the question of nature and accumulation in the rural west of the USA. A region famed for its "wilderness", it is also one whose environment has been increasingly appropriated by corporate capitalism. In this context of corporate intrusion and the environmentalist critique of it in the name of a "wild nature" which needs preservation, McCarthy considers the complicated case of the Wise Use movement. An influential and vocal populist organisation, Wise Use has been seen as both a corporate lackey and as a pro-environment but anti-environmentalist baton. McCarthy rejects these simple antinomies in order to show that Wise Use, its ambiguties notwithstanding, in fact embodies a sober recognition of nature's sociality, one which is deployed in quotidien struggles to reclaim the rural in ways that do not rest easily on the nature–society dualism. While hardly a model for new environmental movements in general, it does, he argues, show the potential within civil society for concerted action in the name of more progressive social natures.

Finally, ending the section on as broad a canvas as Katz begins, **Allan Pred** (Chapter 7) takes a critical look at everyday commodity consumption practices in *fin-de-siècle* societies. Reworking Marx's critique of commodity fetishism, Pred constructs a performative account in which his disruption of narrative continuity and completeness is intended to disrupt the apparent self-sufficiency of the commodity form. In our "extended moment of danger", as he calls it, Pred uses the work of Walter Benjamin to construct a profane illumination of the present in which the creative destruction of nature is daily facilitated and reproduced by distanciated consumption practices, be they of food, furniture, fuel or fur coats.

REFERENCES

Eder, K. (1996) *The Social Construction of Nature*, London: Sage.
Ross, A. (1994) *The Chicago Gangster Theory of Life*, London: Verso.

2
WHOSE NATURE, WHOSE CULTURE?

Private productions of space and the "preservation" of nature

Cindi Katz

INTRODUCTION

Nature changed in the 1970s. However we "value" nature, our conventions and practical engagements with the external world – "the environment" or "nature" – under capitalism have operated as if nature were given, a free good or source of wealth, an unlimited bounty awaiting only the "hand of man" to turn it into a bundle of resources. With decolonization and the environmental movements of the 1960s and 1970s coupled with the oil shock of 1973, the utilitarian presumptions that undergirded so much of the relationship to nature under capitalism hit their limits. Capitalist actors could no longer be sure that "natural resources" would be everywhere and eternally available to them. The very grounds of capitalism's global ambition – environmental as much as spatial – had been altered. Yet at the same moment that recognition of environmental exploitation increasingly scripted capitalists as the enemy of nature, those exploitative practices, indeed nature itself, was remade for capitalism. In less than two decades, corporate capitalism reversed its dismissive opposition to environmental movements and gleefully embraced various brands of environmentalism as its own. In the course of this shift, and central to it, nature became an accumulation strategy for capital.[1]

Nature is no longer an "open frontier" for capitalism in the sense of an absolute arena of economic expansion. Rather, nature has undergone an "involution" (cf. Geertz 1963), much as space did in the first few years of the twentieth century when planetary expansion was effectively at an end (Kern 1983; Smith 1984; cf. Lefebvre 1991). In this period, productions of space no longer pushed the borders of the unknown so much as reworked its internal subdivisions. Conceptions of space changed dramatically as a result; cubism was a case in point but so too was Einstein's relativity theory which promised the recombination of space and matter through the ascendence of relative over absolute space (Smith 1984: 72). Today, biodiversity prospecting – to take just one example – illustrates the case of a similar and equally consequential redefinition and involution of nature at the end of the millenium.

In summary, after the limits imposed by decolonization, the environmental movement, and the so-called oil shock, culminating in the early 1970s, the contours of nature produced and conceived under capitalism were reworked in ways that are continuous with and analogous to those of space in the early years of the century. The time–space of nature has changed irrevocably. Rejecting any limits to its own growth, corporate capitalism quickly morphed into a green version of itself by the 1980s, while science, an always ready accomplice and increasingly indistinguishable partner, had embraced startling new objects of inquiry and practice – from mapping the human genome, through documenting the earth's biodiversity, to cloning. Faced with the loss of extensive nature, capital regrouped to plumb an everyday more intensive nature. The process was facilitated by the appearance (in both senses of the term) of corporate environmentalism in the 1980s.

This shift promulgated and was propelled by the conversion of nature into an accumulation strategy. To be sure, the traditional entanglement of environmental and human exploitation remains, but productions of nature under capitalism now also reflect a different spatio-temporality. Without absolute control over the mineral and vegetation resources of the former colonies and other parts of Africa, Asia, and Latin America, or the security of cheap access to the seemingly bottomless fuel reserves of the oil exporting states, Western capital no longer found nature so unproblematically there for exploitation. The environmental language of nature as an "investment" in the future took on an explicitly capitalist meaning with increasing privatization, whether in the form of "preserves" or as a component of intellectual property rights, and as a result, nature was scrutinized and "mapped" in wholly new ways. The entailments of this rescripting of nature are witnessed in, as much as driven by, the rise of corporate environmentalism, the re-ascendence of "preservation" and "restoration" *as* environmental politics, and the increasing privatization of public environments. These are obviously connected, and come into play in interesting and troubling ways in the practices and policies of, among others, The Nature Conservancy, which operates on an increasingly global scale, and in the operations of the Central Park Conservancy in New York City, which obviously works on a much smaller scale.

In this chapter I want to address the new enclosure movement, witnessed in the growing acreage worldwide commanded by "Park Enhancement Districts," "World Wildlife Zones," biosphere reserves and the like, and their special significance to poor people and poor regions in remote areas of the so-called Third World. Driven by a common impulse and portending increased privatization of the public environment, the intent of these natural set-asides is to cordon off discrete patches of nature in ways that efface their own historical geographies while simultaneously serving up these preserves for "bio-accumulation." Underwriting these strategies are deeply problematic constructions of nature that turn around peculiar and problematic tropes of wild and wilderness; a class-based, racialized, and imperially inflected notion of the "public" and its "commons"; and a paradoxical understanding of material social practices as somehow outside of nature. By drawing transnational parallels between preservation attempts at different geographical scales, this chapter will examine the new ideological commitment to "preserve" nature, the contested boundaries of what is to be preserved, and the significance of the privatization of nature via which this occurs. I will argue for a political ecology that is rooted in productions of nature that hold environmental concerns in tension with social, cultural, and political economic considerations.

NATURE AS AN ACCUMULATION STRATEGY

An instrumentalist view of nature as a source of value or a "resource base" has been a feature of enlightenment thinking and capitalist social relations of production since the eighteenth century. Marxist and eco-feminist theorists, among others, have exposed and analyzed the common threads between the exploitation of nature and the exploitation of people. However, the notion of nature as an accumulation strategy introduces an altered and broader purview for capital's interests in nature. First, nature *qua* nature has become an "investment" in the future. Second, to secure that investment nature has been commodified and privatized at all scales. Finally, with the traditional means of access to nature for metropolitan capital no longer ensured by colonialism, patron–client state relations, or the acquiescence of trusting or environmentally unaware publics, capital's need for clear channels of access to control nature and environmental resources has been refashioned and reasserted strongly in recent years. Each of these related bundles of practices is associated in distinct ways with the reproduction of nature as an accumulation strategy.

The environmentalist literature – corporate and otherwise – is so full of the metaphors of investment, saving, and future gain that it often reads like board room script. This is no accident, given the conservatism of so much of the conservation movement. As biodiversity prospecting has taken off, much of the rhetoric advocating the salvation of particular habitats or restoration of ecological "balance" stresses the potential uses of "as yet unknown" species and organisms. This logic pushes instrumentalism to the vanishing point; apparently *nothing* should be allowed to become extinct, let alone destroyed, because it might one day prove useful (and profitable) to humankind. Darwin be damned. Rooted in an homologous rhetoric of care or biocentrism to that espoused by many environmentalists, such preservation agendas mooted by corporations, foundations, non-governmental organizations (NGOs), and various governments are directed to a much more instrumentalist cause.

While nature "preserves," and the more specialized biosphere reserves, appear to, and indeed do, protect particular environments from a range of "inappropriate" uses and thus from damage, they also invite and encourage scientific documentation and analysis of endemic flora and fauna with the explicit intent of facilitating future expropriation. To all appearances, the preserved landscape is secure; but in the world of action, mediated by particular axes of knowledge, power, and wealth, its conversion to resource in some global accounting ledger has fundamentally altered its status and temporality. The preserve becomes, in current lingo, "a biodiversity bank." Deferred consumption coupled with various investments in money and scientific inquiry are expected to pay off in the future. All of which begs the question of who has rights to determine the "appropriate" use of preserved land; of how the altered temporalities of nature bias future social access to the landscape. Like any corporate investment, the biodiversity bank exists for its investors, and access is strictly controlled. In these ways and others, biosphere reserves and nature preserves come to represent a peculiar form of fetishized nature.

Of course most nature preserves are more than banks of biodiversity. Many of them are also sites of nature appreciation and learning – destinations for eco-tourists or those looking to experience nature in their local environment. Yet there is an interesting difference between the properties set aside for preservation in the contemporary landscape and those made into parks in the nineteenth and early twentieth centuries. Where preserves or national parks were once a means of ennobling wastelands or landscapes that promised little in the way of potential resource extraction – it was easy to make a park of a landscape as starkly beautiful *and* unyielding as Yellowstone, for instance – contemporary preserves are immediately recognized as productive or potentially productive sites. Indeed, that is why many of them are selected for preservation initiatives. Putting such properties aside in the name of some global citizenship is actually a form of luxury consumption requiring considerable reserves of money and power. In an era of ascendant neo-liberalism, preservation and privatization are mutually implicated.

There is, of course, a geography to the process of preservation, and it is uneven. The transnational and class aspects of modern preservation practices are particularly troubling. Thanks to dramatic reductions in biodiversity resulting from industrial development in Europe and North America, as well as, more directly, the enhanced geographical power of capital to dictate conditions of exploitation at the margins, many of the biodiversity battles are being waged in the underdeveloped countries of Latin America, Asia, and Africa. Environmentalists and governments of the "north" (often in collaboration with scientists and policy-makers from the "south") have determined that, for the good of the global environment, substantial parts of such regions rich in biodiversity must be preserved rather than squandered. But here, the attentiveness to diversity in terms of plants and animals is strangely contrasted with a one-dimensional treatment of vastly different human constituencies who use the protected environment. Poor people are thereby constituted as poachers or intruders on their own land (Peluso 1992; Brandon and Wells 1992; Colchester 1994). Preservation has too often been deployed like a blunt instrument when a more fine-grained approach – ecologically and socially sensitive – might have done better. As a growing percentage of land in biodiverse regions is locked in preserves,[2] those with lower mobility and fewer economic options are constrained more by the restrictions on land use than large-scale users such as multinational timber, pharmaceutical, and ranching companies whose production practices are in any case more damaging. Equivalent restrictions on land use by corporate resource exploiters and small-scale agriculturalists and resource users have radically unequal results; for one group, extractive exploitation can be pursued elsewhere, while the other is exiled from the means of their existence. There is no metric of equivalences in universal preservation, equally but blindly applied.

The privatization of nature is also witnessed in such recent innovations as "debt-for-nature" swaps. When the debt of a poor nation is assumed (after renegotiation at a much lower rate) by a non-governmental organization or one of the northern industrialized states, in exchange for the "preservation" of an area they deem to have particular environmental value, the economic valuation of preserved nature is made explicit. This value is clearly intended to accrue primarily to the investors, and to require the long-term accommodation of the debtors (the language is instructive) to the necessary conditions of investment. This is especially true of those who live or lived in or use the environment in question even though the national debt was surely an abstraction to them. This arrangement – of growing popularity in recent years – exemplifies imperialism redux via nature (Mahony 1992).

The presumption that there is impending environmental disaster, especially in the Third World, has become such orthodoxy, that it is hardly remarked that such strategies as debt-for-nature swaps, environmental preservation projects, or the creation of buffer zones to protect various threatened ecologies, represent a sea change in north–south relations which were premised for years in an extractive relationship so intense that many of the current environmental problems can be traced to them. More commonly, the new environmental policies and practices are touted as evidence of global environmental concern contra imperial ambition. While this dissembling might be expected from the purveyors of such projects, liberal environmentalists and radical ecocentrics not infrequently toe the same line (Bonner 1993; Colchester 1994; Neumann 1995, 1996). Yet these policies and practices betoken a whole new regime of imperial exploitation camouflaged as environmentalism. There is big money to be made from "preserving" nature, and the current transnational political ecological relations by and large ensure that the eventual profits will flow north.

The role of science in these preservation efforts and other nature investment schemes is substantial. While science – especially agronomy, biochemistry, and biotechnology – contributed in important ways to earlier "development" endeavors, helping to shape the contours of the "green revolution" and to develop appropriate conservation measures for tropical soils, for instance (Blaikie and Brookfield 1987; Shiva 1991), contemporary scientists are increasingly concerned to patent knowledge and nature prior to any social use (Blum 1993, *Africa* 1996, Martinez-Alier 1996). Ian Wilmut, the Roslin scientist who cloned the sheep "Dolly" (as in Parton) from a cell in her original's mammary gland, secured the patent for his reproduction before announcing the results in a scientific journal in early 1997. Financed largely by a US-based pharmaceutical firm interested in genetically engineered animals that will produce drugs in their milk, Wilmut remained beholden to the investment interests of his backers six months after his accomplishment (Kolata 1997). Science might here be compared to a true wolf in sheep's clothing – in the age of bio-mechanical reproduction cum neo-liberalism, science has become as much a means of production as Herbert Marcuse anticipated. The "productivizing" practices of many contemporary scientists provide a new ironic twist to the claim that science is "value" neutral. An ever-shrinking number of contemporary university scientists work on projects that do not directly commodify nature for the corporate benefit of their employers and/or funders. So called "intellectual property rights" are the latest profit frontier in the privatization of nature. This may be the true "tragedy of the commons."

CORPORATE ENVIRONMENTALISM

The commodification of ever smaller bits of biodiversity, the reproduction of nature as product – whether whole sheep or bit of DNA – and the privatization of common property resources of particular instrumental or aesthetic value, suggest the myriad ways that investments in nature pay. These interests underlie (and have underwritten) much of the surge in US corporate environmentalism in the last two decades. The cache of environmental awareness, focused on consumption practices or gushing sympathy for "charismatic megafauna" in distant places, has not been lost on US corporate leaders. For a relatively small price, corporate capitalists buy the good will, averted glance, and forgiveness, as well as patronage, of much of the population, with changes in packaging and tokenistic "green" gestures. With substantial financial support of various environmental causes, they have bought off much of the environmental movement (Cockburn and Silverstein 1996). Environmentalism is now a pillar of establishment orthodoxy, its own cash cow. As Neil Smith (1996) notes, it was only a few years from Reagan's retro tree bashing in the 1980s – recall his memorable, "if you've seen one redwood . . . " or his quirky notion that trees cause pollution – to Bush's declaration that he would be the "environmental President." In a parallel shift, large corporations have discovered the currency of environmentalism. And like the Republicans, who laid the groundwork for gutting federal environmental protection legislation in the USA and for increased (subsidized) access to federal lands for commercial timber, mining and ranching interests since the "environmental President" took office, most corporations have camouflaged more than changed their environmentally destructive practices.

Advances over older forms of environmental destruction have been real enough, but the "racket in nature," as Horkheimer and Adorno (1987) called capitalism more than half a century ago, has moved to a different level.

Corporate environmentalism sells as well as buys "nature" – whether eco-tourism outfits, or shops such as The Nature Company, The Body Shop, or the recently opened Evolution in downtown Manhattan, which sell *inter alia* bones, fossils, natural elixirs, and dead bugs (Kaplan 1995; Smith 1996; Luke 1997). What I like to call "greenateering," has become an unabashed marketing strategy. Greenateers pander to and assuage consumers' environmental concerns by making it part of their sales pitch that their products are packaged in "environmentally friendly" containers, they use recycled materials, and they only use goods produced in ecologically sensitive ways. Green sensitivity pays tremendous dividends with a public that has itself become consumed with the environment. But religious recycling and the consumption of "green" goods in "green" packaging in the USA have become little more than consciousness cleansing, although they pass for politics.

With so much green, I start to see red. In the tide of "win–win" bonhomie politics for which these gestures pass, it should not be forgotten that corporations – both those that trade in nature directly and those that use it to trade other goods – make a great deal of money in the process. Perhaps there is nothing wrong with that – clean capitalism is better than dirty to be sure – but other issues are at stake. Politics as consumption (and vice versa) works to individualize environmental problems and their solutions in ways that repeatedly forestall and mystify any meaningful ways of dealing with them. As many others have noted, focus on the scale of individual recycling or consumption practices often serves to efface the much broader realm wherein environmental problems are produced and to lull people into a problematic sense of security. So focused on individual solutions is contemporary recycling policy, for example, that almost no one marks the astonishing rate of growth in waste production. Martin Melosi (1981, cited in Horton 1995), indicates that US per capita waste production went from 2.75 pounds per day in 1920 to 8 pounds a day in 1980. While we recycle frantically (and I do), plastics manufacturers, paper companies, and the metal industry, among others, produce, profit, and can pollute – or at least dissemble on the question of their pollution – with abandon. Moreover, while recycling may reduce the inputs of various raw materials to production, it also represents a net economic transfer from individuals to business. Corporate responsibility for dirty production is individualized.

Other corporations curry favor with the public by funding various environmental projects from biodiversity protection efforts through watershed preservation to wildlife conservation. These companies may be among the world's biggest polluters or habitat destroyers, but their environmentalism buys a protective if not mystifying shield for their actions. How else to explain the prominent role of the big oil, chemical, and timber companies on list after list of "environmental" donors. Of course, the conservation movement in the USA and elsewhere has a long history of self-serving connections with sport hunters, forestry interests, and others interested in "resource managerialism" (Luke 1994; cf. Bonner 1993). But what I am describing here as characteristic of the contemporary era of corporate environmentalism is different. It has at least as much in common with "blood or hush money" as it does with the assurance of sustained yields. Thus, Cockburn and Silverstein (1996) describe how environmental opposition has been blunted under pressure from corporate donors, and expose the ways that environmentally inclined foundations are directly underwritten by profits gained from environmentally destructive practices. Environmentalism may be good for business, but environmental destruction – new forms are preferable to old – is still better.

Of course there is a geography to these practices too, and it is most easily uncovered in the Janus face of so many corporate environmentalists. These companies destroy or wreak havoc on environments at one scale or location, and present an entirely different face at another scale or place. For instance, the Ordway family, heirs to the once notoriously polluting Minnesota Mining and Manufacturing Company (3M Company), is one of the darlings of The Nature Conservancy. They and their well-heeled neighbors along the Brule River in an exclusive neck of northern Wisconsin formed a property association in the 1950s to protect "their" river and its environs, their estates of thousands of acres each. Not trusting their own heirs to continue to protect their property, they enlisted The Nature Conservancy to protect their property and prevent future development except for "inoffensive construction . . . a guest cottage or another outbuilding" (Krasemann and Grove 1992: 84). In keeping with their "focus," the Conservancy seems willing to overlook whatever environmental problems companies like 3M might cause elsewhere in order to help such powerful donors as the Ordway family keep the Brule pristine and protected, even from such class offenses as "boisterous, destructive fun-seekers," such as "tubers [who] *crawled* out on the banks to leave defecation and litter."[3] The combined clout of the ruling-class landowners and The Nature Conservancy resulted in a law – believed to be the first of its kind in the nation – banning tubing on the Brule, and thereby eliminating the unsavory "splashing and yelling of *beered-up* groups" from the Ordways' otherwise peaceful world (Krasemann and Grove 1992: 84, emphasis added).

Such dissonant geographical practices – and here we could include the conservation of biomass or rerouting the carbon cycle at a global scale through such mechanisms as "pollution exchange credits" – suggest the ways that corporate environmentalists manipulate geographic scale and produce nature quite differently at different scales. The Montreal Protocol of 1987 ushered in a novel approach to coping with increasingly restrictive environmental legislation in various localities by advocating "mitigation banking" according to which "pollution credits" are traded on a world scale. With mitigation banking, a company polluting in one location can continue unpenalized by, for example, reforesting in another. The logic of such exchanges is that at a global scale biomass or the carbon balance is maintained. Too bad for those ("beered-up"?) people who live near the still belching source of toxic emissions or those whose forests are cleared while saplings take root elsewhere.

Probing environmental politics through the lens of scale can bring even more sinister contradictions to light. For instance, despite my insinuations above, the 3M Company in recent years has embraced the notion that environmental responsibility will enhance profitability,[4] and among other things, reduced its emissions substantially even where not required to by law. While at a global and national scale, 3M has become a leader among corporate environmentalists, its productions of nature at other scales are more equivocal. One of the biggest law suits facing 3M, according to its 1996 *Annual Report*, concerns its role in the production of breast implants. Thus, it is across scale as much as across space that the contradictions of corporate environmentalism need to be monitored.

ENVIRONMENTAL PRESERVATION

Environmental restoration and preservation projects combine imperatives of absolution and accumulation. Preservation turns on an intrinsic contradiction. It requires that a particular patch of nature – ecological niche, biome, or park – be cordoned off as an island in space and time. Preservation represents an attempt both to delineate and maintain a boundary in space and to arrest time in the interests of a supposedly pristine nature which, of course, is neither bounded nor static. As such, preservation is quite unecological, defying natural history and the vibrancy of the borders – physical, temporal, spatial – where evolution, change, and challenge are negotiated and worked out in nature as in culture. "Preservation" is most commonly accomplished by a physical and textual exclusion of sedimented layers of social activity and actors, past and present.

Several authors have therefore noted the problematic relationship between the preserved and not preserved. They suggest that the preservation of certain sites often legitimates and mystifies the continued or even heightened destructive use of all that is outside the preserves' borders. Timothy Luke (1995), for example, refers to land preserved by The Nature Conservancy as a nature cemetery, suggesting that as certain areas are preserved from a more general environmental exploitation, which continues outside of them, "preserves" will not actually preserve any sort of viable ecologies, but, more accurately, will serve as memorials to what once was. Drawing on strikingly similar metaphors, Wes Jackson (1991: 51) suggests that wilderness – "an artifact of civilization" – has become a kind of "saint" in the USA. Like Luke, he notes that people pay homage to the saint, enshrined in preserves, to commemorate what has been lost, to assuage their sense of loss, and to absolve their guilt in not treating the lost object with all due respect. Such soothing shrines to nature actually enable business elsewhere to go on as usual. But nature preserves can also work to license further depredation elsewhere. Jackson notes poignantly that it is dangerous and deeply flawed in environmental politics to separate "the holy" from "the rest." If we do not care for cities, farmland, and other "pieces of nonwilderness" – East Saint Louis, Harlem, Iowa, and Kansas – the "pristine" wilderness will, he cautions, be "doomed" (ibid.).

Of course, it is precisely this separation between the wild and non-wild that defines The Nature Conservancy's environmental strategy. Their vision and their attractiveness is founded in the insistence that nature can be located, fixed, and preserved outside of culture. The social doubly removed: first, their work perpetuates and hardens the boundaries Jackson and others dissolve between urban, agricultural, and wilderness landscapes, to valorize only the latter as the vestiges of pure nature; second, they read generations of social actors out of the "nature" they preserve, denying any social history of landscape.

At another scale, preservation efforts often partition landscapes in a fairly arbitrary way, rendering parts of inhabited environments off limits to future habitation. Apart from the forced displacement of usually marginalized peoples, such a strategy intensifies the stress on non-preserved environments. One of the responses to this problem has been to create buffer zones around protected areas, but predictably, mixed success has led in turn to demands for buffering the buffers, creating a sort of bull's eye of preservation ringed by buffers of decreasingly restricted use (Brandon and Wells 1992). These dartboards of nature are often constructed and overseen by non-residents whose livelihood is not dependent on the preserved environment. Marcus Colchester (1994) among others (Hecht and Cockburn 1990; Peluso 1992; Nepal and Weber 1995; Wood 1995; Neumann 1997), has addressed the political fallout from these sorts of divisions. He notes that when biodiversity or some other perceived environmental resource is "locked up" in a particular place without regard to the broader social, economic, cultural, and political context of resource use, it not only leads to compensatory exploitation elsewhere, often quite proximate, but ultimately is ineffective even within the site itself. Long-term local land users and residents removed from a preserve have no stake in its abandonment (International Alliance 1992, cited in Colchester 1994). Here again, history is sideswiped and landscape misrecognized as a solely natural artifact. The result is a continued pairing of preservation and plunder, with deeply problematic implications for "local people."

The environmental conservation literature and the literature on "development" are both adept at producing (and often patronizing) the figure of the local. Resident populations in conservation areas or preservation initiatives are generally scripted homogeneously as "local people" by those who presumably see themselves, and their interests, as translocal, and thus more important. As with all strategies of "othering," the creation of "local people" enables planners, policy-makers, and practitioners to romanticize as they exclude, exploit, and marginalize those with most at stake.

Finally the whole notion of preservation is pregnant with Malthusian assumption. Malthusian and neo-Malthusian presumptions are rarely more than a heartbeat away from environmental politics, except perhaps among some socialist and feminist environmentalists, and Malthusianism has made a comeback in the 1990s. Part of what drives the impulse to "preserve" is the notion that resources are running out, that people are destroying the environment, and that these problems are exacerbated by unchecked population growth. It is a classic rationale for blaming the victim (Harvey 1974). Malthusian scenarios demonize especially the poor, implying that population growth among poor people must be checked if collective resources are not to be jeopardized. The stubborn attractiveness of this logic, and the nasty environmental politics it engenders, speak volumes about a range of global agendas that consistently deny the broader ecologic importance of people's self-control over the relations of production and reproduction in their community. The purpose of such denials is less to create a viable program of conservation than to make particular claims on global resources. Such practices ensure that conservation will falter or succeed only through coercion (Peluso 1993).

RESTORATION AND ITS LIMITS

The politics of ecological restoration, most eloquently espoused by the late Alexander Wilson (1992), is, by contrast, built upon the recognition that landscape is a social activity, a social text. In *The Culture of Nature*, Wilson offers restoration as an explicit alternative to preservation. Rather than "saving what's left," he suggests that environmental politics center on "repairing" ruptures in the landscape and "reconnecting" its parts (1992: 17). Recognizing that landscapes are by definition disturbed – "worked, lived on, meddled with, developed" – Wilson calls for greater intervention and "care." His plea to "make intelligible our connections" (ibid.) with one another and our environments via active work in the landscape, resonates strongly with that of Wes Jackson at the Land Institute in Kansas. Restoration ecology is intended to "reproduce, or at least mimic, natural systems," and is envisioned to take place at all scales, from habitat to biosphere. Unlike preservation, restoration is not an "elegiac exercise" for Wilson; it offers rather, an environmental ethic that "nurtures a new appreciation of working landscape" (p. 115).

Restoration ecology offers a more promising environmental politics than preservation. Rather than enshrining nature, restoration works it; rather than ignoring, eclipsing, defacing, or erasing environmental knowledge, restoration is premised on its ongoing production and exchange. In reconnecting nature and culture, restoration offers a politics that is much more ecological than the politics that drive preservation. Taken seriously, restoration ecology would undermine preservationists' and other environmentalists' exclusion of people from the environment, and make impossible the narrow gauge, anti-social politics of biosphere preserves and strict nature reserves.

But restoration ecology also has its limits. It operates at a smaller scale than that in which many environmental problems are generated; it can still be driven by deeply romantic notions of nature; and it has a tendency to privilege certain landscapes and land use practices. Despite a rhetoric that covers all scales, restoration ecology is very much locally focused. It fails to "jump scale" as Neil Smith (1992) put it in a different context, and this limits the viability of restoration as an environmental politics at the transnational scale. In other words, restoration and repair at the grassroots level, however important, are not enough either to fashion a hybrid, liveable world out of a multiply troubled landscape or to cut through and undo the tiresome and moralistic narratives of scarcity, ends, and limits that pervade environmentalist discourse (Katz 1992). Given the scope of problems diagnosed by Wilson and other environmental activists, a refashioning is needed as much as a rehabilitation or restoration; the production of wholly new political ecologies is inevitable.

If the best restoration ecology is appealing insofar as it denies the separation between nature and culture and incorporates histories of environmental knowledge, it nevertheless romanticizes particular historical geographies. It privileges certain landscapes over others, which begs the question who determines what a "good landscape" is. To which period is the political ecology to be restored? Restoration ecologists appeal to "nature" for the answers, and inevitably advocate, valorize, and fix a specific historical landscape as idealized and ahistorical, somewhat antithetical to the living, socialized ecology they set out to remake. Rather than building upon Raymond Williams's idea of "livelihood" as an active practice within a mediated physical world as Wilson (1992) advocated, restoration ecology also tends to naturalize the produced and produce the natural. The subliminal appeal of such neatness makes restoration ecology that much more seductive. But once everything is set aright, "active practice" can easily be jettisoned for authenticity and "livelihood" sacrificed to lawful use. Nature as measure and arbiter of what is good and right has had a long and powerful history, as appealing to fascists as to those who would "heal" the land (Bramwell 1989). The ease of this appeal should trouble anyone interested in the radical project of interrogating nature as a social construction and producing new political ecologies. Recuperation smuggles in the real danger of stopping nature dead.

PRESERVATION AND RESTORATION AS GATEWAYS TO PRIVATIZATION

The politics of preservation and restoration short circuit the radical possibilities of producing nature, authorizing instead, a privatized rescripting of nature. The social is excluded as a redemptive prelude to the resocialization of nature in a very particular guise. The doctrine of "wise use" (see McCarthy, this volume) operates for example, as if wisdom and use were entirely separable from questions of history, geography, or power, while claiming nature for some social and economic interests over others. Nature indeed becomes an accumulation strategy, and provides simultaneously the new material for present and future production and an in-built justification of the naturalness of exploitation.

As a scratch almost anywhere on the transnational landscape will reveal, preservation and restoration facilitate the privatization of nature and space that have become the hallmark of global neo-liberalism. The operations of the transnational Nature Conservancy, and the local Central Park Conservancy, are illustrative. These two conservancies do their work with the zeal and self-righteousness of missionaries. Indeed, they are capitalists with a mission – saving and protecting nature – which, in the spirit of global capitalism, they see as everywhere theirs. Like religious missionaries, both conservancies presume that the larger "good" of their endeavor will immunize them from charges of self-interest, or from the erasures of history that their projects require.[5]

The unquestioned assumptions that drive the work of both organizations and their overweening self-approbation are formidable. Yet what is actually going on in urban "park enhancement districts" such as New York's Central Park, and the productions of nature authorized by The Nature Conservancy? The Central Park Conservancy (CPC) promulgates restoration ecology, while The Nature Conservancy, in the main, promotes preservation. Each exemplifies issues raised in this chapter.

Devoted since 1980 to a rehabilitation of the park, the CPC has nearly completed an extraordinary project. This achievement has incurred many of the political and social costs that dog restoration ecology. First, it has valorized a particular moment of the Park's history, choosing the halcyon days of its architects, Frederick Law Olmsted and Calvert Vaux, as the authentic moment. Little regard is given for intervening histories or the landscapes that were erased when the Park was developed.[6] Second, the CPC has drawn a rigid boundary around the Park. Since 1980 the Conservancy has invested more than $110 million for restoration and maintenance. As of 1995 it had assumed 70 per cent of the Park's operating costs and by 1998 had taken over the day to day running of the Park (Central Park Conservancy 1995; Martin 1998). Thus privatized, the experience of Central Park has led to a diminution of resources for other city parks, both by enabling the City Government to reduce its Parks budget drastically without penalty to New York's flagship park, enjoyed by its wealthiest citizens and by tourists, and also by claiming the attention (and money) of powerful social actors concerned with parks and public space. The ahistoricity and social bias of a revitalized Central Park is explicit. For instance, CPC's 1995 *Annual Report* boasted that it replaced an "abandoned playground" with a "grassy glade." Unmentioned was the historical geography of the playground that might have revealed why it was abandoned, and which would have raised the possibility of its creative rehabilitation rather than removal. While the Conservancy has redesigned most of the Park's twenty-one playgrounds, the preference in this case for a "grassy glade" reveals the power of their naturalized restoration language, even at the expense of children.

The Nature Conservancy (TNC) is also concerned with restoration as well as preservation, but its agenda is dominated by the latter. It operates "the largest private system of nature sanctuaries in the world" (The Nature Conservancy 1996), and is fond of pointing out, mimicking an advertisement for a large brokerage house: "We protect land the old-fashioned way, we buy it." Since 1980 it has worked with "partner organizations" internationally to establish nature preserves in Latin America, the Caribbean, the Pacific, and Asia. Many of these become destinations for eco-tourists who are exhorted during their visits that these sites are there "for the plants and animals, not for humans." The same logic applied to resident populations leads to a far crueller fate rippling with Malthusian sentiments. The Nature Conservancy's preservation efforts insistently evict people from nature. So adept at recognizing diversity in the biosphere, TNC cannot seem to distinguish between different kinds of hunting or varieties of human occupance and use, and thus works in partnership with national environmental organizations to block all "human interference" with the environments it protects.

Oblivious to the problem of constructing and preserving nature apart from people, TNC's overseas partnerships under the banner of environmentalism reinstate imperial and neo-colonial relationships to land and other environmental resources. To take just one case, TNC has joined with the Foundation for the Sustainable Development of the Chaco (Paraguay), an organization founded and directed largely by Mennonite settlers in the area, to protect more than over 250,000 sq km of ecologically diverse terrain. Together they have established a "conservation buyer program" whereby parcels of "ecologically significant land" are purchased by "nature lovers all over the world" (Thigpen 1996). The contradictions of nature lovers from afar owning land that may be the livelihood and means of existence of people in the Paraguayan Chaco are transparent.

A final almost hallucinogenic event highlights the role of both conservancies in serving up nature as accumulation strategy. In May 1996 The Nature Conservancy of New York collaborated with the Central Park Conservancy to host the "Second Great Party to Save the Last Great Places." Held in Central Park with "entertainment, corporate, and media" leadership, complete with awards to Ted Turner and Charles Kuralt, the party raised hundreds of thousands of dollars for The Nature Conservancy of New York, which in turn helped to support a restoration project in Central Park. The draw was five "eco-tents" offering the "opportunity to travel to one of the Last Great Places in the world: the Desert Southwest of Utah and Arizona, the Peconic Bioreserve on eastern Long Island; Alaska [period]; the Pantanal in Brazil; and the Lore Lindu Park on the Island of Sulawesi in Indonesia." Each tent featured "authentic decoration; food and drink, dance and music, and crafts, and costumed characters who . . . inform[ed] and entertain[ed] our guests' (The Nature Conservancy of New York 1996). Consistent with its theme park-within-a-park approach, the conservancies managed to conflate a nature reserve at the end of New York's Long Island, the entire state of Alaska, and the largest wetlands in the world.

4 3M produces, among other things, chemical coatings and abrasives, two notoriously toxic products. L. D. DeSimone, the Chairman of 3M, indicated upon receipt of a 1996 Presidential Award for Sustainable Development for 3M's pollution prevention program, "We are convinced that, in the future the most environmentally responsible companies will also be the most competitive companies." Indeed, their pollution program was called "3P" for "Pollution Prevention Pays" (3M 1996, 1997).

5 I must confess (since we're talking about missionaries, why not get religious) that I feel squeamish sometimes about taking on either of these conservancies. They are on one level "good guys" in a sea of far worse political operators, so why (I hear my father asking me) do I need to go after them? Precisely because they trade on being "good guys," to evade scrutiny. Not only do they operate politically with very little external accountability, but their funding strategies explicitly remove not only tax dollars but public environmental responsibility from the state. Privatizing nature and space, as these conservancies do, reduces the tax base for less noble environments (viz. Jackson 1991), siphons off the pressure for safe, engaging, healthy public environments else-where, and eclipses the environmental interests of non-dominant populations (cf. Katz 1995).

6 One of the nice exceptions to privileging the Olmsted landscape has been the Conservancy's "legitimization" of footpaths made by Park visitors heading to popular areas. By constructing these "recreational pathways" the Conservancy recognizes contemporary social practice as it protects other areas from being trodden.

REFERENCES

Africa (1996) 66, 1, (special issue), "The invention of biodiversity".

Blaikie, P. and Brookfield, H. (1987) *Land Degradation and Society*, London: Methuen.

Blum, E. (1993) "Making biodiversity conservation profitable," *Environment* 35: 16–20, 38–45.

Bonner, R. (1993) *At the Hand of Man: Perils and Hope for Africa's Wildlife*, New York: Vintage.

Bramwell, A. (1989) *Ecology in the 20th Century*, New Haven: Yale University Press.

Brandon, K. E. and Wells, M. (1992) "Planning for people and parks: design dilemmas," *World Development* 20: 557–70.

Central Park Conservancy (1995) *Annual Report*, New York: Central Park Conservancy.

Cockburn, A. and Silverstein, J. (1996) *Washington Babylon*, London: Verso.

Colchester, M. (1994) *Salvaging Nature: Indigenous Peoples, Protected Areas and Biodiversity Conservation*, Geneva: United Nations Research Institute for Social Development, Discussion Paper 55.

Geertz, C. (1963) *Agricultural Involution: The Processes of Ecological Change in Indonesia*, Berkeley: University of California Press.

Harvey, D. (1974) "Population, resources, and the ideology of science," *Economic Geography* 50: 256–77.

—— (1998) "The body as an accumulation strategy," *Society and Space* (forthcoming).

Hecht, S. and Cockburn, A. (1990) *The Fate of the Forest*, New York: Harper Collins.

Horkheimer, M. and Adorno, T. W. (1987) *The Dialectic of Enlightenment*, trans. John Cumming, New York: Continuum.

Horton, S. (1995) "Rethinking recycling: the politics of the waste crisis," *Capitalism Nature Socialism* 6: 1–19.

International Alliance (1992) *Charter of the Indigenous-Tribal Peoples of the Tropical Forests*, Penang: International Alliance.

Jackson, W. (1991) "Nature as the measure for a sustainable agriculture," in F. H. Bormann and S. R. Kellert (eds) *Ecology, Economics, Ethics*, New Haven: Yale University Press.

Kaplan, C. (1995) "'A world without boundaries': the Body Shop's trans/national geographics," *Social Text* 43: 45–66.

Katz, C. (1992) "Review of Alexander Wilson, *The Culture of Nature: North American Landscape from Disney to the Exxon Valdez*," *Voice Literary Supplement* April: 20.

—— (1995) "Power, space and terror: social reproduction and the public environment," paper given at Landscape Architecture, Social Ideology and the Politics of Place Conference, Harvard University, Cambridge, March.

Kern, S. (1983) *The Culture of Time and Space 1880–1918*, Cambridge: Harvard University Press.

Kolata, G. (1997) "Rush is on for cloning of animals," *New York Times* 3 June: C8.

Krasemann, S. J. and Grove, N. (1992) *Preserving Eden: The Nature Conservancy*, New York: Harry N. Abrams.

Lefebvre, H. (1991) *The Production of Space*, trans. D. Nicholson-Smith, Oxford: Blackwell.

Luke, T. W. (1994) "Worldwatching at the limits of growth," *Capitalism Nature Socialism* 5: 43–63.

—— (1995) "The Nature Conservancy or the nature cemetery: buying and selling 'perpetual care' as environmental resistance," *Capitalism Nature Socialism* 6: 1–20.

—— (1997) "Nature protection or nature projection: a cultural critique of the Sierra Club," *Capitalism Nature Socialism* 8: 37–63.

Mahony, R. (1992) "Debt-for-Nature swaps who really benefits?," *The Ecologist* 22: 97–103.

Martin, D. (1988) "Private group signs Central Park deal to be its manager," *New York Times* 12 February: A1, B5.

Martinez-Alier, J. (1996) "The merchandising of biodiversity," *Capitalism Nature Socialism* 7: 37–54.

Melosi, M. (1981) *Garbage in the Cities: Refuse, Reform and the Environment 1880–1980*, Chicago: Dorsey Press.

Nature Conservancy (1996) *Reporter*, Summer.

Nature Conservancy of New York (1996) *Newsletter*, Spring.

Nepal, S. K. and Weber, K. E. (1995) "Managing resources and resolving conflicts: national parks and local people," *International Journal of Sustainable Development* 2: 11–25.

Neumann, R. P. (1995) "Ways of seeing Africa: colonial recasting of African society and landscape in Serengeti National Park," *Ecumene* 2: 149–69.

—— (1996) "Dukes, earls and ersatz Edens: aristocratic nature preservationists in colonial Africa," Environment and Planning D *Society and Space* 14: 79–98.

—— (1997) "Primitive ideas: protected area buffer zones and the politics of land in Africa," paper presented at Politics of Poverty and Environmental Interventions Conference, Nordiska Afrikainstitutet, Stockholm, May.

Nygren, A. (1995) "Deforestation in Costa Rica," *Forest and Conservation History* 39: 27–35.

Peluso, N. L. (1992) *Rich Forests, Poor People*, Berkeley and Los Angeles: University of California Press.

—— (1993) "Coercing conservation? The politics of state resource control," *Global Environmental Change* 3: 199–217.

Shiva, V. (1991) *The Violence of the Green Revolution,* London: Zed Books.

Smith, N. (1984) *Uneven Development,* 2nd edn 1990, Oxford: Basil Blackwell.

—— (1992) "Contours of a spatialized politics: homeless vehicles and the production of geographical scale," *Social Text* 33: 54–81.

—— (1996) "The production of nature," in G. Robertson, M. Mash, L. Tickner, J. Bird, B. Curtis and T. Putnam (eds) *Future Natural,* London: Routledge.

Thigpen, J. (1996) "The savior of 'Green Hell'," *Nature Conservancy* September/October: 11–15.

3M (1996) *Annual Report,* St. Paul: 3M.

3M (1997) http://www.mmm.com ("environment"), 14 April.

Wilson, A. (1992) *The Culture of Nature,* Oxford: Basil Blackwell.

Wood, D. (1995) "Conserved to death," *Land Use Policy* 12: 115–135.

3

FLUID BODIES, MANAGED NATURE

Emily Martin

INTRODUCTION

In Euro-American culture during the modern era, "nature" was typically viewed as a domain separate from humans, one that could be acted upon by human agents, for survival, accumulation, or aesthetic pleasure. This view of nature depended on a way of separating human agents from the world, delineating one from the other in a definitive way. In her comparison between Melanesian and Euro-American cultural assumptions, Marilyn Strathern (1988: 89) depicts the latter this way: "Industry and culture are conceived of as a break away from nature and suppose domination over it. Within these terms, to be a full person one must be culturally creative." It follows that women are less full persons than men in this scheme, because at least part of their creativity is in "nature." It also follows that nature is seen as a domain that through passivity or recalcitrance, offers resistance to domination by human agency. It must be transformed through conscious action.[1] "To the Western European view, culture is production, it makes things; it is artifice, it builds on an underlying nature; and it is an agent, a manifestation of power and efficacy, for in what should power and efficacy be shown but in the taming of the natural world and the products of the created one?" (p. 55). The agent called an individual is similarly "a source of action, an embodiment of sentiment and emotion, author of ideas, and one who reveals the imprint of culture . . . Since the individual as an agent is also conceived as a single entity, many of her or his problems are presented existentially, as boundary ones" (ibid.).

Strathern focuses on how agents, persons, or individuals are situated in the world: in this chapter I will consider how the bodies of these persons (the body is treated in Euro-American culture as a kind of real property that belongs to a person) have begun to be reconceptualized in relation to the external world. Elsewhere I have argued that in the USA at present there is a shift in emphasis from one model (the body as a sharply bounded machine) to another (the body as a blurrily bounded complex system).[2] Here I will extend this argument to show

64

how the separation between human agents and nature that Strathern described is being replaced by a quite different relationship, one in which humans are embedded in nature. Problems associated with the separation of humans from nature, such as ("male") humans' exploitation or domination over ("female") nature, are being joined by a new set of problems. I will suggest that persons with bodies that flow easily into spaces beyond the skin create the potential for disturbingly labile forms of association; these associations, whatever their merits, are unmoderated by the brakes of categorically opposed divisions between the human and natural worlds. In other words, new conceptions of the body which may undermine cultural discourses that have legitimated certain forms of exploitation and domination in the past should not be embraced uncritically, especially because they are often experienced as providing an escape from familiar forms of oppression. This chapter contributes in a limited way to the urgent task of examining what new forms of domination and control might be brought by concepts that seem to provide nothing but liberation.

DEFENDED BORDERS:
THE BODY AS MACHINE

In a modernist mode, medical imagery of the first half of the century and beyond often depicts a body suited to the machine age, specifically the postwar era of mass production (the 1950s), engaged in orderly, periodically regular, assembly line production on a rigid time schedule. This imagery had particular consequences for how hierarchical rankings of people according to gender and race affected medical concepts and treatment. For example, the standard medical accounts of a woman going through menstruation, birth and menopause depict her body as metaphorically engaged in various forms of production on an industrial model: when she menstruates instead of getting pregnant it is interpreted as a result of failed production. (Menstrual fluids, are described by one author of a standard text used in medical schools as "the uterus crying for want of a baby." They are seen negatively, as the result of breakdown, decay, necrosis or death of tissue.) When she gives birth it is regarded as successful production, but this production is often held to a rather strict timetable reminiscent of assembly line production. When a woman reaches menopause, the main headquarters governing her body's reproductive system is thought to undergo a devastating breakdown leading to loss of the centralized control necessary to keep order (Martin 1987).

In the biology of fertilization, the courtship drama of the egg and sperm is rigidly patterned, on a post-World War II model, machine-like in its inexorable conclusion in heterosexual union and fecundity. The standard medical (as well as popular science) accounts of fertilization (at a metaphorical level) see the egg as a damsel in waiting, or a damsel in need of rescue, and the sperm as her seducer, or rescuer, depending on whose account you read (Martin 1990). In all these functions, regular periodicity between well-defined limits is considered normal – estrogen, progesterone and other hormones are produced (if all is normal) with machine-like regularity; menstruation occurs (if all is normal) with the periodicity of a metronome. Disease produces irregularity, and shifts between stages of maturation (puberty and early menopause) produce irregularity. Regularity is normal, good, and valued, irregularity is abnormal and negatively valued.[3]

In its machine-like solidity and concreteness, this body has definite, clear-cut edges. Where the body ends and the outside world begins is marked with precision, and is emphasized because this border is where the battle for health goes on. The drawing, "The Lilliputian Hoards," from *Life Magazine* in 1950 shows the preoccupation at the time with body surfaces. All the action is taking place on the skin of a prone, passive body, where germs are depicted as hoards of little devils trying to puncture the barrier of the skin with sharp objects like drill bits, swords, and needles (Coughlan 1955: 122).

Bodies with clear boundaries and machine-tooled parts may have had their heyday in the 1950s, but in fact, they are very much with us still. For example, contemporary popular and scientific accounts of the immune system see it as maintaining the health of the body through continuous warfare against the foreign enemy. In medical models of HIV and the immune system, macho heterosexualized T-cells, wielding weapons, graduate from "technical college" (the thymus), and guard the healthy body, like intelligent Rambos, killing invading non-self cells. Flexible, creative, innovative B cells, producers of anti-bodies, provide additional ever-changing protection, like talented females. Lowly, dumb, primitive macrophages, marked as low-ranked females or racial others, are the infantry/garbage collectors who die in great numbers on the body's battlefield or live to cannibalize the dead bodies or dispose of the garbage. This is the body as nation-state, isolated, precariously defended at its borders, attempting to maintain purity within, and to guard from contamination without: "self" is sharply divided from and defended against "non-self."[4] Waging war against hostile invaders, the job of the immune system is to protect good "self" from bad "non-self."

Such a picture presupposes an absolute difference between what is friendly "self" and what is unfriendly "non-self." Imagery of aggressive immuno-warfare against the foreign focuses on a body that is all of one kind, all purely self. A body made up only of self is the "normal" body, the desirable body to have. It is as if the body were a castle and its ramparts held stalwartly against anything foreign ever entering.

Not only the bodies of persons, but also the bodies of other legal "personages" (who have, in the USA, a legal standing much like a person), corporations, wish to believe they have strongly defended boundaries. The illustrations that financial firms choose to represent themselves convey the idea: the strong, well-defended corporation proclaims itself through ubiquitous ads in popular magazines that feature photographs of huge, monolithic, and massively defended castles hewn of solid stone, the corporate logo modestly superimposed in one corner. As Donna Haraway (1992: 320) puts it, "The perfection of the fully defended, 'victorious' self is a chilling fantasy." She asks, "when is a self enough of a self that its boundaries become central to institutionalized discourses in biomedicine, war, and business?" (p. 320).

Looking at the cover of a popular book about the immune system, *The Body Victorious* by Lennart Nilsson (1985), gives us very specific clues about the identity of the defended self, in terms of both race and gender. The cover shows a nude, muscular male figure, back lit so that the edges of his body are sharply outlined. But enough stray light escapes to show us that his facial features and hair type identify him as white. The identification is tacit and because it is unmarked in linguistic terms, it contributes to the powerful and silent cultural construction of whiteness, the unmarked racial category. Superimposed across his abdomen is a greatly enlarged, color photograph of an immune system cell, a macrophage, in the act of eating up bacteria. This white male body has boundaries that are defined extremely clearly. Inside is only self; outside is only non-self. Should any foreign matter enter, it will be swiftly dispatched by the roving armies of the man's immune system.

People who belong to other racial or gender categories cannot compete with the internal purity of this self. Women, for example, fall far short. When they are pregnant, they are truly hybrid, uneasily "tolerating" the foreign fetus. Since in immunological terms every cell of the fetus is marked as non-self, technical immunological articles wonder how any fetus ever survives to be born. The puzzle is often phrased: "Why does the mother not mount an attack against the fetus?"[5] A headline in an article in *The Economist* (1985) wonders "Why does the body allow fetuses to live?"

In addition to the "mixing" of self and other in pregnancy, women are statistically more prone to auto-immune diseases. These diseases are conceptualized as caused by the immune system mistakenly attacking self. In one newspaper illustration of a woman with Lupus, an auto-immune disease, she is shown lying inside the ramparts of a castle. But instead of protecting her against threats to her health, the sharp spikes on top of the castle walls are turned inward: the castle of her body has literally turned against her.[6] The depiction of a woman in this illustration is no accident either: the American Auto-immune Related Disease Association estimates there are 50 million Americans affected by 80 known auto-immune diseases, and most of them are women (Brody 1994).

Non-whites are at no advantage when it comes to auto-immunity, either. Although statistics that relate health to race and class are almost nonexistent in the USA (our analyses are prisoners of the categories used to count cases), one form of auto-immunity, asthma, is documented as 26 per cent more common among African Americans than among whites (NAID 1995). And African Americans are three times more likely than whites to die from asthma in the USA (Sly 1994).[7] So it might be argued that the model of the body as a fortress defended behind its ramparts in some ways casts women and minorities, with their greater tendency toward auto-immunity and allergy, in a disadvantaged position.[8]

BLURRED BOUNDARIES: THE BODY AS A COMPLEX SYSTEM

A set of metaphors with implications quite different from "the body as a machine" or as a defended fortress is currently exercising a significant amount of influence in some medical specialties. Some scientists are exploring metaphors derived from chaos theory, also known as non-linear dynamics or complexity theory. Cardiologists, for example, are coming to see the heart, not as the quintessential mechanical body part, the pump, but as a self-organizing system that only beats with a mechanical regularity when the body is near death.

> Until recently, it was widely held that sudden cardiac death represented an abrupt change from the apparently periodic state of the normal heartbeat to one in which chaotic arrhythmias occur. Work from several sources has suggested that under normal conditions the heart has chaotic dynamics and that fatal disturbances of the cardiac rhythm are often preceded by a decrease in the degree of physiological chaos. This represents a reversal in the conventional usage of the term "chaos" when applied to the injured heart.
>
> (Skinner *et al.* 1990: 1019; see also Denton *et al.* 1990)

In the machine model, regularity is a sign of health, irregularity a sign of disease or impending death. In the chaos model, it is just the opposite. Quoting again from the heart researchers:

> In the past five years or so we and our colleagues have discovered that the heart and other physiological systems may behave most erratically when they are young and healthy. Counter intuitively, increasingly regular behavior sometimes accompanies aging and disease. Irregularity and unpredictability, then, are important features of health.
>
> (Goldberger *et al.* 1990: 43–4)

The reason chaotic organization might be an advantage to the heart is this:

> Chaotic systems operate under a wide range of conditions and are therefore adaptable and flexible. This plasticity allows systems to cope with the exigencies of an unpredictable and changing environment.
>
> (Goldberger *et al.* 1990: 49)

Looking through the lens of a complex systems model, women's reproductive functions appear transformed. In the logic of a chaotic system, "irregularity" becomes an adaptive response to a changing internal and external environment. The young woman whose menstrual cycle is affected by exercise, by stress, or by puberty, could then think what a good job her endocrine system is doing flexibly adjusting to her life, rather than worry unduly about a pathological "irregularity." Epidemiological studies that have shown greater menstrual irregularity in women who work at night (Miyauchi *et al.* 1992) and women who are vegetarians (Lloyd *et al.* 1991) could be taken to reveal the responsiveness of these women's physiological systems to their particular environment, rather than a pathological deviation from a putative norm of machine like periodicity. The older woman whose menstrual cycle is affected by approaching menopause or other aspects of her life could see irregularity as a sign of vigor and health instead of impending disease and death.

Whether or not non-lineal models are applied more widely in medicine, my ethnographic fieldwork in a wide variety of settings – an immunology lab, clinics, a hospice for AIDS patients, support groups, urban neighborhoods and work places – has shown that people in all walks of life, quite commonly see their bodies as complex non-linear systems (Martin 1994).

This has happened in part by way of an enormous cultural emphasis on the immune system, which has moved to the very center of the way ordinary people now think of health. Let me illustrate with some examples of how people talk about the immune system [two men are talking, Bill Walters and Peter Herman]:

Bill: I don't even think about the heart anymore, I think about the immune system as being the major thing that's keeping the heart going in the first place, and now that I think about it I would have to say yeah the immune system is really . . . important . . . and the immune system isn't even a vital organ, it's just an act, you know?

Peter: It's like a complete network . . . if one thing fails, I mean if . . .

Bill: If something goes wrong, the immune system fixes it, it's like a back up system. It's a perfect balance.

Steven Baker The immune system is the whole body, it's not just the lungs or the abdomen, it's, I mean if I cut myself, doesn't my immune system start to work right away to prevent infection? So it's in your finger, I mean it's everywhere.[9]

Instead of a mechanical body made up of simple components with different functions, these people are operating as if the body were a dispersed, fluid system. Something like a fractal, the protective ordering functions of the immune system are present in every part of the body no matter how different each appears.

Not only men and women who are not professional scientists are working with a conception of a fluid, flexible, ever-changing body. In the 1970s, immunologists began devising models that treat the immune system as a complex system embedded in a set of interactions with many related complex systems. Although they are not yet in the mainstream, a subset of immunologists argue that the immune system can be described as a network with a kind of distributed intelligence. In their technical articles, imagery of the dance replaces traditional imagery of the immune system in battle against external foes.[10] The body positively reaches out into the world and takes it in, continuously changing its state to adjust to a constantly changing environment:

> The dance of the immune system and the body is the key to the alternative view proposed here, since it is this dance that allows the body to have a changing and plastic identity throughout its life and its multiple encounters. Now the establishment of the system's identity is a positive task and not a reaction against antigens (foreign substances).
>
> (Varela and Coutinho 1991: 251)

In mainstream science, too, there is a shift of interest to newly open and porous edges of the body. We are beginning to hear that macrophages, cells once glossed as primitive garbage eaters, are major players in the maintenance of the immune system and once sabotaged, an able porter for disease organisms. Headline news as early as June 1991 proclaimed that "AIDS virus can get into body through mucous membranes" (Garrett 1991). At the Seventh International Conference on AIDS, researchers reported that HIV can enter the body through dendritic cells (a type of macrophage) and mucosal cells (sticky cells in the mucous membranes that can transport cells much the way macrophages do):

> This finding helps explain HIV transmission among heterosexuals and people who have no sores or cuts that would allow HIV to enter the bloodstream directly. Everyone has mucous membranes in mouth, anus, or vagina across which HIV can travel via these cells. "Look", Dr. Haseltine said, "let's be candid. AIDS is a venereal disease . . . it's time people stopped thinking there was something special about somebody else that put them at risk for HIV. Mucosa is mucosa: VD is VD."[11]
>
> (Garrett 1991: 12A)

Simultaneously in a variety of contexts attention is shifting to the mucous membranes, which provide a route through which anyone, even a baby, could be susceptible to HIV.

It is too soon to tell whether the macrophage might be rehabilitated. The macrophage could continue to be depicted as lowly and primitive, with overtones of race and gender ranking, while more negative traits might be added on top: macrophage as traitorous Trojan horse, hiding a lethal agent within. Or the conceptual categories could radically shift ground: distinctions between self and other could blur, the master T cell might become just another member of the team, the lowly macrophage might become just as important as any other component of the system.

Mucus membranes of nose, mouth, eyes, vagina, anus are where we come off on each other through snot, saliva, semen, secretions, tears. These slimy substances, of and not of the body, dramatically blur self–non-self lines. With respect to HIV transmission, there is a hint of a shift in science from the current focus on the body as citadel, vulnerable only when its defenses are penetrated through a wound, sore, puncture, or thrust, to the body as blurrily linked to others through its many surface secretions. This shift hints at a change in the previously common perception that the body is an autonomous, bounded entity that ends at the skin.

This possibility seems confirmed by the recent rise in activity in the sub-specialty called mucosal immunology. A biologist who works in this area commented:

> I find it intriguing that immunologists have devoted far more attention to the systemic immune system, the system that monitors the blood and lymph, than to the mucosal immune system, the system that monitors the mucus on the surfaces of the gut, lungs, eyes, and reproductive tracts ... Mucosal immunology has emerged as a separate discipline only recently, in part because it is easier to study immune activities in the blood than in mucus. It seems that studying the blood is somehow more noble or "important" than studying mucus. The research in my laboratory is aimed at developing better methods for preventing sexually transmitted diseases and unwanted pregnancy and as a result students must work with semen, cervical mucus, and other unmentionable secretions. They joke, somewhat defensively, about being "masters of mucus."

It is surely no accident that mucosal immunology is relatively undeveloped. Much has been written by anthropologists like Mary Douglas (1966) and Victor Turner (1967) about the deep and complex cultural meanings that often become attached to bodily substances like blood or mucus and bodily orifices like the mouth or anus. Especially strong and troubled significance seems to be attached to substances that are "betwixt and between" (neither solid nor liquid, for example, but a sticky glop like mucus). Such substances arouse horror, but not because of "lack of cleanliness or health," rather because they "disturb identity, system, order. [They are] what does not respect borders, positions, rules. The in-between, the ambiguous, the composite" (Kristeva 1982: 4). In Kristeva's phenomenological account, horror is aroused by these substances because they attest to the impossible task of maintaining a clear bodily boundary between the self and the world: "the fragile border ... where identities (subject/object, etc.) do not exist or only barely so – double, fuzzy, heterogeneous, animal, metamorphosed, altered, abject" (p. 207).[12] And of course, female: the female in much western thought represents disorder both at the level of the body and at the level of her social roles. She threatens to be unruly, turbulent; she poses

71

the problem of how she can be contained in her allotted role, her body is uncontained, messily secreting unmentionable substances (Irigaray 1985: 206–7).

Luce Irigaray explicitly juxtaposes the solidity of the "ruling symbolics" of the phallus with the fluidity of the female:[13]

> Woman never speaks the same way. What she emits is flowing, fluctuating. *Blurring*, And she is not listened to, unless proper meaning (meaning of the proper) is lost. Whence the resistances to that voice that overflows the "subject." Which the "subject" then congeals, freezes, in its categories until it paralyzes the voice in its flow.
>
> (Irigaray 1985: 112)

A woman's speech is:

> Continuous, compressible, dilatable, viscous, conductible, diffusable . . . That it is unending, potent and impotent owing to its resistance to the countable; that it enjoys and suffers from a greater sensitivity to pressures; that it changes – in volume or in force, for example – according to the degree of heat; . . . that it mixes with bodies of a like state, sometimes dilutes itself in them in an almost homogenous manner, which makes the distinction between the one and the other problematical; and furthermore that it is already diffuse "in itself," which disconcerts any attempt at static identification.
>
> (Irigaray 1985: 11)

Scientists' disinclination to study mucus, and the lab members' joking about being "masters of mucus" could well be related to these cultural and phenomenological matters. Might not responses like Sartre's to "the viscous" come into play, its overtones despising the female? "The slime is like a liquid seen in a nightmare, where all its properties are animated by a sort of life and turn back against me . . . the slimy offers a horrible image; it is horrible in itself for a consciousness to *become slimy*" (1981: 138, 140).

To recapitulate, some images, in which the body is a sharp-edged monolithic entity, are related to models of production from the Fordist, mass production era. They value such things as incessant, massive production of male sperm and they denigrate the cyclic, limited and often failed production of the female. Beginning in the 1970s, in scientific fields such as immunology, and in the popular imagination, very different models of the body came into being. In these models, the body is seen as a single, internally related field, a system in which all parts are in complex mutual interaction. When such a model is taken seriously, there is no longer an obvious hierarchy of parts or processes, for all play an equally important part in the functioning of the whole.[14] In addition, the female body, with its cyclic changes, adaptation to different phases of life, and considerable irregularity, even begins to provide a closer approximation to the most desirable bodily states.

How these various bodies live out their days is not determined only by where they sit in a particular cultural world view or how they are construed by a set of metaphors. A variety of concrete practices, material forms of action, go along with any way of imaging the body. When the female body is seen as a mechanical production device, irregularity is understood as pathology, and efforts are made to restore regularity. For the menstrual cycle, these efforts might include administering hormones such as progesterone orally or removing tissue surgically to restore regular periodicity. For the stages of labor, they might include administering hormones such as Pitocin or using surgical techniques such as caesarean section or forceps delivery to ensure labor adheres to a proper schedule of production. When the body comes to be seen as part of a complex system, new practices can potentially take shape. The focus in treating allergies could shift from moderating the immune response within the body to moderating what crosses into the body from outside. Medical agencies might find they had to "treat" the complex of institutions involved in the production of industrial pollutants, for example.[15] As mucosal surfaces of the body attract research interest, the way may be opened for new forms of protection against sexually transmitted diseases that will not perturb the body's internal defense system as a vaccine would. Instead the body's interface with the outside world could be redesigned selectively to prevent pathogens like HIV or Chlamydia from crossing it.[16]

MANAGING "OUR ROOTS IN THE WORLD"

Reflecting on the changes that complex systems models, derived from chaos theory, signal for conceptions of nature as something external to and the object of human agency, Carolyn Merchant comments:

> Chaos theory . . . fundamentally destabilizes the very concept of nature as a standard or referent. It disrupts the idea of the "balance of nature," of nature as resilient actor or mother who will repair the errors of human actors and continue as fecund garden (Eve as mother). It questions the possibility that humans as agents can control and master nature through science and technology, undermining the myth of nature as virgin female to be developed (Eve as virgin). Chaos is the reemergence of nature as power over humans, nature as active, dark, wild, turbulent, and uncontrollable (fallen Eve).
>
> (Merchant 1995: 156–7)

There is, in the valorization of unruly, uncontrollable, and unknowable forces and slippery, slimy, and uncategorizable substances, at the very least, a re-alignment of the oppositions between "male" human as active agent and "female" nature as passive resource I discussed earlier.

But it is not clear that there will simply be a reversal of the ascendant male-as-culture and the subordinate female-as-nature. In some ways the categories themselves seem to be collapsing.

Boundary blurring images from mainstream science find commonality with imagery from alternative health discourse, as in this quote from *Green Lifestyle*, a collection of writings from the alternative health movement and the environmental movement:

> Although each of us seems to be bounded by his or her skin, this is a sheer illusion. When we view our physical boundaries with pinpoint accuracy, they are so fuzzy as to be nonexistent. With each bodily movement, we trail such a haze of chemicals, vapors, and gases behind us that we resemble out-of-focus images.
>
> Not only are we constantly blending physically into the world and our environment, we are blending into each other. Quite literally, we are sharing bodies. How? As writer Guy Murchie has shown, each breath of air we inhale contains a quadrillion or 10^{15} atoms that have been breathed by the rest of mankind within the past few weeks, and more than a million atoms breathed by each and every person on Earth. These atoms don't just shuttle in and out of our lungs, they enter our blood and tissue and make up the actual stuff of our bodies. This means that human bodies are constantly being interchanged with those of any and all things that breathe – not just the bodies of humans but those of cows, crocodiles, serpents, birds, fish, etc. These exhaled "pieces" of our bodies remain after we die to be taken in by other bodies. Yet *our roots in the world go even deeper, even to the stars themselves*. Many of the elements that comprise our bodies were not born on Earth but were recycled through lifetimes of several stars before becoming localized on our planet. Thus, not only are our roots in each other, they are also in the stars. We are, literally, star stuff.
>
> (Dossey 1990: 79, emphasis added)[17]

In this new kind of body, attuned to its environment, what counts for maintaining health is what goes on in interactions between the body (inside) and nature (outside). Accordingly, in striking contrast to the sharp-edged, closed images of the body in the 1950s and earlier, contemporary images more and more show us a body in motion, not in repose, a body in action, not reclining passively, and a body with no skin at all, exposing the inner workings of its protective system, opening the body to its environment. In this light we can understand the cover of a recent issue of *Science* magazine (vol. 260, 14 May 1993). An anatomical figure in the classical tradition stands alert and tense, posed like an athlete or dancer. His body is skinless, allowing the viewer to see directly inside his body, where the lymph nodes under his armpits and in his groin (key sites of immune function) glow with luminescent ink.

The loss of the skin as a kind of secure packaging material for the body/person can be experienced as frightening. As one woman felt it:

> In the past year and a half I've gotten rashes, just all these weird things that I've never gotten before. I was really sick at one point and I decided that it was really a bizarre kind of thing. I thought I had rheumatic fever or something. It started in all my joints and this stuff, I was in the hospital for a day. I remember thinking, I got a rash in the spring and then I got poison ivy and my *skin had just decided it didn't want to protect me anymore. It's given up. It's just saying forget it.* Anything that touches me affects me now. Which is kind of a funny thing, I mean, because your skin is your first defense. It can be considered part of your immune system. Really it's your first defense against anything.

Instead of a castle wall or moat, her skin has become a membrane, an interface with everything in the outside world.

An acupuncturist explains her view of how the skin has receded in importance.

> I don't think our immune systems probably stop at our skin, although I think our skin is a big part of our immune system . . . I'm sure it does extend . . . the person who's working in [an industrial area] is going to have lungs that are not quite as good, as clear, as someone who would be working up in the country . . . there's no barrier there.

She goes on to explain that environmental toxins, radiation of all kinds, use of antibiotics, diet, rest, stress, personal relationships, work, and every other aspect of life directly involve immune function and therefore health. All of these areas must be managed to improve their role in how the body intercalates with the world.

Management is key here, because maintaining the health of this kind of body is an elaborate process: virtually everything that a person experiences comes to seem relevant to health. Stress, emotions of all kinds, the environment, personal relationships, diet, rest, work, and so on: all these areas must be scanned and their good effects maximized. Since what is sought is a complex interaction among a multitude of factors, only "management" that continuously monitors the body's broad field of health and facilitates the flow of information among all the linked parts of the field will do. Management using rigid, top-down control would be as inappropriate for the contemporary body as for the contemporary corporation – both are ideally fluid, ever-changing bodies containing turbulence and instability, in delicate relationship to their environments, complex systems nested in an infinite series of other complex systems. Donna Haraway once commented that "the body is an accumulation strategy, in the deepest sense" (Harvey and Haraway 1995: 510). When people come to see the body as a system in a field of systems, what they most want to accumulate in its vicinity is watchful, continuous, and wise management – as much as their means and circumstances allow.

Apart from images of the body used by immunologists and by non-scientists, my research has also shown that flexibility, adaptability, and the ability rapidly to alter in response to an ever-changing environment with agility and grace are ideals that are "breeding" widely in the culture, bubbling up in all kinds of different contexts. For example, they are now well entrenched as ideal characteristics of work organizations, the government, and educational institutions. Flexibility is used to characterize the most desirable personality, the highest form of intelligence, and the species most likely to survive. It labels countless products and concepts, from Nordic Flex Gold, an exercise machine, through flex space, an architectural design.[18]

To return to the theme of the corporation as a castle, some advertising agencies have tackled head on how to recast the solid immobility of the corporation as castle for a time that extols the agile and flexible. How do you make a massive castle appear to become nimble and mobile? If you are the New York based Republic National Bank, you put it on a boat, on top of a turbulent sea and draw the threatening waters in the style of a Japanese artist (*Business Week* 19 June 1995: 25). Corporate images featuring boats are exceedingly frequent. Many have gone one better than Republic National Bank's image of a massive edifice mounted on top of a boat: usually they feature many small boats, one per employee. Individual initiative, individual fates: each person is alone and works alone. In a recent Chemical Bank Company advertisement with the heading "Be Innovative or Begone", the lone woman in a group of oarsmen has innovatively designed a sail for her row boat and is shown outsailing all the men who fall behind as they doggedly ply their oars (*Business Week* 28 December–4 January 1993: 76–7). Or, in a People Soft ad, a lone employee is pictured with his row boat pulled up to an isolated island, a tiny dot of rock in the wilderness. Although he is surrounded by the beauties of nature, he has his laptop turned on and his head buried in it, concentrated on working (*Business Week* 4 March 1996: 54 E5). When employees are exhorted to loose their creativity to release their initiative, they are also exhorted to become "self-managed" workers whose managers reside inside them wherever they are.

In connection with these lone oarsmen, increasingly in the 1990s there has been a rapid increase in the "You, Inc." phenomenon. According to a 1996 *US News and World Report* poll, 19 per cent of Americans are self-employed, freelance, or sequential temporary workers (Saltzman 1996: 71). Because workers can no longer count on job security, they can only hope that business consultant Tom Peters' advice is taken seriously: "workers, not corporations, should be the primary 'carrier' of adequate benefit packages" (1992: 762). Just as people invest in management of their bodies, "people need to invest in their development as if they *were* a corporation," says Anthony Carnevale, Chairman of the National Commission for Employment Policy at the Department of Labor

(Saltzman 1996: 71). People are, so to speak, coming to see themselves as mini-corporations, collections of assets that each person must continually invest in, nurture, manage, and develop.

CORRELATIONS

In sum, at one level there is a shift from the body as a defended castle poised like a hostile nation-state against its foes, to the body as blurrily becoming part of its environment, reaching out with the sticky fingers of its mucosal surfaces, flexibly and nimbly changing to meet a continuous stream of challenges from without and within. At another level, in the larger realm of political and economic organization, there is a shift from Fordist industrial organization to post-Fordist, with a premium on rapid movement of people and resources in space, swiftly changing markets, production, and products. A number of theorists have observed that the nation-state will, under these emerging circumstances, either cease to exist in the same form or cease to exist altogether (Harvey 1989; Balibar 1991; Gupta 1996). To some theorists, it seems likely that the multinational corporation will replace the nation-state as the preeminent actor on the global stage (Gupta 1996). To the extent that the multinational corporation comes to hold sway over the globe, orchestrating fluid flows of people, resources, capital, and jobs, the emergence of a fluid body, moving liquidly in its field, accumulating assets like a miniature corporation, seems disturbingly in touch.

CAUTIONS

The widespread appeal of these new images of the ideal warns us, veterans of other ideological projects, to be cautious. Carolyn Merchant remarks optimistically:

> What would a chaotic, nonlinear, nongendered history with a different plot look like? Would it be as compelling as the linear version, even if that linear version were extremely nuanced and complicated? A post-modern history might posit characteristics other than those identified with modernism, such as a multiplicity of real actors; acausal, nonsequential events; nonessentialized symbols and meanings; many authorial voices, rather than one; dialectical action and process, rather than the imposed logos of form; situated and contextualized, rather than universal, knowledge. It would be a story (or multiplicity of stories) that perhaps can only be acted and lived, not written at all.
>
> (Merchant 1995: 157–8)

Merchant finds these new horizons appealing in many ways, as do I, but there is one aspect of them that I find particularly worrisome. In the aftermath of a weakening of the bounded body and self, what new kinds of allegiances among people will come into being? I have already mentioned that people are beginning to treat the person/body as a set of assets. In MacPherson's description of the individual, which has been central in Anglo-American culture since at least seventeenth-century liberal democratic theory, the individual was "essentially the proprietor of his own person or capacities, owing nothing to society for them. The individual was seen neither as a moral whole, nor as part of a larger social whole, but as an owner of himself" (1962: 3). My question would be: can liquid bodies blurred across time and space belong to "individuals" in this sense? If parts of the person/body are spread across time and space, which part is the owner? Can there be joint ownership – as in the now familiar form of corporate employee ownership? Or is the link MacPherson describes involving ownership of the self giving way to something else? If the boundaries around the individual are becoming more porous in some ways, are we seeing some form of what Marilyn Strathern (1988) called "dividuation" – some kind of relational selfhood? Numerous observers are remarking on ways in which concepts of self and personhood are becoming multiple: Petchesky heralds the pluralistic meanings of self-ownership, involving permeability, interdependence, and communality and Franklin notes in reproductive technology the breaking of boundaries in terms of individuality or personhood (Ginsburg and Rapp 1995).

At the least there is a broadening here in the forms that personhood and embodiment have recently taken. On the one hand, obliterating old barriers and divisions among individuals might also remove some forms of oppression and discrimination; obliterating old barriers between (female, exploited) nature and (male, exploiting) human agency might increase the likelihood that the "natural" parts of the world would satisfy a greater range of interests in the "human" parts of the world. On the other hand, bodies/persons opened to the world might facilitate forms of mass social movements in which there is a direct, unmediated link between a person and a (putatively pure) whole. Disaggregated workplaces, families, communities and nation-states (Stacey 1990), fleeting ties to serial co-workers, partners, neighbors and support group members, might enhance the appeal of abstract ideals of oneness and purity.[19] Such ideals can involve a participatory mentalité, a version of which was described very negatively at the turn of the century by Le Bon: "Isolated, a person may be a cultivated individual; in a crowd, he is a barbarian – that is, a creature acting by instinct. He possesses the spontaneity, the violence, the ferocity, and also the enthusiasm and heroism of primitive beings" (quoted in Giddens 1979: 124–5). Of course, Le Bon participated in the Eurocentrism and racism of his day in likening crowd behavior to "primitive" behavior. But he was also pointing to what apparently can happen when the borders of the controlled, rational, cultivated individual break down: "When caught up in crowd action . . . they become highly suggestible to influences which, outside of the charged atmosphere of the mob, they would be prone to appraise in a more reasoned fashion" (p. 124).

Clearly it would be simplistic to conflate the loss of individuality in mass movements with a loss of individual borders accompanying complex systems thinking. But it is important to ask whether the rampant social Darwinism in the USA today together with an attenuation of the strength of intermediate social groups such as the family, work team, and community are fertile ground for the suggestion that certain people, linked intimately to certain higher authorities, are valuable and worthy, while others, not being so worthy, deserve expulsion and destruction. There are extreme groups, neo-fascist in form, who are able to champion "the good of the planet" as a reason for their purifying measures (Laqueur 1996: 95).[20] But any mass movement contains some notion of exclusion of those who are not chosen, saved, or enlightened. In their farewell videotapes, members of the Heaven's Gate cult implored: "Come join us, the time is now, the window is small" (Brooke 1997: 16). Press articles and letters to the editor following their mass suicide blamed this event specifically on the current trend toward more fluid concepts of the person and on the fluidity of concepts in general. A theologian commented: "In our postmodern situation, where words can mean anything, it is not surprising a cult like Heaven's Gate can reemerge" (Marquand 1997: 1). A letter writer to *The New York Times* reminded readers that democracy depends on the "participation of free-thinking individuals" and the protection of individuality. He is alarmed by the cult's failure to protect the individuality of its members, which he attributes to "a growing tendency in modern society towards the abandonment of the self in favor of a larger collectivity" (Neufeld 1997).

A number of recent developments seems to link people directly to authorities promoting "right conduct," bypassing the interests of others who prefer not to be so linked. There are wildly popular television shows ("America's Most Wanted") in which people are encouraged to (and do) turn in suspected criminals who could be hiding as a neighbor, kinsman or church member. There are home testing kits that pharmaceutical corporations are marketing for parents to use on their children's urine to detect drug use. There are increasing numbers of mass movements (religious and otherwise) that involve "transcendent meaning and deep public distrust of established authority, whether secular or ecclesiastical" (Niebuhr 1997: 10A). And there are studies on the shop floor of how workers can become part of the surveillance apparatus that exposes the misdeeds of other workers (Sewell and Wilkinson 1992). To the extent that fluid links among individuals can feed the life of a "pure" super-individual entity such as the corporation or the planet, our enthusiasm for such links might well be curtailed. We might even begin to feel nostalgic for the blockages to such collective body/persons that were provided by liberal democratic notions of the clearly demarcated individual and modernist notions of the separation between the natural and the human.

NOTES

1 See Mellor (1987) for a discussion of how "the aggressive, virile male scientist legitimately captures and enslaves a passive, fertile female nature" in the imagery of the seventeenth-century scientific revolution, and in later literature, such as *Frankenstein*.

2 The contrast I posit here and below between models of the body, forms of economic organization, and phases of modernity are overly simple and overly drawn. I intend the contrast as a heuristic device that allows me to throw certain differences into light for discussion. I do not intend to imply a unitary, uniform ladder of development that all societies must go through in the same way. As I make clear later in the chapter, models of the body, like forms of economic organization, coexist in contradictory, complex ways.

3 Menstrual irregularity is often regarded medically as a pathology related to some organic dysfunction. The dysfunction is variously attributed to a "presumed" malfunction of the ovaries, or such problems as hyperandrogenism (Arai and Chrousas 1994), diabetes (Adcock *et al.* 1994), PMS (Khella 1992), or anorexia nervosa (Whitaker 1992). I do not mean to suggest that these correlations are spurious. Rather I want to call attention to the unexamined assumptions that normal equals periodically regular. There is a sharp contrast set up in the typology of the normal and regular vs. the abnormal and irregular. To begin to move toward a different view, one might ask, how regular are most women?

4 This depiction of the body as defended nation-state allows the medical description of HIV+ gay men suffering from AIDS to entail a loss of heterosexual potency, in the guise of their progressive loss of virile T cells (Martin 1992).

5 See Billington (1992) for an example of the genre.

6 Cover of *To Your Health*, October 3, 1989, *The Baltimore Sun*.

7 Ironically "management" of asthma, if successful, results in better tolerance of the air we have to breathe. Seeing the body as a fortress focuses attention on the integrity of the castle walls, not the content of the outside environment.

8 The question of why women and African Americans have higher rates of these maladies would lead me too far afield here, but a place to start would be new research that ties the increasing incidence of allergy and auto-immunity world wide to the rise in the amount of airborne toxins from cars and industry.

9 These quotes are taken from an ethnographic study carried out in Baltimore, MD and described in more detail in Martin (1994).

10 Tauber (1995) traces a shift within immunology from a modernist program, which followed a reductionist paradigm to reach genetic control of immune function, to a postmodern program, which uses a complex systems approach.

11 I would speculate that one social grounding on which this shift in science lies, is a change in our perceptions of which populations are affected by HIV infection. In particular, as HIV+ women and their babies become more apparent, some other model of vulnerability to disease that does not rely on loss of heterosexual virility seems called for.

12 Grosz (1994: 192ff) cogently discusses Kristeva, Douglas and Sartre on the in-between and the viscous.

13 Elizabeth Grosz (1994: 199) notes that the male body is represented as hard, solid and impermeable. Male substances, such as semen, that could bring qualities of seepage, fluidity, or formlessness to the male, are only seen as causal agents, and hence as things.

14 See Haraway (1991) and Treichler (1992) on this transition.

15 See Cone and Martin (forthcoming) for a discussion of the growing evidence

that industrial pollutants increase the allergic effect of pollen and other airborne particles.

16 Cone and Martin (forthcoming) discuss the imminent prospects for such a development.

17 This passage was read to members of a class I took on psychoneuroimmunology in a university setting, as a way of emphasizing the claims of this sub-discipline striving for credibility in research science.

18 See Martin (1994).

19 My current research concerns forms of sociality in interstitial sites such as support groups (see Martin forthcoming). Dobrzynski (1997: 1) describes how marriage is increasingly treated as a corporate partnership, especially by the wealthy. A former corporate wife contested a divorce settlement: "Gary wanted to buy out my partnership, and I didn't want to be bought out," she said, using language she learned in her role. "It's like a hostile takeover – he offered me a very small percentage, and I said that's not the price of the buyout."

20 Some form of nationalism is generally taken to be one of the central tenets of fascist movements (Laqueur 1996: 93). But Laqueur details the movement of neo-fascists today away from nationalist allegiance to any particular country and toward allegiance to "the idea of Europe" (p. 94). This extension of allegiance to political units that are not exactly nation-states, together with these movements' emphasis on social Darwinism, racialism, a new aristocracy, and obedience (p. 96) seems to point to groups formed on the conviction that their members are pure, worthy, and valuable, while others are not – the basis of which need not be the nation-state.

REFERENCES

Adcock, C. J., Perry, L. A., Lindsell, D. R., Taylor, A. M., Jones, J. and Dunger, D. B. (1994) "Menstrual irregularities are more common in adolescents with type 1 diabetes," *Diabetic Medicine* 1: 465–70.

Arai, K. and Chrousas, G. P. (1994) "Glucocorticoid resistance," *Baillière's Clinical Endocrinology and Metabolism* 8: 317–31.

Balibar, E. (1991) "Is there a 'neo-racism'?," in E. Balibar and I. Wallerstein (eds) *Race, Nation, Class: Ambiguous Identities*, London: Verso.

Billington, W. D. (1992) "The normal fetomaternal immune relationship," *Baillière's Clinical Obstetrics and Gynaecology* 6: 417–38.

Brody, J. E. (1994) "'Hair of dog' tried as cure for autoimmune disease," *The New York Times* 18, October: C1

Brooke, J. (1997) "Former cultists warn of believers now adrift," *The New York Times*, 1 April: 16.

Cone, R. A. and Martin, E. (forthcoming) "Corporeal flows: the immune system, global economies of food and new implications for health," in P. Treichler and C. Penley (eds) *The Visible Woman*, New York: New York University Press.

Coughlan, R. (1955) "Science moves in on viruses," *Life* 38: 122–36.

Denton, T. A., Diamond, G. A., Helfant, R. H., Khan, S. and Karagueuzian, H. (1990) "Fascinating rhythm: a primer on chaos theory and its application to cardiology," *American Heart Journal*: 1419–40.

Dobrzynski, J. H. (1997) "Divorce executive style, revisited," *The New York Times* 24 January: 1.

Dossey, L. (1990) "Personal health and the environment," in J. Rifkin (ed.) *The Green*

Lifestyle Handbook: 1001 Ways You Can Heal the Earth, New York: Henry Holt & Company.

Douglas, M. 1966. *Purity and Danger: An Analysis of Concepts of Pollution and Taboo*, Praeger: New York.

The Economist (1985) "Why does the body allow fetuses to live?," *The Economist* 296: 89.

Garrett, L. (1991) "AIDS virus can get into body through mucous membranes, new studies show", *The Sun*, 20 June: 12.

Giddens A. (1979) *Central Problems in Social Theory: Action, Structure and Contradiction in Social Analysis*, Berkeley: University of California Press.

Ginsburg F. and Rapp R. (eds) (1995) *Conceiving the New World Order*, Berkeley: University of California Press.

Goldberger, A. L., Rigney, D. R. and West, B. J. (1990) "Chaos and fractals in human physiology," *Scientific American* 262: 42–9.

Grosz, E. (1994) *Volatile Bodies: Toward a Corporeal Feminism*, Bloomington: Indiana University Press.

Gupta, A. (1996) Lecture at Humanities Research Institute, University of California, Irvine.

Haraway, D. (1991) "The politics of postmodern bodies: constitutions of self in immune system discourse," in *Simians, Cyborgs, and Women*, New York: Routledge.

—— (1992) "The promises of monsters: a regenerative politics for inappropriate/d others," in L. Grossberg, C. Nelson, and P. A. Treichler (eds) *Cultural Studies*, New York: Routledge.

Harvey, D. (1989) *The Condition of Postmodernity: An Enquiry into the Origins of Social Change*, Oxford: Basil Blackwell.

Harvey, D. and Haraway, D. (1995) "Nature, politics, and possibilities: a debate and discussion with David Harvey and Donna Haraway," *Environment and Planning D: Society and Space* 13: 507–27.

Irigaray, L. (1985) *This Sex Which Is Not One*, Ithaca: Cornell University Press.

Khella, A. K. (1992) "Epidemiologic study of premenstrual symptoms," *Journal of the Egyptian Public Health Association* 67: 109–18.

Kristeva, J. (1982) *Powers of Horror: An Essay on Abjection*, New York: Columbia University Press.

Laqueur, W. (1996) *Fascism: Past, Present and Future*, New York: Oxford University Press.

Lloyd, T., Schaiffer, J. M. and Demers, L. M. (1991) "Urinary hormonal concentrations and spinal bone densities of premenopausal vegetarian and nonvegetarian women," *American Journal of Clinical Nutrition* 54: 1005–10.

Macpherson, C. B. (1962) *The Political Theory of Possessive Individualism: Hobbes to Locke*, Oxford: Oxford University Press.

Marquand, R. (1997) "Mainstream culture embraces – but redefines – meaning of 'spirituality'," *The Christian Science Monitor* 31 March: 1.

Martin, E. (1987) *The Woman in the Body: A Cultural Analysis of Reproduction*, Boston: Beacon Press.

—— (1990) "The egg and the sperm: how science has constructed a romance based on stereotypical male–female roles," *Signs* 16: 485–501.

—— (1992) "The end of the body?", *American Ethnologist* 19: 120–38.

—— (1994) *Flexible Bodies: Tracking Immunity in America from the Days of Polio to the Age of AIDS*, Boston: Beacon Press.

—— (1996) "Managing Americans: policy, and changes in the meanings of work and the self," in C. Shore and S. Wright (eds) *The Anthropology of Policy*, New York: Routledge.

Mellor, A. K. (1987) "*Frankenstein*: a feminist critique of science," in G. Levine (ed.) *One Culture: Essays in Science and Literature*, Madison: University of Wisconsin Press.

Merchant, C. (1995) "Reinventing Eden: Western culture as a recovery narrative," in W. Cronon (ed.) *Uncommon Ground: Toward Reinventing Nature*, New York: W. W. Norton.

Miyauchi, F., Nanjo, K. and Otsuka, K. (1992) "Effects of night shift on plasma concentrations of melatonin, LH, FSH and prolactin, and menstrual irregularity," *Sangyo Igaku (Japanese Journal of Industrial Health)* 34: 545–50.

National Institute of Allergy and Infectious Diseases (NAID) (1995) Washington, DC: National Insititutes of Health, Public Health Service.

Neufeld, K. G. (1997) "Cultists' tragedy defies our understanding; abandoning the self," *The New York Times* 5 April: 20.

Niebuhr, G. (1997) "Lay women of faith seek joy on a journey," *The New York Times* 27 January: 10.

Nilsson, L. (1985) *The Body Victorious*, New York: Delacorte Press.

Peters, T. (1992) *Liberation Management: Necessary Disorganization for the Nanosecond Nineties*, New York: Alfred Knopf.

Saltzman, A. (1996) "How to prosper in the You, Inc. age," *US News and World Report* 12: 66–79.

Sartre, J.-P. (1981) *Existential Psychoanalysis*, Washington, DC: Regnery Gateway.

Sewell, G. and Wilkinson, B. (1992) "'Someone to watch over me': surveillance, discipline and the just-in-time labour process," *Sociology: Journal of the British Sociological Association* 26: 271–89.

Skinner, J. E., Goldberger, A., Mayer-Kress, G., and Ideker, R. (1990) "Chaos in the heart: implications for clinical cardiology," *Bio/Technology* 8: 1018–1024.

Sly, R. M. (1994) "Changing asthma mortality," *Annals Of Allergy* 73: 259–68.

Stacey, J. (1990) *Brave New Families: Stories of Democratic Upheaval in Late Twentieth-Century America*, New York: Basic Books.

Strathern, M. (1988) *The Gender of the Gift: Problems with Women and Problems with Society in Melanesia*, Berkeley: University of California Press.

Tauber, A. (1995) "Postmodernism and immune selfhood," *Science in Context* 8: 579–607.

Treichler, P. (1992) "AIDS, HIV, and the cultural construction of reality," in S. Lindenbaum and G. Herdt (eds) *The Time of AIDS: Social Analysis, Theory, and Method*, Newbury Park, CA: Sage Publications.

Turner, V. (1967) *The Forest of Symbols*, Ithaca: Cornell University Press.

Varela, F. J. and Coutinho, A. (1991) "Immunoknowledge: the immune system as a learning process of somatic individuation," in J. Brockman (ed.) *Doing Science: The Reality Club*, New York: Prentice-Hall.

Whitaker, A. H. (1992) "An epidemiological study of anorectic and bulimic symptoms in adolescent girls: implications for pediatricians," *Pediatric Annals* 21: 752–9.

4

MOVING ON FROM BOTH STATE AND CONSUMER EUGENICS?

Hilary Rose

INTRODUCTION: EUGENICS AND THE GENOME PROJECT

Thinking about the Human Genome Project (HGP), not least the vaunted possibility of the "perfect baby", seems, as we approach the millennium, to be both near and light years away from the spectre of eugenics which has haunted the twentieth century, above all in the hideous episode of Nazism. This new possibility in which intending parents get to choose eye colour, height, intelligence, looks, and so on, is part of a new consumer culture without limits. If you want it, can pay for it, and someone can provide it, then, whatever "it" is, it is yours. A revitalised economic liberalism enthrones the consumer as king – or even queen. Of course, there will be some moral discourse questioning the desirability of letting the market into parenthood, but the ethicists are themselves weakened by their subscription to the thesis of the importance of the market as the chief arbiter of our futures. Thus while tasteless, absurd, even impossible, the dream of the perfect baby takes its place alongside other consumer fantasies, of the perfect house, suit, job, garden, partner, etc. The epitome of this unrestrained consumerism is, as usual, the USA, and it is important not to dismiss or undermine the institutional structures of social solidarity still evident in European countries which, though weakened, still serve as a constraint against the marketisation of everyday life. It is this deteriorating context which makes me increasingly uncomfortable at the ease with which the risk of a new eugenicism paralleling the growth of the new genetics has been dismissed.

At one level I agree with Evelyn Fox Keller (1992: 299) that: "Of course it is true that in 1990, we have no Nazi conspiracy to fear. All we have to fear today is our own complacency that there are some 'right hands' in which to invest this responsibility – above all, the responsibility for arbitrating normality." But at another level I find myself uneasy as to whether the new eugenicism is sufficiently different from the old. Certainly there are more than traces of eugenics scattered around the HGP literature. Although it is written without the language of hatred for the *untermenschen*, there is a taken for granted eugenicism embedded within the medical technology – the increasing array of diagnostic tests – which Bertrand Jordan (1993: 168) accurately reflects: "The impact of the Genome programme on society as a whole is far from insignificant. The new knowledge thus gained leads to the elimination of embryos through prenatal diagnosis and pregnancy termination."

For that matter, although personal choice about keeping or not keeping a particular pregnancy is stressed within the literature of clinical genetics, in Britain the profession consistently makes clear to the state the economic benefits to be gained if the numbers of genetically seriously impaired infants go down. Pointing to the statistic that genetic impairment only constitutes 3 per cent of all impairment may weaken the cost-benefit effectiveness of such a claim, but what it does not do is to tackle the taken for granted eugenicism in which some "right hands" decide which of the "impaired" foetuses need to have maternal decisions made about their survival (Hubbard and Henifin 1984). Nor does testing, which claims to give reassurance, actually provide it (Green 1990) Scarcely surprisingly, then, that the movement of disabled people is deeply suspicious about the proliferation of genetic testing (Shakespeare: 1995).

Thus the question of "sufficient difference" between the old and new eugenicism nags away underneath this chapter's discussion of the social criticism of the HGP: who are the actors and what are their alliances both within the natural and the social sciences, and within the new social movements? Neither the HGP nor its critics are identical cross-nationally, so how are we to understand these variations? And do the critics seek to block science as many leading geneticists claim, or do they seek to shape science in new directions?

ORIGIN STORIES

The Human Genome Project has two distinctly whiggish and linked histories as told by scientists and science writers: one an internalist account which begins with a number of precursor research findings, culminating in 1953 with Crick and Watson's helical model of DNA which is seen as the crucial launch pad for the new genetics; the second an externalist history which, in the USA, locates the launching of the immense international effort to map and sequence the Human Genome at a meeting of most of the significant figures in the field convened by Robert Sinsheimer in Santa Cruz in May 1985. By contrast with this carefully orchestrated political/technical bid, which energetically corresponds to the actor network analyses of social science accounts (Callon 1987), the UK version celebrates the casual word in the right ear beloved of British elite political storytelling (Bodmer and McKie 1994). But the point, however the story is told by leading actors (and one of the interesting things about science is how frequently its practitioners not only want to construct the knowledge but also the history of the knowledge), is that the HGP was from its inception a highly self-conscious bringing together of different scientific, industrial, and political interests, differing somewhat country by country and reflecting different configurations of interest. Thus, while the French complained about the lack of industrial interest in making laboratory equipment, they rejoiced in national technophilia and so experienced very little opposition (Jordan 1993). By contrast the key US actors, having already experienced a highly public social responsibility debate concerning genetic manipulation, were well aware of potential con-troversy and planned for 5 per cent of the budget to be dedicated to the Ethical Legal and Social Implications (ELSI) of the HGP. With more confidence than the British were to display, the USA placed a philosopher (Eric Juengst at the National Institutes of Health [NIH]) and a jurist (Michael Yesley at the Department of Energy, which was to play a key role) in joint charge.

Global networks also had to be constructed, moving from the standard form of open international scientific unions which agree standards and measures to the establishment of a new international scientific organisation, the Human Genome Organisation, HUGO, which actually selects its members. Such networks had to be made sufficiently stable both to constitute and reflect stable knowledge and technological knowhow. In the early years the self-appointed HUGO played a crucial part in this networking and stabilisation process, not least through the work of its early presidents. The first, US geneticist Victor McKusick, wrote the influential report on the prospects of the new genetics for the US National Academy of Science. The second, UK molecular biologist Sydney Brenner, whose brainchild HUGO was, played a similar role for the Medical Research Council (MRC) and the Royal Society in the UK. Through HUGO this small but immensely influential network of scientists sought to shape the research programme so that mapping and sequencing were to become both the national and international objectives. Initially HUGO seemed stronger on agenda making than delivery, not least on work on the ethical and social dimensions of the project, and the European Science Foundation (1991) wrote a critical report. Arguably it is now more securely founded and certainly has rather less opaque admission procedures.

While modern science makes global claims, the formation of these Human Genome research networks have primarily taken place in the advanced capitalist societies. Despite a track record in molecular biology in the countries of the former Soviet Union, particularly Russia, the research system is in such crisis that currently the lead players are the USA, Britain, France and Japan in that order (according to an analysis carried out by Academia Europea), with the latter three following some way behind the former. Despite Germany's growing scientific strength, it does not play a major part in Genome research, a consequence of the response to the active role played in the Nazi period by biologists who then continued as the directors of leading laboratories in the postwar period, fostering a revulsion against the new genetics. This revulsion was fed by new critical histories of genetics and biomedicine (Muller Hill 1988; Weindling 1989), effectively mobilised by the Greens and taken by them both into Länder and national debates, and internationally into discussions within the European Parliament where they effectively challenged the original proposed name of the European programme, Predictive Medicine. After much bargaining the Parliament gave its support to the Genome programme, providing that 7 per cent of the budget was allocated to ethical, legal and social aspects.

Discussion of the Human Genome programme has been muted in the UK. Even the cloning of Dolly the sheep in Scotland did not raise the kind of fundamentalist religious debate which this technical achievement raised in the USA. Indeed attempts to produce debates between scientists about a number of issues (e.g. John Durant's "Science Frictions" media series footnote) have been conspicuous for their lack of liveliness, with scientists cosily agreeing with one another. Science journalism echoes this cosiness and rarely takes an independent critical stance. In Britain only the joint Wellcome/CIBA meeting in the spring of 1995 on the Genetics of Anti-Social Behaviour, with its perceived kinship to the 1992 US meeting on the Genetics of Violence which had been blocked as racist, has precipitated a public exchange between scientists about the ethics and politics of such research (Hubbard 1995). Unusually critical commentaries were run in most of the broadsheets, notably *The Independent*, where the conference was hostilely reviewed by ethologist Patrick Bateson, a review which in turn attracted a dismissive rejoinder by psychologist Michael Rutter who had chaired the CIBA/Wellcome proceedings.

By contrast, very strong exchanges over Genome have taken place continuously within the USA since the inception of HGP, with the *Science* editorial serving as a much quoted voice of biological determinism, in which the editor Daniel Koshland (1989:189) claimed that the HGP would solve everything from schizophrenia and ageing to homelessness. The debate among US scientists was both public and acknowledged. In a subsequent issue of *Science*, Nobel Prize Winner Salvador Luria robustly challenged the sociotechnical agenda of the HGP, arguing that the Project has been promoted by a "self-serving coterie" and asking:

> Whether the Nazi programme to eradicate Jewish and otherwise "inferior" genes by mass murder be translated here into a gentler kinder programme to "perfect" human individuals by "correcting" their genes in conformity to an ideal "white Judeo-Christian economically successful" genotype?
>
> (Luria 1989: 873)

While it is probably necessary to be a Nobel Prize Winner to get such views published in a major scientific journal, even when there is a recognised controversy, it is the case that feminist, liberal and left critics of science in the USA are ranged against the HGP and have produced powerful critiques. These have gained sufficient cultural purchase that when the historian of eugenics, Daniel

Kevles, and Leroy Hood, a leading figure within the Human Genome Project, jointly edited *The Code of Codes* (1992), it was felt appropriate to invite Evelyn Fox Keller, a distinguished feminist critic of science, to contribute. But neither the claims being made for the HGP nor by its critics have remained constant. Few, following Luria's fundamentalist denunciation of Koshland's fantasy claims, have either advocated or attacked the HGP in quite these unequivocal terms. But while Keller is surely right that history does not repeat itself, it was surely an error to use the term "Nazi conspiracy", as this serves to suggest that a minority of bad people (the Nazis) were involved and that they rather successfully hoodwinked the good (the majority). Yet the immense difficulty of confronting the Nazi period and eugenicism is that Nazi regimes with openly eugenicist programmes were popularly elected in both Germany and Austria, with only the communists, socialists and a handful of Christians – apart from the militant among their proposed victims – politically engaged against them.

EUGENIC ENTHUSIASMS

Further, while the Nazi episode stands as the historical embodiment of state eugenics in all its horror, it is important to recall a parallel enthusiasm for eugenics in most industrial countries – socialist well as capitalist – which was supported by socially progressive scientists and social thinkers as well as racists and reactionaries (Pickens 1968; Weeks 1981; Kevles 1985). Nations were, however, rather different in their levels of implementing their eugenicist enthusiasms. The USA sterilised some 20 thousand "feeble minded" women, while many of the Scandinavian countries had race laws disturbingly similar to those of Nazi Germany. A number of British social policy thinkers and biologists (for instance, William Beveridge, Julian Huxley and J. B. S. Haldane) were very attracted to ideas of improving the race. "Eutelegenis", the progressivists' version of eugenics, Haldane saw as synonymous with socialism (Kevles 1985). That Britain got little farther than a policy of custodial care and sexual segregation for mentally impaired women and men may be a source of some relief. However, the point is that eugenicist ideas were actively discussed and widely endorsed by the intelligentsia in most industrial societies. It was only the advent of the Nazi mass extermination of the mentally impaired and sick in the hospitals, and the mass extermination of the death camps, which silenced such naïve enthusiasms for racial improvement. Such eugenicist histories and enthusiasms are, if not actually hidden, at least distinctly underplayed in national cultural self-accounting, leaving the Nazi episode to stand out as a singular horror story rather than as the monstrous epitome of a widespread current. Consequently the current contrast with "state eugenics" is almost always set against the Nazi horror rather than this much more pervasive state eugenicism which lies uneasily, only half silenced, within the culture. As the repressed rather than the confronted, such histories threaten to return.

Even in 1968, that year of revolutionary hope, eugenicist enthusiasms were openly expressed:

> There should be tattooed on the forehead of every young person a symbol showing possession of the sickle cell gene or whatever other similar gene. . . . It is my opinion that legislation along this line, compulsory testing for defective gene before marriage, and some form of semi-public display of this possession, should be adopted.
>
> (cited in Kay 1993: 276)

The painful thing is that the author of this appalling statement was not some bizarre figure from the racist right, but a hero of the anti-war and alternative health movements, none other than Linus Pauling. Taking Vitamin C in mega doses, whatever else it achieves, clearly offers no protection against disability hating thought.

Nor was eugenicism to be left solely at the level of advocacy. In the early 1970s, ironically at the height of the War on Poverty which seemed to promise a fresh start in social policy, small numbers of "feeble minded" young black women were still being sterilised (Rose and Hanmer 1976). Subsequently President Nixon's drive to screen for sickle cell anaemia, a genetic disease common among African Americans, led to extensive stigmatisation and employment discrimination. Rayna Rapp in her anthropological study of genetic counselling noted that, during the late 1980s, 95 per cent of genetic counsellors were white (1988: 145). African American and sociologist Troy Duster shares Rapp's concerns and is less confident that eugenicism is part of an unreturnable past. His study of the workings of genetics counselling in the lives of black and white Americans is not by chance entitled *Backdoor to Eugenics* (1990). For African Americans it is not only a "Nazi conspiracy" which gives cause for concern, for US liberal democracy has its own negative record. As Duster (1990: 92) observes: 'Once again, whether this new genetic knowledge is an advantage or a cross depends only partly on the how the genes are arranged. It depends as well on where one is located in the social order.'

Yet even for the apparently socially secure the new genetics poses a historically new problem. As Dorothy Nelkin and Lawrence Tancredi document in their book *Dangerous Diagnostics* (1994), the new genetics has proliferated tests but few therapies. Indeed the diagnostic technology is integral to the development of the field, both in the sense of the technics of production of the knowledge and in a more directly commercial sense. While the HGP was launched to the sound of promises of gene therapy and thus secured substantial investment from venture capital, the new diagnostics, even without therapies, offers to provide those promised returns. Insofar as tests are offered to pregnant women the only "medical treatment" that can be proposed is abortion for foetuses deemed to be "imperfect".

The response from ethically sensitive clinical geneticists and from groups such as the King's Fund Consensus Forum (1987) on prenatal screening is to support the concept of choice, privileging the right of the pregnant woman to make the choice, supported by the best quality information and counselling. But this assumes that a consensus supporting a woman's choice is ultimately available, and that "choice" itself is unproblematic. In the USA this consensus is conspicuously not present, not least because of fundamentalist Christianity's success in introducing the concept of personhood for the foetus – a success which has affected abortion rights and more. While feminism was claiming agency for women, this powerful right wing movement was claiming agency for what in another discourse was described as a bunch of cells. The cultural efficacy of claiming agency for the foetus cannot be ignored, even though its implications in law do not necessarily go along with the anti-abortion intentions of the protagonists. US lawyers have demonstrated their willingness to pursue cases of "wrongful life" against doctors failing to provide tests for pregnant women, or against women failing to act on them.

CRITICISING CONSUMER EUGENICISM

Particularly but not only in the US context, the problem, as numbers of feminists have pointed out, is that the concept of "choice" and indeed the entire discourse of "reproductive rights" – which through the 1970s so richly supported the feminist project – has come into difficulties (Petchesky 1985). Two things have been important in this change. First, reproductive science and technology, as relatively low status areas in the past, have undergone tremendous growth and are no longer meaningfully separated as two distinct areas; instead they fuse as a powerful "technoscience" which pervades the lives of women. Second, culture has over the last fifteen years taken a powerfully consumerist twist, at its most intense in the unregulated medical market economy of the USA.

Following this line of analysis, geneticist Paul Billings, one of the critics associated with the US Council for Responsible Genetics, argues that the new danger stems precisely from the contemporary consumerist culture, so that "choice" becomes a highly problematic concept (Billings 1994). In this culture (and I would add for the socially secure, which given the Clinton failure to secure adequate nationwide health care, still excludes 30–40 million Americans) the new consumer eugenics replaces the old state eugenics. Thus different sections of US society, and particularly pregnant women in those different sections, are likely to find themselves challenged by very different kinds of eugenicism.

The Council, in part a rather more respectable offspring of the radical grouping Science for the People, includes a number of distinguished biologists who are both socially and scientifically concerned by recent developments in genetics. Over the years leading figures such as Jonathan Beckwith, Ruth Hubbard, Sheldon Krimsky and Richard Lewontin have successfully punctuated the self-reported success story of the new genetics, whether in its applications to green nature through biotechnology or to human nature. Among the more conspicuous recent examples of such interventions was Lewontin's challenge to the near absolutist truth claims being made for DNA "fingerprinting". The upshot of this intervention was that the claims made by experts advising law enforcement agencies were significantly modified (Lewontin 1991). Usually the intervention is made through a critical public understanding of science approach. Hubbard's contribution as both a feminist and biologist and her involvement with the women's health movement is a particularly good example. In a string of publications she has set out both to demythologise the new genetics and to alert women to the imperialising claims of the new diagnostics (Hubbard and Henibi 1984; Hubbard and Wald 1993; Hubbard 1995). Despite the ambivalences which seem to exist between most new social movements and scientists, the challenge posed by the new genetics has led to a cautious regard for the scientists associated with the Council.

But these robust socially critical voices are not part of the professional world of Science Technology and Society (STS) studies. Instead they are outsiders, their legitimacy lying primarily within their background as molecular or theoretical biologists or biochemists that provides the cultural capital which amplifies their socio-technical criticism of the HGP. While academic STS studies in the USA do not have a strong normative agenda, individual STS scholars such as Dorothy Nelkin, Lawrence Tancredi and Troy Duster have produced highly critical work which has nourished public debate. But the one consistent academic source of self-consciously normative criticism has been Women's Studies, with its peculiar location between the academy and the movement. Here increasing numbers of feminists engaged in STS studies have produced powerful critiques of aspects of the new genetics and include: Anne Fausto Sterling, Evelyn Fox Keller, Donna Haraway, Sandra Harding and Rayna Rapp.

US critics of the Genome Project found little support in the institutionalised critical evaluation of science and technology policy. The US government Office of Technology Assessment (OTA) which many looked to with hope, proved, when the Genome Project was proposed, to turn itself into a facilitating agency (see Robert Cook-Deegan's description of his role at the OTA 1993). None the less despite this handmaiden role to the HGP the counter-revolution of the Republicans has destroyed the OTA.

BRITISH CRITICISM OF THE NEW GENETICS

Voluntary associations in the UK equivalent to the Council for Responsible Genetics have developed more weakly. Although the USA is at the centre of the production system of modern science, UK molecular biological and genetic traditions are strong and in principle there would seem to be a sufficiently self-confident cultural elite to ensure critical debate. But this does not seem to be the case. The Genetics Forum, while energetic within a small budget and a handful of activists and which publishes a monthly *The Splice of Life*, does not command such a distinguished line up of biologists, not least human geneticists. Like the majority of green British groups, but unlike the German Greens, the Forum mainly focuses on biotechnology; human genetics comes in a rather poor second.

Nor are there many individual voices from within the scientific community raising strong socio-technical criticism against the HGP. Social criticism of science from the 1970s onwards has passed over to the muted tones of professional STS studies. To some extent numbers of the current luminaries of STS were natural scientists mobilised by the 1960s and 1970s radical science movement and then were professionalised. The normativity of their earlier radicalism has been transformed into the intellectually radical project of the sociology of scientific knowledge, a project which, in a contradictory way, both intellectually nourishes and politically weakens more openly normative strategies. There are now few survivors of a past socially engaged tradition of radical scientists which extended, with a break for the war and Cold War, from the 1930s to the 1970s (Werskey 1971; Rose and Rose 1976). It is true that neurobiologist Steven Rose (1995) has criticised the US move into neurogenetics and that ethologist Patrick Bateson was openly hostile to the CIBA/Wellcome meeting, but such public displays of unequivocal scientific dissent are increasingly rare in Britain.

Yet the current vigour of the US debates and the UK mild disagreements are in marked contrast to the genetics debate of the 1930s, when British geneticists were at the centre. The shift within Britain during the war was intense, nowhere more marked than in the anti-eugenicist politics of Lionel Penrose and his colleagues at University College. That he persuaded his colleagues to retitle his Chair in Eugenics to Human Genetics symbolised the marginalisation of the Galton/Pearson eugenic tradition.

Today Steve Jones, also of University College, is the most eloquent and widely read British geneticist who positions himself on the left. None the less his Reith lecture book *The Language of the Genes* (1994) shows little awareness of the dangers of a situation in which the new genetics can diagnose but not treat. He seems to believe that biomedical research is automatically in the best interest of the patients, and is clearly unaware of a substantial research literature which might challenge this. His second book and accompanying television programme *In the Blood* (1996) displays a conspicuous social insensitivity. He shows no critical reflection on the rapid geneticisation of culture in which human diseases, disorders and (mis)behaviours are explained in terms of genes. The 1995 MRC exhibition on Euston Station "Genes Are Us" said it all. They, and by implication we, exist outside the social order.

The one unambiguous oppositional voice throughout the 1980s was that of the Feminist International Network against New Reproductive Technologies and Genetic Engineering (Finnrage). Finnrage has had considerable success in raising internationally the issue of assisted reproduction and genetic engineering, not least in Europe in alliance with the German Greens (Yoxen 1983; Zipper and Sevenhuijsen 1987). Currently it is less an audible voice. Finnrage's problems have lain in its total opposition to both reproductive technology and genetic engineering, at the same moment that many women, including numbers of feminists, were turning to the new technologies for help in overcoming problems of infertility or in their concern about grievous inherited conditions. Thus Finnrage was to become one of those organisations that recruit powerfully because of the radicalism of their critique, but which because of the incompatibility of women's lived lives and the critique was to become a staging post. Numbers of feminists joined, but moved on, on their way to still critical, but rather more complex political positions frequently entering feminist STS studies. Here, because of the successful institutionalisation of Women's Studies as a critical approach to the knowledge, academic feminists have been able to maintain a normative agenda which has been largely lost by mainstream STS (Stacey 1992; Rose 1994).

THE PECULIARITY OF BRITISH SCIENCE FRICTIONS?

But if HGP is not central to national science frictions, what is? Unquestionably the big British debate about nature is the issue of animal rights and welfare, usually versus the experimental biologists, though until the emergence of BSE (mad cow disease) the livestock exporters were demonized over and above them. Defending animals against cruelty is not new but has bubbled through the national culture, emerging from time to time, not least since the birth of animal experimentation within physiology in the 1870s. Unquestionably this is now a strong cultural current, reflected, for example, in the advertised commitment of the Co-operative Bank not to invest either in repressive regimes – or, like the Body Shop, in animal experimentation for cosmetics. Increasingly it is assumed that no sensitive caring person can be anything other than into animal rights. For feminism the history of opposition to vivisectionism has been more constant, though by no means commanding a consensus among either past or present feminists. But where in the nineteenth century violence against women was commonly linked to cruelty to animals, today's struggles are culturally and politically distinct.

Despite this popularity, for some the animal rights movement has uncomfortable links with Nazism. The animal welfare legislation still current in Germany was initially introduced by the Nazis, and a concern with nature an important strand of Nazi ideology. Even today the November the Ninth Group (the Nazi Crystal Night) has been active in Britain within animal rights campaigns, although not without protest from democrats within the movement. Contemporary deep ecology's near exclusion of human animals from the concept of nature and the attack on "speciesism" raise similar difficulties for those all too aware of the history of National Socialism's nature politics.

If the epicentre of today's UK nature politics is animal rights with green nature in second place, it becomes difficult for the British debate to share the energy which comes off the US scene. Instead, the danger is that British and perhaps European social science becomes acceptance science devoted to ironing out the Ethical Legal and Social Aspects (ELSA) problems of genetics as an innovatory technoscience. Should we be more cautious about and therefore deconstruct the concepts of "acceptability" and "demand" which scatter both the ELSI and the ELSA literature, not least the research programmes within which social science has to couch its requests?

MANAGING UK CRITICISM

In the UK we have no OTA, with a clear (if on, the HGP, disregarded) brief, although parliamentarians in the House of Commons have the Parliamentary Office of Science and Technology (POST), and the EU Parliament has Science and Technology Assessment (STOA). While both POST and STOA offer scientific support, how far this is informational and how far it assists with critical technology assessment is unresolved. The belief that what Parliamentarians need are the scientific facts, and that these will somehow speak for themselves, dies hard.

Within Britain the Medical Research Council, like the NIH in America, is managing the British ELSA programme for the Human Genome Project and also acts as the link for the European Framework programme. But there are crucial differences, for in Britain not only is the committee considering applications for ELSA funds chaired by a biologist, thus not reproducing the disciplinary independence of the US ELSI, but this is the manifestly partisan figure of the embryologist Lewis Wolpert, who has continually engaged in public polemic against the sociology of science since the British Association meeting of 1994.

It is important to say something about Wolpert's views because of his key location within both ELSA and the Royal Society's Committee for the Public Understanding of Science (COPUS). He homogenises the diverse epistemological positions among the sociologists of science as if they were all relativists who inevitably fostered anti-science. That his reading misses most of the differences between sociologists, and of course ignores feminist STS work where some (Birke, Cockburn, Harding, Haraway, Hubbard, Keller, Rose) cling to a constrained version of objectivity in a way he should find quite endearingly conservative, is beside the point. But what is peculiar to the British situation is that someone with no expertise in ethics, law or the social sciences, who is on public record as hostile to the last, was felt to be appropriate to chair the British committee to manage the ELSA programme.

Wolpert took over the chair of COPUS from the geneticist and erstwhile president of HUGO, Sir Walter Bodmer. Both hold the view that if the experts tell the public about the facts of science, then science will move back into public esteem, and the public will be able to make better informed decisions about science. However, "understanding" in this model equals "knowledge of scientific facts". Despite the Economic and Social Research Council (ESRC) programme on the public understanding of science which showed with both qualitative and quantitative evidence that with increased scientific knowledge came increased willingness to evaluate and criticise, this ideological commitment to the equation between information and trust has remained unchanged on the part of these influential actors.

For very few social scientists is the concept of "understanding" interchangeable with factual knowledge; in a non-positivist epistemology understanding is profoundly linked to meaning and to history. Indeed the ESRC qualitative studies of understanding science pointed to the ways that different publics made sense of scientific knowledge alongside other equally important knowledges of everyday life (Irwin and Wynne 1996). At a time when the wider culture increasingly recognises the limits of science and that science itself is socially shaped, this defence of science as being independent from society feels like a nostalgic throwback to a time when it was able to claim much greater authority.

The dirigistic British climate of eighteen years of right wing conservativism provided a supportive context for this move from within science to contain and control social criticism of science. There is a still a widely held belief in the necessary subordination of the social to the natural under the banner of positivism. The MRC's longstanding commitment to quantitative research and its inclination towards preferably quantified social psychology over and above sociology is a continuing expression of this handmaiden view of the social sciences. This is not to suggest that the MRC supported research in this area is socially uncritical, although I have considerable reservations about the MRC's recruitment of the American Robert Plomin and his attempt to reinfuse the old biologistic claims of psychometry with ever fancier statistical manipulations. It is gloomily fitting that he shares the old Institute of Psychiatry's address of Hans Eysenck to re-establish the apparatus of twin registers, etc. through which he claims to be able numerically to allocate nature and nurture. It is interesting that Plomin's programme, which was supposed to cost the MRC £1.8 million, has caused such internal criticism within biomedical research circles that the MRC has felt it necessary to consult more widely.

Despite this, we do need the detailed empirical accounts, which the MRC also supports, of how diagnostic tests work out for different groups, not only for people at different points in their lives, such as having a baby, trying to get a job or take out insurance, but also for very different genetic conditions not least whether immediate or late onset. The question is whether and how far such micro social research is able to influence in a positive way the already in place actor network of the molecular biologists, the geneticists and industry. Such a network presently finds expression and support in such measures as the government Technology Foresight exercise devoted to technological innovation and wealth creation, and which pays rather little attention to the relevance of research to the quality of social life.

MOVING ON

The British response to the HGP is part of a deep-rooted cultural difficulty in coping with democratic debate about matters which involve high levels of technical expertise. Instead of a willingness to be open with the public and accept that specific technological developments may be rejected, such for example as is the case in Denmark, there is a peculiar British desire to manage the outcomes. This means controlling who, and holding which values, is to be permitted to enter the actor networks. Such an institutionalised anxiety to control was demonstrated in the (for Britain) pioneering consensus forum on biotechnology, held at Imperial College in and funded by BBSRC. Participants were, however, informed that they were not permitted to write a minority report. By contrast, in a similar technology assessment exercise in Germany which deliberately recruited a wider range of people and positions, the Greens walked out. This was not seen as a failure but as a fair expression within the task of technology assessment (van den Daele 1994).

It is precisely the outsider groups, in Britain particularly the disability movement which currently offers an energetic opposition to the new genetics, which need to be brought into debates. The movement's challenge to the twin assumption that all would-be parents want a "normal" baby, and that bio-medicine is able to define this "normality", disrupts the quiet and naturalised eugenicism of the dominant culture of the very late twentieth century. To say this is not to subscribe to everything proposed by the disability movement, which at times has come dangerously close to an anti-abortion politics, nor is it necessary to ignore that there is popular support for the new genetics from many families with children suffering severe genetic diseases, as offering the most likely means of finding new and effective therapies. Nonetheless foregrounding this quiet and naturalised eugenicism, provides a crucial additional location from which the HGP requires critical assessment.

It is time, though difficult for the reasons I have tried to set out, to put the gains from the social studies of science and technology, both mainstream and feminist, together with the urgent social concerns of anti-eugenicists, to develop the new institutions which could help make the HGP socially accountable. Back in the 1960s, the radical demand was for a Science for the People. The rhetoric was fine but we had no very clear idea of how the science got to be quite so anti-social in the first place, thus the goal to reinfuse science with new values was unrealisable. Today, as we approach the millennium, there is less political energy directed towards the old goal of radicalising science, but a better understanding of how science and technology might be reshaped in socially responsible ways. STS scholars have spent two, almost three, decades empirically mapping the social practices which have shaped both scientific knowledges and technological artefacts. Gradually the very idea of science and technology as socially shaped has won through, and has received cultural recognition well beyond the academic boundaries of STS. It is this recognition which has opened the door to a normative project of "reshaping", to the possibility of different sciences and technologies.

To realise these alternative possibilities, to move away from both state and consumer eugenicism, requires new institutions. The task of such new institutions, of consensus fora, citizens' juries and the like, is technological assessment of the products of the Human Genome Programme. Technological assessment by new institutions offers the possibility of restoring social trust in the technological innovation which is integral to a highly scientific and technological culture. The alternative is to let the defeatist nostalgia for a past which never existed expand, so that fashionable commentators like Fay Weldon or Brian Appleyard demogogically denounce science, ignoring the highly technical media which make their communications possible. Retro romanticism or unthinking technophilia is no choice.

Creating new institutions does not of itself offer any guarantees of easy consensus. What it does is to bring the democratic possibility of conversations between many differently situated people and groups, not least between potential end users and providers of the new technosciences. There is some modest evidence that industry is more open to such ideas than the old secretive traditions of state scientific paternalism. Industry and the Ministry of Agriculture (MAFF) were the joint funders of Dolly, but where industry was willing and prepared to talk about the ethical issues, MAFF was (as usual) silent. Overcoming this "Daddy knows best" tradition is a huge problem for British science cultural struggles. Consumer resistance to genetic tests, notably in the case of Cystic Fibrosis, is compelling both UK and US industry to recognise that their market model of the "consumer" does not quite fit this world of women and their partners deciding to have a baby. Such modest auguries, which serve to limit and constrain the much bruited construction of a consumer eugenic society, are precious to both providers and potential users. Moving on, out of the old state eugenicism and out of the new consumer eugenicism, requires the patient building of social trust through a democratic inclusiveness.

REFERENCES

Billings, P. (1994) "Contribution to an International Seminar on Technology Assessment of Neurogenetics", January, Hamburg University.

Bodmer, W. and McKie, R. (1994) *The Book of Man*, New York: Little Brown.

Callon, M. (1987) "Society in the making: the study of technology as a tool for sociological analysis", in W. Bijker, T. Hughes and T. Pinch (eds) *The Social Construction of Technological Systems*, Cambridge, MA: MIT Press.

Deegan, R. C. (1994) *The Gene Wars*, 4th edn, New York: W. W. Norton.

Duster, T. (1990) *Back Door to Eugenics*, New York: Routledge.

European Science Foundation (1991) *Report on Genome Research*, Strasbourg: European Science Foundation.

Green, J. (1990) *Calming or Harming? A Critical Overview of Psychological Effects of Foetal Diagnosis on Pregnant Women*, vol. 2, London: Galton Institute Occasional Papers.

Hubbard, R. (1995) *Profitable Promises: Essays on Women, Science and Health*, Monroe, ME: Common Courage.

Hubbard, R. and Henifin, R. (1984) "Prenatal diagnosis and eugenic ideology", *Women's Studies International Forum* 8: 567–76.

Hubbard, R. and Wald, E. (1993) *Exploding the Gene Myth*, Boston: Beacon.

Irwin, A. and Wynne, B. (eds) (1996) *Mis-Understanding Science? The Public Construction of Science and Technology*, Cambridge: Cambridge University Press.

Jones, S. (1994) *The Language of the Genes*, London: HarperCollins.

—— (1996) *In the Blood: God, Genes and Destiny*, London: HarperCollins.

Jordan, B. (1993) *Travelling Around the Human Genome*, Paris: Inserm.

Kay, L. (1993) *The Molecular Vision of Life: Caltech, the Rockefeller Foundation and the Rise of the New Biology*, New York: Oxford University Press.

Keller, E. F. (1992) "Nature, nurture and the human genome project", in D. Kevles and L. Hood (eds) *The Code of Codes: Scientific and Social Issues in the Human Genome Project*, Cambridge, MA: Harvard University Press.

Kevles, D. (1985) *In the Name of Eugenics: Genetics and the Uses of Human Heredity*, New York: Alfred Knopf.

Kevles, D. and Hood, L. (eds) (1992) *The Code of Codes: Scientific and Social Issues in the Human Genome Project*, Cambridge, MA: Harvard University Press.

King's Fund (1987) *Screening for Foetal and Genetic Abnormality: Consensus Statement*, London: King's Fund Centre.

Koshland, D. (1989) Editorial, *Science* 13 October, 246: 189.

Lewontin, R. C. (1991) *Biology as Ideology: The Doctrine of DNA*, New York: Harper.

Luria, S. (1989) Letter, *Science* 17 November, 246: 873.

Muller Hill, B. (1988) *Murderous Science: Elimination by Scientific Selection Jews, Gypsies and Others, Germany 1933–45*, Oxford: Clarendon.

Nelkin, D. and Tancredi, L. (1994) *Dangerous Diagnostics*, New York: Basic Books.

Petchesky, R. (1985) *Abortion and Woman's Choice: The State, Sexuality and Reproductive Freedom*, Boston: Northeastern University Press.

Pickens, D. K. (1968) *Eugenics and the Progressives*, Nashville: Vanderbilt University Press.

Rapp, R. (1988) "Chromosomes and communications: the discourse of genetic counsellors", *Medical Anthropology Quarterly* 2: 143–57.

Rose, H. (1994) *Love, Power and Knowledge: Towards a Feminist Transformation of the Sciences*, Cambridge: Polity Press.

Rose, H. and Hanmer, J. (1976) "Women's liberation, reproduction and the technological fix" in D. L. Barker and S. Allen (eds) *Sexual Divisions and Society*, London: Tavistock.

Rose, H. and Rose, S. (eds) (1976) *The Radicalisation of Science*, London: Macmillan.

Rose, S. (1995) *Nature* 353, February: 280–1.

Shakespeare, T. (1995) "Eugenics by the backdoor? The disability movement's concerns with the new genetics", paper given at the Edinburgh International Science Festival, April.

Stacey, M. (ed.) (1992) *Changing Human Reproduction: Social Science Perspectives*, London: Sage.

Van den Daele, W. (1994) *Technology Assessment as a Political Experiment*, Berlin: Wissenschaft Zentrum für Social Forschung (WZB).

Weeks, J. (1981) *Sex, Politics and Society*, London: Tavistock.

Weindling, P. (1989) *Health, Race and German Politics between National Unification and Nazism: 1870–1945*, Cambridge: Cambridge University Press.

Werskey, G. (1971) *The Invisible College*, Harmondsworth: Penguin.

Yoxen, E. (1983) *The Gene Business: Who Should Control Biotechnology?* London: Pan Books.

Zipper, J. and Sevenhuijsen, S. (1987) "Surrogacy: feminist notions of motherhood reconsidered", in M. Stanworth (ed.) *Reproductive Technologies: Gender, Motherhood, and Medicine*, Cambridge, Polity Press.

5

REASSERTING NATURE

Constructing urban environments after Fordism

Roger Keil and John Graham

INTRODUCTION

This [urban] space is the seat of a specific contradiction. The city expands extensively; it explodes. When urbanization of society takes place, and when the city consequently absorbs the countryside, there is simultaneously a ruralization of the city. The urban expansions (suburbs, near or distant peripheries) are subject to landed property and to its consequences: ground rent; speculation; the natural or provoked scarcity of land.

(Lefebvre 1975: 115)

During the final days of the summer of 1996, residents of the eastern inner suburbs of metropolitan Toronto encountered just a bit more "nature" than they had bargained for. Two workers from the Metropolitan Zoo – located in a wide expanse in Scarborough's Rouge Valley – and an off-duty police officer confirmed what people in the area had been reporting for over a year: a 45kg cougar was prowling the edges of Canada's largest city. The animal – which had not escaped from the nearby zoo – was about 2,500km away from its "natural" habitat in the western mountains of North America. The thing to note, though, was not the cougar's distance from its nature, but the beast's proximity to the front lawns of suburban humanity. The unexpected intrusion of wildness added a chapter to the story of Toronto as "zoöpolis": spaces inhabited commonly by animals as well as people.[1]

Such returns of nature to the city are puzzling events. Historically, urbanization has proceeded as a progressive distancing from nature through the production of second nature (Lefebvre 1976; Gregory 1994). Indeed, cities have been sites of assumed (in the double sense of the word) control over ecological processes, and, as we sketch below, this distancing and control peaked in the

100

Fordist period of urbanization, to which Toronto, like other North American cities, owes most of its growth. In such a context, cougars appear resolutely "out of place." But for our purposes, the wild cat's presence usefully draws attention to something larger: the emergence of new rhetorical and material modes of "producing" urban natures after Fordism. This chapter argues that in contrast to past practices of building cities against nature, today, nature – although rarely in the form of a roaming cougar – is being reinserted into the urbanization process in new, although often contradictory, ways. We will attempt to isolate and re-connect some of the major aspects of cities' natures at the end of the millennium using current trends and events in the Toronto urban region.

Changes in the production of urban natures must be evaluated in relation to larger urban restructuring processes. Toronto, like many other western, industrial cities, has been under the spell of the secular crisis of Fordism and rampant globalization. Most accounts of the Fordist and post-Fordist city have focused on the social and political crisis engendered by these processes, and the institutional responses and spatial strategies taken by urban political regimes (Goodwin *et al.* 1993; Amin 1994). But Fordism and globalization have affected natural and built environments as well as social relations and political processes. The reinsertion of nature into urbanization, then, is specifically related to the crisis of societal relations with nature engendered by the social and spatial dynamics of Fordism and globalization.

The argument we make is that "nature" has become central to how urban spatial regulation and urban political regimes are reconstituted. Today, urban growth and development occurs increasingly through and not against nature. This does not mean that the Fordist city was "unnatural" or that cities after Fordism are more "ecologically sustainable." Rather, it signals an important transformation: in European and North American cities at the close of the twentieth century, forms of urban growth are being articulated in new ways with discourses on environment, nature and sustainability. We argue that these discourses underwrite processes of eco-modernization (Hajer 1995; Kipfer *et al.* 1996) which become expressed materially in the (re)organization and regulation of urban space.[2]

This chapter is divided into two halves. In the first we explore the general and historical conditions which we believe have governed societal relations with nature in the urbanization process. In the second we examine recent articulations of nature and urbanism through a triptych of examples drawn from Toronto in the 1990s: urban planning reform processes in the Province of Ontario; exurban developments on the city's northern edge; and, the marketing of residential spaces in the era of reflexive urbanization in Southern Ontario.

THE "NATURE" OF URBANIZATION

Environmental and social theorists have found it notoriously difficult to come to terms with the relationship of the city to nature (Williams 1973; Harvey 1993).

Cities are built in nature, with nature, through nature, yet so often appear to be external and opposed to nature. "Nature" has become viewed as that which is outside cities, yet it is everywhere in cities (as parks, electricity or even automobiles). Nature is continuously remade in cities at the same time as it is taken to be destroyed by cities. On the one hand, then, we follow David Harvey, who has argued:

> In a fundamental sense, there is in the final analysis nothing *unnatural* about New York City and sustaining such an ecosystem even in transition entails an inevitable compromise with the forms of social organization and social relations that produced it. . . . It is fundamentally mistaken, therefore, to speak of the impact of society on the ecosystem as if these are two separate systems in interaction with each other. The typical manner of depicting the world in terms of a box called "society" in interaction with a box "labeled" environment not only makes little intuitive sense . . . but it also has just as little fundamental theoretical and historical justification.
>
> (Harvey 1993: 28)

In short, cities are part of nature (they are the site of complex, socially organized relationships between "social" and "natural" processes), but it is precisely its ecologies that are often most difficult to see (since urbanization distances people both spatially and perceptually from the larger bio-physical processes in which cities occur). Thus, people in cities are natural beings at the same time as they are social beings. Their reproduction is achieved through a combination of biological and societal processes – through a complex urban metabolism. As Harvey notes, seen this way, cities are not "anti-ecological": "we must recognize that the distinction between environment as commonly understood and the *built environment* is artificial and that the urban and everything that goes into it is as much a part of the solution as it is a contributing factor to ecological difficulties" (1996: 60).

To speak of nature in the context of urbanization invites the writing of environmental histories of urbanization that understand nature and the city to be continuously remade in and through the other (Davis 1990; Cronon 1991). From such histories it becomes clear that the modern metropolis grew in a process of "destructive creation," as it were, inverting the famous label attached to capitalist production by Schumpeter, Harvey and others. Indeed, it is perhaps due to the specific dialectics involved in urbanization – killing living things so that other beings may live and grow – that has led writers on cities to use crudely biologistic idiom for the expansion of cities: as if they were alive, "consuming" the countryside.

While often being naturalized (as in the case of the Chicago School) and subjectified (as in the frequent colloquial reference to the city as an actor), cities have rarely been seen as "natural" in the material or physical sense. Instead, urbanization has commonly been viewed as a process which compromises natural environments to a degree that they become perverted, distorted and destroyed. In cities, built environments are seen to have replaced "natural environments" to such an extent that only domesticated forms of "nature" such as parks and backyards (and their diverse and changing animal and plant populations) as well as depleted watersheds appear to remain of the natural basis of the city.[3]

Histories of urban natures

The foundation of every division of labour that is well developed, and brought about by the exchange of commodities, is the separation between town and country. It may be said, that the whole economic history of society is summed up in the movement of this antithesis.

(Marx 1887: 333)

For centuries, the city has been perceived, understood and judged in relation to the land and through the land in relation to nature. Now the situation has been turning around for the past century. The land is perceived and understood through the reference to the city. It recedes from the city which overgrows it. . . . In this moment when the city becomes the point of reference, it disappears as a tangible certainty.

(Lefebvre 1972: 162–3, our translation)

Building cities means building "thresholds" (Menninghaas 1986), "transitory zones" (Prigge 1991), or "liminal spaces" (Zukin 1991). Liminal spaces are those spaces "in between" where one frame of reference has been left behind and a new frame has not entirely gained shape. Apart from Prigge (1991), most contributors to the Benjamin induced debate on transitions as an urban condition, have not taken the specific liminality of the urban–natural interface into account. Yet, much of today's urban liminality is constituted in a redefinition of exactly this concatenation. Historically, the creation of cities and natures has meant the continuous redefinition of inside and outside, of centrality and periphery, of town and countryside. With each new phase of urbanization has come a reorganization of societal relationships with nature, and these in turn are related to the reorganization at any given point of the relationships of production and consumption. As artifacts of production and consumption, cities and natures are constantly repositioned toward each other.

In the medieval and the baroque city in Europe, for instance, nature was kept outside, in front of the gates and the trenches. Old birds-eye renditions of the ring of defensive structures show neatly clustered houses inside, with gardens and fields along with undesirable human populations outside the walls. While the separation seems unbridgeable, we know that the simple commodity production and the market activity in the city depended on the agricultural and mining activities in the countryside. As social ecologist Dieter Hassenpflug has pointed

103

out, during the medieval (agricole) period of societal relationships with nature, "the land always forms the cultural centre, the city always the periphery . . . The agricole state of nature marks the time of the relationship of country and town when landed property exerts the determining authority" (1993: 105).

The physical appearance of the medieval city was thus one of clear separation of urbanity and nature, yet the social–ecological processes that sustained town–countryside relations during the period were complex and involved continuous movement across this divide. Today, the medieval city has become somewhat of a historical model for current discussions of bioregionalism and sustainability:

> The complete ring wall of Dinkelsbühl, Germany, from above resembles a stone collar, keeping the town separate from the world outside. The centrepiece of the town is the Gothic cathedral. Crafts guilds made the town prosperous – cloth-making, beer-brewing, tanning, corn-milling, and furniture manufacture. Most of these trades depended on the fertility of the surrounding countryside to supply the raw materials. The land usually belonged to farmers who lived within the city walls. Even today cows are herded from the fields through the town gates into sheds for the night. Dung heaps are still stacked against the wall and manure is returned to the land. And so the cycle of plant growth, harvest, and decay is closed again and again, making medieval towns sustainable.
>
> (Girardet 1992: 44–5)

In short, the medieval city appears to mirror nature, tied closely to the land and its cycles. Yet the appeal of the medieval city to contemporary proponents of urban sustainability – such as bioregionalists – conveniently overlooks both the considerable tension stemming from any urban region's integration into larger, even world markets and global ecosystems and the downsides of a less than hygienic process of everyday urban living. As Harvey has noted, the medieval city was "forced to be sustainable . . . because they had to be" (1996: 44).

The early industrial city or the city in the period of "formal urbanization" (Hassenpflug 1993) seems to reverse this relationship. It pretends to shed its dependency on natural metabolisms at the base of the medieval city and appropriates human and non-human nature at an unprecedented pace. From the English Midlands of the nineteenth century to the East German industrial areas and the newly industrializing centers of Brazil, Mexico or Korea, industrialization and capitalization have externalized and objectified nature, environment and landscape to such a degree that they appear as mere fodder for the industrial machine. This is the period of extensive accumulation.[4]

The machine tendentially destroyed the garden, while the reproduction of its operators continued to be achieved through non-capitalist societal relationships with nature, e.g. subsistence production and extended families. Where parks are proposed, and garden cities are planned during this period, this happens decidedly as acts of socio-spatial hygiene and countermeasures to the pathetic destruction of every living thing in and around the city.

The Fordist city during the period of "absolute urbanization" (Hassenpflug 1993), which (in North America and in Western Europe) lasted for the better part of this century, was characterized by intensive accumulation which introduced a virtual cycle of mass production and mass consumption regulated by the macroeconomic policies of the Keynesian state. During this time, societal relationships with nature in the city continued to be brought under the hegemony of capital. In the age of the manicured front lawn and astro-turf, nature becomes materially unimportant and humans' material interaction with non-human nature is replaced by a more ideational and aesthetic interaction (Fowler 1992). Indeed, this is the basis for the reconstruction of nature as a separate entity and is arguably also when the separation of the city and the countryside became most deeply implicated in the destructive trajectory of capitalist urbanization. The separation of town and country establishes nature as an entity unto itself, a place and an idea which can be exploited, trampled on and admired.

In the Fordist city, the physical conditions of survival of humans appear segregated from life in cities. At least this is how it seems. Agriculture happens elsewhere (and becomes insignificant in purely macroeconomic terms: Hassenpflug 1993), tourism opens up nature in foreign and exotic places and children think that milk comes from rectangular cartons. Indeed, as we discuss later, the separation of nature and city in Fordism creates nature as a cultural artifact ready to be reinserted into the urban by way of symbolic action.

The automobile as the key commodity of Fordist production and consumption enables the urban field to be expanded, and villages and hamlets to be overrun by endless subdivisions. Socially, absolute urbanization "*completes the transformation from traditionality to informality.* From the earlier integral oikistic life conditions and nature relationships only systems-compatible elements of shadow- and self-employment, of self-help and self-supply in recreation, family and local communities remain" (Hassenpflug 1993: 171f., original emphasis). The natural base of the Fordist city is appropriately fossilist. How people eat, what their food contains, where the waste and the sewage go are questions of rather minor technical detail – as long as the stream of oil from the well into the gasoline tank of our automobile is not interrupted. The ersatz nature of the suburbs has, despite its constant invocation of the open countryside, established lifelines to industrial natures on far distant continents. From the oilfields of the Middle-East to the vineyards of Chile, from the high-tech agriculture of the Canadian plains to the garbage dumps in West Africa, the Fordist city has established an imperialist urban–nature relationship that spans the globe. It allows nature only to reproduce itself in capitalist and industrial terms.

How, then, is nature produced in the post-Fordist city – an age of "flexible accumulation" and "reflexive urbanization" (Hassenpflug 1993: 172–4)? As we have argued elsewhere, today's cities "appear as multicentered, nodal, flexible, and global" (Keil and Ronneberger 1994: 138).[5] Further, after having been made all but expendable, nature has been creeping back in, albeit in a form which Hassenpflug terms "staged" (*inszeniert* in its German original) or which one could call – with Baudrillard – a "simulacrum." This is true for two reasons. First, post-Fordist urbanization more than ever before occurs outside of traditional urban centers both in the regional sense (Silicon Valley rather than in San Francisco, in Oakland County, Michigan, rather than in Detroit; see Bloch 1994) and in the intraregional sense (edge cities beyond the previously recognized boundaries of the metropolitan region). Cities grow where there were no cities before. And in contrast to their Fordist suburbanization, much of this ex-urbanization has brought towns and cities, rather than just workplaces or dormitory satellites, to the countryside – while not necessarily closer to nature.[6]

Nature, in the post-Fordist city, is re-established, recreated and reborn as an immanent element of the new urban order. As if the builders and planners of today's cities wanted to orchestrate and synthesize all the societal relationships with nature of previous periods of urbanization, the post-Fordist city pretends to have it all, but with myriad contradictions: organic and diverse agriculture inside and outside the urban centers conflicts with the continued neo-industrialization of urban environments; smokestack industries are replaced by so-called "clean industries"; "greenfields" host office parks that handle countless toxic chemicals; suburbanization continues unabatedly, gobbling up prime farmland around cities; and food production has entered the uncharted territory of bio-engineering.

In North America, the European perception of an abundance of available land and of "a landscape untouched by history – nature unmixed with art" (Leo Marx, quoted in Garreau 1991: 389) created what perhaps is an illusion: that urban-ization and natural environments can be integrated beyond the trajectories of historical patterns of urbanization – urbanizing the land without destroying its natural character. The reintegration of nature and urbanization is today en vogue in popular literature on new urban forms in metropolitan peripheries. Joel Garreau expresses this position most succinctly in his trademark *Edge City*:

> The question is whether the forces of preservation and the forces of growth can ever somehow be resolved. Can the Garden-like aspects of what we are doing be encouraged while the hellish scourge restrained that despoils the land, crowds the schools, devours open space, jams traffic, and leaves nothing but a fast-food crisis for the soul?
>
> (Garreau 1991: 390)

Garreau states correctly that Edge City has left the suburban solution (Walker 1981) behind as insufficient, because suburbia was "designed to get away from the environmental and social dislocations – the factories and the slums – of the first hundred years of the Industrial Revolution" (Garreau 1991: 399). Yet his trust in the integration of city and nature in Edge City, the "reintegration of all our functions – including the urban ones of working, marketing, learning, and creating – into those once-suburban landscapes that, after all, are among the most affable we have built this century" (ibid.) is unconvincing in terms of ecological demands. It is not at all clear that Edge City (and other exurban experiments such as New Urbanism) presents a simple way out of the dilemma of making urbanization work in nature (on New Urbanism see Lehrer and Milgrom 1996).

In the Fordist period, industrial production was wedded to its logical counterpart, mass consumption. Progress through industry found its pervasive spatial form during this period, expressed in corporate downtowns surrounded by grids of suburban production clusters and consumer modules tied together by freeways and other technological infrastructures. In post-Fordism, despite its postindustrial narrative, the industrialism of the previous era is taken to its hyperindustrial extreme, where anything and everything is being produced mechanically and in mass. Only its geographies have changed. While industrial and traditional working-class landscapes have been ravaged everywhere between Los Angeles and Detroit, Newcastle and the Ruhr, industrialization and proletarianization have continued to rise worldwide and in capitalism's globalizing trajectory nothing is meant to be left to the accidental realm of simple commodity production and precapitalist subsistence. Even craft production and subsistence farming (and through these, specific non-capitalist productions of nature), long kept out of the reach of the mass producing Fordist complex, have become integrated into the industrial landscape of global capitalism. And thus so has nature, worldwide.

It is no coincidence, then, that just when nature globally has become "artifactual" city builders and urbanists in North America have discovered "nature" as the real matter through which and by which to construct the urban. The domesticized nature of the Fordist suburb is no longer enough, though. In current debates around the city and its environs, for instance, one often encounters the argument that planning and design need to be related more closely to local topography, ecology and climate, such that urbanization is now conceptualized and achieved through nature and ecology rather than against it. "The discourse on growth is now being reborn as a discourse that is naturally *critical* of growth in order for local politicians and planners to be able to sustain it" (Keil and Ronneberger 1994: 164). In short, the usual contraposition of the city and environment only tells part of the story. "Nature" is no longer exiled from the city, but becomes the key element of the current era of urban growth.

In these appeals to "nature," however, it is rarely recognized that what is local is constantly being redefined by the processes of urbanization. While it is true that local topographic, climatic and ecological conditions need to be the basis on which any design or redesign of cities must take place, we need also to take into account the globalization process which is transforming our cities. Globalization changes the meaning of the local and this extends to perceptions and realities of local natures. We need to realize, then, that while climate and topography impose

107

certain restrictions on local natures, any narrow and culturally overdetermined concept of local natures is misleading.

Equally, landscape itself needs to be problematized. Is landscape a congealed form of human practices or a pregiven quality of place with which ecological planning should attempt to align the urban process? If we agree that landscapes are products to a large degree of human practices, we need to look at perceptions and practices currently found in cities. Social collectives and individuals bring a large variety of perceptions and practices to urban nature. They range from hard-core conservationism to consumerist recreation, resulting in "diverse perceptions of space which often contradict each other and call into question each other's legitimacy" (Keil and Ronneberger 1991: 46). Current restructuring processes are leading to a dramatic rewriting of urban landscapes and urban–nature relationships. Post-Fordism has introduced new industrial and consumption landscapes. The emergence of globalized and world city economies and societies has brought fragments of distant landscapes and ecologies to local landscapes and natures.

RESTRUCTURING AND MODERNIZING NATURE AND URBANITY IN THE TORONTO URBAN REGION

Ecology is now central to the institutional and symbolic modernization of capital. So, too, are discourses of urban sustainability and ecological planning, which seek to reorder and rationalize capitalist urbanization in the future. The politics and discursive strategies not only of politically dominant forces but also of a range of reformist efforts are connecting the "city" with "ecology" and both to the production of urban space.

(Kipfer *et al.* 1996: 5)

In the following sections we focus on three dimensions of the current rearticulation of nature and the urban in Toronto. We have chosen to present these dimensions using a version of Henri Lefebvre's triad of spatial practices, representations of space, and representational spaces.[7] We begin by looking at the spatial practices found in the reform of urban and regional planning in Southern Ontario. We then proceed to examine the representations of space expressed in current planning for the municipality of Vaughan located just north of Metropolitan Toronto. Finally, we will focus on the "lived" spaces of representation implicit in the marketing of new residential subdivisions in suburban Southern Ontario.

Modernizing urban planning after Fordism

During the last decade of regional restructuring in Toronto, ecology and nature have been incorporated, through a flurry of institutional and legislative reforms over local spatial practices, as essential elements in local urban development.[8] Urban reform processes occur within specific state and institutional relations in Canada, where municipal authority is delegated by provincial governments (a relation underlined during the recent, forced amalgamation by the provincial government of the six cities of Metropolitan Toronto into a single jurisdiction).[9] The province also determines the administrative and legal framework for local planning, all which is overseen by the Ontario Municipal Board.

These local state institutional contexts formed the arenas within which ongoing regional restructuring occurred (during conditions of severe economic crisis that affected much of the province in the late 1980s). To some this period represented an opportunity to adapt city building practices to "post-Fordist urban forms," or to enlarge modes of socio-spatial regulation that could condition "networks of average-sized cities in a metropolis built on a human scale" (Lipietz 1992: 103, 108). To others such as Pierre Filion (1995: 61), who draws on the Canadian example of Kitchener, Ontario, new fads in civic spatial practices come out mostly "as Fordist urbanization in crisis," a perspective that turns on the simultaneity of concretized Fordist and post-Fordist urban structures and forms, especially historic transportation and built environments. Indeed, it can be argued that new modes of regulation in the Toronto region are best characterized as "neo-Fordist" (Painter 1995: 278–80), a description that explicitly rejects their adequacy as less destructive urban forms.

The rise of an eco-modernized regime of accumulation in the Toronto urban region can be briefly traced from the area's most recent period of civic reform, particularly the work of the Royal Commission on the Future of Toronto's Waterfront,[10] a harbinger of a new urban order that programmatically makes growth and accumulation dependent on, rather than independent of, the reinsertion of nature. The commission's forays into eco-modernization, however, were an outgrowth of its 1988 task to make recommendations over federal lands and agencies on Toronto's waterfront (significant lands in this area were controlled by the federally mandated Toronto Harbour Commission), a mandate made possible by the privatization program of Prime Minister Brian Mulroney.[11] Federal waterfront lands had been the subject of local state and civic contestations over contaminated site development, land use, built forms and public access, especially the valorizing activities on downtown waterfront lands owned by federal agencies (see Desfor *et al.* 1989). The commission was headed by David Crombie, a former mayor of the City of Toronto and minister in Mulroney's cabinet. After the commission's first report (which among other things opened debate over the integrity of the regional watershed) its terms of inquiry were expanded and rolled into a joint federal–provincial royal commission by Ontario Liberal Premier, David Peterson. At the same time, on 17 October 1989, Peterson enacted a development moratorium in some centres of the region "to prevent any major development until it can be determined what is appropriate for the people and the environment, and identify ways of protecting forever the headwaters and river valleys from the Oak Ridges Moraine to Lake

Ontario" (RCFTW 1992: 11), an area that essentially corresponds to the boundaries of the Greater Toronto Area (see Figures 5.1 and 5.2).

With an extended mandate and moratorium on exurban development in proximity to the moraine, the commission assumed a leading role in public consultation, articulating normative, modernized approaches to comprehensive planning, and brokering the regional local state institutional network. Through this process its ecosystem planning approach was tacitly constructed as a critique of the socio-ecological crisis and the byproducts of "conventional planning." Furthermore, the approach identified and applied important new methods and strategies over site analysis (RCFTW 1991a: 69–96), and offered a bare manifesto of design principles for contemporary urban management and development.[12] The commission integrated these principles while navigating the local state network, effecting waterfront partnership agreements, and making coherent policies among land authorities over the "regeneration" of the regional shoreline (RCFTW 1991b). These widely supported initiatives were aimed at a limited revision of shoreline spatial practices, including: water quality and wastewater Remedial Action Plans; a waterfront trail with links to river valleys and ravines; the protection of wetlands and other environmentally significant areas; public open spaces; public transit, road, pedestrian, and bicycle access to and along the waterfront; preservation policies; and, a renewed emphasis on provincial affordable housing policy requirements.

Crombie's tacit critiques of incremental urbanization advanced the provincially mandated Commission on Planning and Development in Ontario, which figured as a modest effort by the New Democratic Party (NDP) government of Bob Rae to inscribe principles of ecological modernization as a mode of regulating spatial practices. Led by John Sewell, another (although far more controversial) former City of Toronto mayor, this stage of the reform process legislated codes aimed at promoting compact development, and attempted to address the problematic cultural issues surrounding "natural heritage," including the preservation of agricultural land, significant landscapes, vistas, ridgelines, green corridors and watershed systems.[13] In the remnant precincts of the Fordist city, this discourse of contemporary planning complemented restructuring processes, driving a green line (Crombie's waterfront trail parallels the edge of the western basin of Lake Ontario) through an ever-increasing inventory of rusting, industrial land in Metropolitan Toronto and the lakeshore corridor. Nature, in the mold of the region's eco-modernizers and the current regime of accumulation, was effectively used to mobilize consent for an unprecedented series of land use conversions across these historic districts, even as more expressways and urbanization occurred on the periphery of Toronto. This added a "natural touch" to the concept of "flexspace" (Lehrer 1994), which marks an increased fluidity of spatial uses and limited dedifferentiation of urban functions in the after-Fordist city.

These controversial legislative initiatives over contemporary spatial practices, while mired in significant questions over their potential to effect environmental protection or limits on regional urban form, gave its proponents a greater

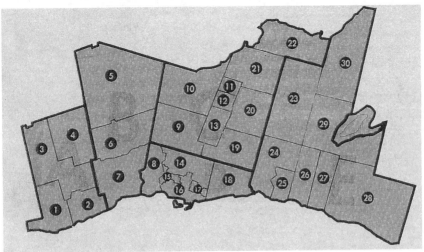

1. Burlington	19. Markham
2. Oakville	20. Whitchurch-Stouffville
3. Milton	21. Esast Gwillimbury
4. Halton Hills	22. Georgina
5. Caledon	23. Uxbridge
6. Brampton	24. Pickering
7. Mississauga	25. Ajax
8. Etobicoke	26. Whitby
9. Vaughan	27. Oshawa
10. King	28. Newcastle
11. Newmarket	29. Scugog
12. Aurora	30. Brock
13. Richmond Hill	
14. North York	**A** Halton Region
15. York	**B** Peel Region
16. Toronto	**C** York Region
17. East York	**D** Metropolitan Toronto
18. Scarborugh	**E** Durham Region

Figure 5.1 Toronto urban region: political boundaries
Source: Office for the Greater Toronto Area 1992, *Shaping Growth in the GTA*, Toronto: Berridge, Lewinberg, Greenberg Ltd.

number of tools to make planning and development in the Toronto region more coherent. However, the eight-year undertaking legislatively to adjust incremental development patterns and "naturalize" the tensions of world city formation only constituted a brief (fourteen-month) interlude in the production of neo-Fordist Toronto. The cardboard reforms were tossed out on 20 May 1996, almost one year after the "commonsense revolution" began under Ontario's Progressive Conservative government. Their streamlined mode of regulating spatial practices in local planning and development, encumbered by the merest signs and symbols of nature in city building, affirmed the historic power of rentiers and their local

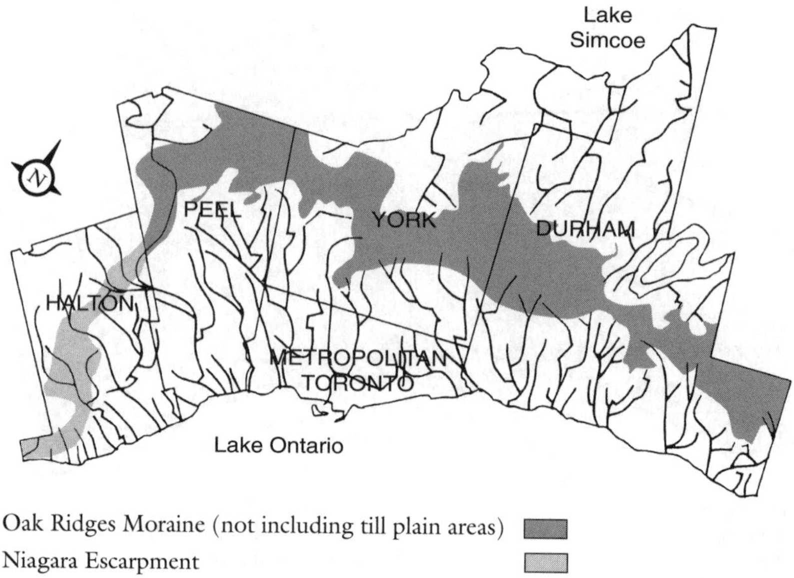
Lake
Simcoe
PEEL
YORK
DURHAM
HALTON
METROPOLITAN
TORONTO
Lake Ontario
Oak Ridges Moraine (not including till plain areas)
Niagara Escarpment

Figure 5.2 Toronto urban region: major watershed features
Source: Office for the Greater Toronto Area 1992, *Shaping Growth in the GTA*, Toronto: Berridge, Lewinberg, Greenberg Ltd.

growth coalitions to determine what counts as ecological modernization, discursive practices that have remained a primary condition for sustaining growth and accumulation. These contested versions of planning and development reform attest to the different paths of eco-modernization and belie any simplistic notions of technocratically induced urban sustainability.

Flexcity: urbanization and the new urban order

Since the storms from Ontario's fractious process to reorient local spatial practices have settled, eco-modernized modes of regulation have figured centrally – rhetorically, if not materially – in the current development boom across Toronto's edge centres. In places like the "flexcentre" of Vaughan, this new urban order has taken up discourses of nature and urbanization both as an urban self-conception and tool of urban management (see Jahn 1996: 97–8). Here we focus on Vaughan's new millennial, "comprehensive" plan – *Official Plan Amendment 400* (OPA 400) – which is an especially important example of Toronto's eco-modernized regime of accumulation. Because the plan was in the loop of provincial approvals prior to the institution of the NDP's legislation of March 1995, which strictly regulated exurban development, it effectively evaded these moderate development controls. Thus, OPA 400's already "streamlined" version of planning and development reform was presciently cast for a post-social democratic planning era and stands as a benchmark for other rapidly urbanizing flexcentres on the periphery of Metropolitan Toronto.

OPA 400 marked an effort to "clean up" local planning and development practices, and was tied to Vaughan's rapid ascendancy as a leading manufacturing, distribution, warehouse and construction centre in the regional economy, a thirty-year process propelled through flexibilization, globalization and a rapacious local growth coalition.[14] Vaughan's intermodal and automobile freight terminals represent a key portal in the regional manufacturing and distribution complex, which sustains a combined 77 per cent of Toronto's traded economy. While Vaughan's largest employer is Paramount Canada's Wonderland Inc. (the theme park provides 3,200 part time jobs), manufacturing firms represent a total of 36.2 per cent of all local enterprises, employing a total of 18,060 workers (Regional Municipality of York 1995). The derivative autoparts sector is the largest cluster in this complex with 38 flex-spec firms employing a total of 7,144 workers (City of Vaughan Business and Office Directory 1993–4). This sector includes Magna International, the world automobile part giant and largest manufacturing grouping (34 plants and 8,000 workers) on Metropolitan Toronto's fringe.

During its thirty-year growth, the rural image of the amalgamated township was reconfigured by the imposition of 3,326 hectares of flexspace and massive suburbanization (area population went from 25,000 in 1981 to 111,000 by 1991), all of which occurred on the hydrogeologically sensitive "south slope till plain" of the Oak Ridges Moraine Complex (ORMC). The till plain is a significant moraine recharge and headwater area that feeds major rivers in the watershed of the larger urban region.[15] After two centuries of agricultural use, aggregate extraction, water channelization and urbanization, hydrogeological conditions (water quantity and quality) in the 160 sq km glacial deposit have reached "crisis levels" (Howard *et al.* 1997). Thus, Vaughan's plans to urbanize the "till plain" and moraine areas of the ORMC are unprecedented since the rhetorics of its spatial practices emphasize the compatibility of their urban development strategies with environmental protection.

At the very least, it would seem, such claims should support an immoderate departure from conventional spatial practices. But this is far from the case. Vaughan's growth coalition defined their "ecosystem approach" for suburbanizing the south slope till plain through OPA 400, which allocated 2,611 hectares of land for mainly low density housing projects (see Figure 5.3). Normative conceptions of ecosystem planning remain elusive, although serious efforts to elaborate on this metaphor of urban management can be found in scientific discourses, resource management, parks planning, and public forums like the Royal Commission on the Future of Toronto's Waterfront (Tomalty *et al.* 1994). Ecosystem based planning has also been identified with utopian planning initiatives, from green cities, bioregionalism and healthy cities to eco-cities, eco-towns and eco-villages. The one element common to these "new" approaches to planning is their emphasis on comprehensiveness, taking holistic, systems based

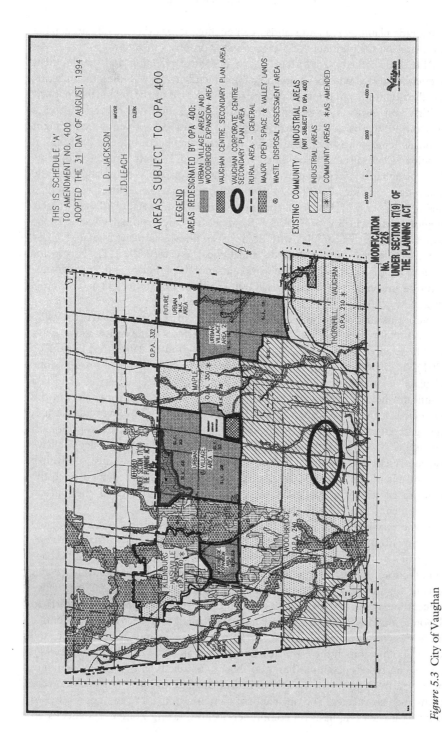

Figure 5.3 City of Vaughan
Source: City of Vaughan 1995, Official Plan Amendment 400, Appendix A.

approaches to environmental planning. In Vaughan's case, its growth coalition linked their ecosystem approach to some of the more moderate currents of "new urbanism," such as "transit oriented development" and "traditional neighbour-hood development." Yet Vaughan's machined version of ecosystem planning remains a hybrid of Fordist and post-Fordist regimes of eco-modernization, resulting in an amalgam of practices: Garden City meets Levittown; Ford meets Calthorpe; Geddes meets Moses; Le Corbusier meets Duany; and Berg meets the Chicago School.

Vaughan's application of "ecosystem boundaries" and goals to "design with nature," for instance, are highly contradictory. The plan's boundaries and limited urban design controls appear simply to mirror the growth desires of its political leaders. Moreover, OPA 400 lacks comprehensiveness – it applies only to the non-urban areas of the city. Secondary plans over Vaughan's 26 industrial and commercial precincts, and existing community plans remain untouched by the policies of the amendment. In addition, other boundaries defined by OPA 400 in relation to the south slope till plain and the rest of the Moraine complex remain highly contested by hydrogeologists, residents and watershed activists. The distribution of prime ecosystem goodies is mainly to low density, future high-rent districts. While their new Environment Management Guidelines (EMG) established a number of progressive growth controls,[16] policies to "augment" the remnant surface water system are just as likely to compound problems of area hydrogeological health, particularly over siltification of water systems and car-related contaminates (Howard et al. 1993, 1997). New policies over valley and stream corridors limit built forms in proximity to riparian and valley edges, but this policy may be seriously jeopardized by the entrepreneurial city's "greenways systems" strategy, which places equal emphasis on linking these corridors with "commercial centres, compatible public and private institutional uses," and other human flows. The "tableland wood lot protection" program contains one of the few new "exactive" policies in the plan, by which developers must contribute to the acquisition of these highground, remnant agricultural resource areas.

The plan identifies a number of criteria for long-term monitoring and assessing cumulative impacts, and has recognized that a significant amount of land in the Urban Village areas "may subsequently be determined to be inappropriate for urban development due to the environmental impacts on some of the lands." However, block planning and implementation procedures, which are "expected to pave the way for expeditious review and approval of subdivision plans" (Robinson 1995: 24), will preclude meaningful adaptations of the plan. This process integrates all land owners (and approvals for individual plans of subdivision) through master environmental/servicing plans in 1,000-acre concessions. Block plans, which "reflect a level of detail approaching that of a sub-division plan" (OPA 400: 26), set the infrastructural stage for the fast approval and implementation of individual plans of subdivision. Public input through the Block plan stage has been relegated to the end of the process.

Another large component of Vaughan's contradictory design for environmental protection is its sweeping community design concepts, which represent little more than a "repackaging of sprawl" (Lehrer and Milgrom 1996). Vaughan's resistance to criticisms of sprawl and the earnest attempts to modify the codes of spatial practices is evident through their way of describing their community development plan. *Transit-Supportive Land Use/Development Pattern* and the *Urban Village Areas*, when measured against their supporting policies and codes, are simply empty rhetoric. Far from producing a new "ecological" city, Vaughan's density prescriptions represent yet another way of producing neo-Fordist urban natures and spatial practices in exurban Toronto. Like earlier suburban developments, in the newly designated *Urban Village Area 1*, 48 per cent of developable land is devoted to low density development, 32 per cent to medium density, and only 20 per cent for high density residential development. Similarly, in *Urban Village Area 2*, 8,100 low density housing units account for 45 per cent of the total, with 6,120 units (34 per cent) of medium density and 3,710 high density units (21 per cent). OPA 400 has planned for 14.4 housing units per hectare of the land use total, a long way from the 30–37 units per hectare the province wanted during its regional reform process and further still from a "transit oriented land use pattern." Indeed, despite its rhetorics, the plan has no exaction policies to finance mass transit systems which, given the current regime of accumulation, will keep Vaughan's "regional mass transit plan" firmly in the symbolic realm. Nor are there any urban design codes over built forms. Instead OPA 400 has emphasized (on less than one page of the plan) a few goals and objectives, leaving urban design details to the block planning process and the hands of property entrepreneurs.[17]

Vaughan's millennial land plan, with its array of urbanizing approaches, can hardly be understood as a plan to protect nature – especially in relation to the south slope till plain of the moraine and the ongoing regional crisis of water quality and quantity. Rather the plan merely captures and redefines, as little more than an urban design feature, elements of remnant "nature" as particles of a technologically defined exurban landscape. The neo-Fordist form and function of "flexcity" neatly dovetails with the eco-modernization of urban planning in the larger Toronto region.

New residential spaces

The production of nature on the after-Fordist frontier of urban Toronto proceeds through the framework of spatial practices in the new planning regulations of the Province of Ontario and the representations of space manufactured by planners and developers in Vaughan. In the lived contradictions of the new suburban subdivisions that now blanket the northern landscape, representational spaces of a new kind have emerged which also redefine the ontological divide of nature and space. Here, our object is the sale of residential real estate.

Lefebvre describes his third aspect of space as "directly *lived* through its associated images and symbols, and hence the space of 'inhabitants' and 'users'" (Lefebvre 1991: 39). Building on the perceived space of the urban–suburban metropolitan structure and on the conceived spaces of new planning in Ontario, as well as trivializing the potentially utopian or change-oriented character of lived spaces through their subsumption to consumer culture, local place entrepreneurs have started to sell the (sub)urbanization of the "green aurora" around Toronto through symbolically invoking "nature." While the destruction of prime farmland proceeds apace, and while sprawl eats into ecologically valuable (and perhaps irreplaceable) wetlands at the foot of the post-glacial Oakridges Moraine, symbolic natures cement the foundations of a pervasive sales pitch in which fragments of "green" environments are articulated with urban post-Fordist lifestyle needs.

A combination of record low interest rates and a lack of institutional and regulatory development controls in post-social democratic Ontario has led to surging exurbanization in the north of Toronto. The lived space of exurbanization entails a contradiction: the eradication of ecological systems by way of development investment in nature (*Toronto Star* 19 October 1996: G1). Combining what Whatmore and Boucher (1993) have called the commodity and ecology narratives, nature has become the symbolic matter of the development discourse at the same time as fears of material extinction of ecological systems abound. Assisted by privatized road, sewer and water infrastructure, which has long left behind the constraints posed by a downtown Toronto-centered development logic, exurbanized residential projects have become possible everywhere in the Greater Toronto Area.[18] A sales map of Mattamy, a local development company, depicts the natural spaces of Toronto's western suburbs as mere embroidery of a seamless web of roads and residences (see Figure 5.4).

Another advertising by the same company is yet more explicit and illustrative (Figure 5.5). Captioned "Mattamy. It's the Trees!," the ad takes us on a home-buying tour guided by Duchess, an imaginary family dog. Exurban developers are outdoing each other in offering potential consumers of residential space "parkside living," waterfronts, coziness under trees, forests, creeks, woods, an enchanted forest, heaven on earth, daffodils (courtesy of William Wordsworth), rolling countryside, "naturally" right prices, the great outdoors in Brampton's White Spruce Park – "a nature lover's delight with lakes, ravines and woodlands" – a backyard as big as your dreams and the ubiquitous total lifestyle golf course community, all in no particular order.[19] Lifestyle is matched with lifecycle with specific communities for retirement, for kids, or for adults: treeplantings, high art (the McMichael Art Gallery in Kleinburg), high-tech (everywhere), and country festivalization (including country fairs with hay rides and creamy butter tarts). The lure of home ownership is paired with traditional family values, community, and heritage (including Native history). Interspersed with the traditional language of family values is the suggestive imagery of eroticism which, in one case, presents a male-dominated nature (golf-course!) and a half-clad woman

117

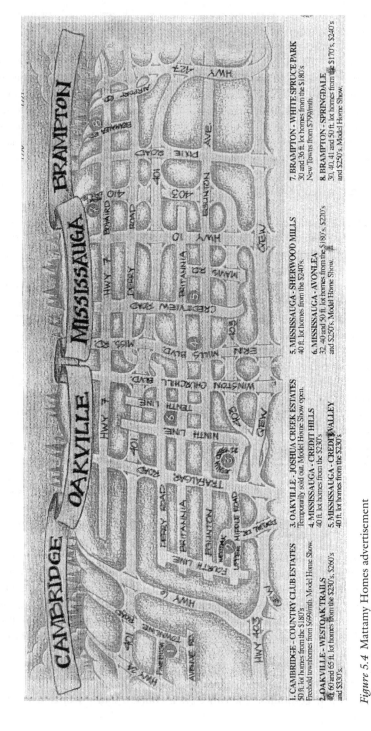

Figure 5.4 Mattamy Homes advertisement

Source: Reproduced with permission of Mattamy Homes.

"home-maker" observing a golfing husband through the gauze of a curtained window. The articulation of exurbanity with nature rests firmly on the reintegration of traditional gender roles considered "natural" by conservatives all along.

The construction of exurbia produces nature as more than just the greenfield site of suburban sprawl typical of the anti-city Fordist urbanism; nature, much more, becomes the symbolic terrain for the realization of urban lifestyle projections. The more vividly nature is evoked as a destination of the minivan armada with its accompanying throng of high-tech urbanist wants, the more viciously it is attacked in the form of filled-in wetlands, paved ravine bottoms and felled trees. While the emergence of the edge city and other exurban phenomena seem to suggest an end of the city, they rather signify the end of nature and the rebirth of nature© in the process.

Whereas planning has often helped formalize "the ontological separation of nature and space" (Whatmore and Boucher 1993: 167), the lived spaces of the political economy of exurbanization around Toronto assist in breaking down the discursive barriers of nature and (ex)urbanity.

CONCLUSION

We have argued in this chapter that today nature is rearticulated with urbanization in new ways. Using Toronto as an example, we have shown how planning practice, plans, and lived spaces reconfigure nature and environment discursively and physically. While the examples used – new planning practice in Ontario, ecosystem plans in Vaughan, and the marketing of residential spaces – were chosen somewhat arbitrarily, they show typical ways in which nature and environment are reinserted into urbanization in western cities at the end of the millennium. Specifically, planning is forced to recognize certain environmental management features in its practice; plans depicting (ex)urban expansion and sprawl internalize ecological concerns in their representations, while images of exurban residential spaces link humans with natural habitat. In the post-Fordist city, nature has become a major discursive element of the production of urban space.

Having shown ways in which nature and environment become part of urbanization in this period, we must also state that there are ways in which it does not. None of the three cases analyzed offers a radical, fundamental ecological alternative to the "nature" of urbanism during the past decades.[20] The specific crisis of societal relationships with nature encountered today in western cities – closely imbricated with the crisis of the unsustainable fossilist Fordist regime of accumulation – is not addressed in the cases discussed above. The reinsertion of nature after Fordism in Toronto, for instance, occurs with a blatant disregard for the material processes that sustain life – urban and non-urban, human and non-human. Cities embody ecological relations, but these are still subsumed under,

Figure 5.5 Mattamy Homes advertisement
Source: Reproduced with permission of Mattamy Homes.

rather than consciously articulated with, social relations, and more specifically, the production of urban space. The processes of material production and reproduction, the physical metabolisms of urban–nature relations in Toronto after Fordism, remain stuck largely in a linear, exploitative, fossilist mode (Girardet 1992; Jahn 1996).

Along similar lines, the symbolic relations of urbanity have not been altered in ways that would allow a drastic change in the material ecological base of urban life. Despite attempts at reinserting nature into the urban, we continue to inhabit a symbolic world in which city and countryside, core and suburb, humans and animals are clearly separated. The spaces visited in this chapter – the Toronto regional ecosystem and residential exurbia – resemble nature only in the pastiche sense of post-modern practice; they subordinate nature once again as simply the raw material from which cities are made.

Finally, the spaces reviewed in this chapter signify an emerging mode of spatial regulation that appears more as a rough-and-tumble reaction to the crisis of Fordism in the Toronto region and less as the beginning of a new era. The secular crisis of Fordism has brought into sharp relief the millennial crisis of urban–societal relationships with nature. However, the current redrawing of meanings and boundaries of regulation and governance have not been a match for this millennial problematic. The hegemonic regional governance structure and process has few tools and even fewer ideas to cope with the disarticulations of an urban humanity from non-human nature. The reinsertions of nature into the city as discussed in this chapter, while noteworthy as a new trend in the building of cities, remain ill-equipped to offset the dramatic damages experienced under Fordist urbanization.

ACKNOWLEDGEMENTS

We gratefully acknowledge the editorial comments and contributions to this chapter by Gene Desfor and the editors of this volume.

NOTES

1 Wolch identifies zoöpolis as a discursive space in which urban relationships are viewed in a comprehensive ecological manner and which "brings animals back in": "Zoöpolis invites a critique of contemporary urbanization from the standpoint of animals, but also from the perspective of people who together with animals suffer from urban pollution and habitat degradation, and who are denied the experience of animal kinship and otherness so vital to their well-being. Rejecting alienated theme park models of human interaction with animals in the city, zoöpolis instead asks for a future in which animals and nature would no longer be incarcerated beyond the reach of our everyday lives" (Wolch, 1996: 47).
2 Whatmore and Boucher (1993) have observed a similar tendency in what they have termed the commodity discourse in British environmental planning.

3 We disregard, at this point, the complex and pervasive metabolisms that sustain urban living (Girardet 1992).

4 Following Aglietta (1979), extensive accumulation refers to the pre-Fordist social formation of capitalism characterized by mostly agrarian social relationships, consumption habits and life forms. Work processes were little mechanized and the accumulation of capital was concentrated in the industries producing the means of production. The basis of this regime was the unlimited reservoir of cheap, extensively exploitable, labour.

5 "Multiple centers are the spatial expression of the division of labor in each city; nodality refers to the restructuring of the pyramidal hierarchy of these centers into a network of nodes; flexibilization describes the changing internal and external organization of the Fordist accumulation regime; and global means the internationalization of urban economies, labor markets, cultures etc." (Keil and Ronneberger 1994: 138).

6 We do not claim here that post-Fordist urbanization is more "natural" or that there is a "real" essential "nature" in the countryside but that locationally and symbolically post-Fordist urbanism expresses itself through evocations of nature.

7 Lefebvre has explained this triad most explicitly in his *Production of Space* (1991: 38–9). Harvey provides an accurate and lucid reformulation in *The Condition of Postmodernity* (1989: 218–19) where he defines spatial practices as referring to "the physical and material flows, transfers, and interactions that occur in and across space in such a way as to assure production and social reproduction"; representations of space encompass "all of the signs and significations, codes and knowledge, that allow such material practices to be talked about and understood" (e.g. plans of all sorts); while spaces of representation are "mental inventions . . . that imagine new meanings or possibilities for spatial practices."

8 For summaries of development in the Toronto area see Todd (1995); Caulfield (1994, particularly chapters 2 and 4); Sewell (1993).

9 See Figure 5.1, area D. At point of writing a democratic citizens rebellion continues against this plan. Since December 1996, residents of Toronto have organized in unprecedented numbers to resist the provincial incursion into local governance. This has taken the form of weekly meetings of Citizens for Local Democracy, countless neighbourhood town hall meetings, mass demonstrations, and committee hearings on amalgamation (Bill 103). Although in a referendum in March 1997, Toronto residents rejected the province's plans, the Conservative government of Mike Harris proceeded with a slightly modified version of the original legislation.

10 Royal Commissions have been a favoured device of governments in Canada to analyze, evaluate and advocate public policy. Frequently these have only served as a catalyst for public debate and social learning. The Royal Commission on the Future of Toronto's Waterfront (RCFTW) was only the second joint federal–provincial royal commission in Canadian history. Following the commission's final report in 1992, it was institutionalized as the Waterfront Regeneration Trust.

11 At the same time Toronto was competing for the 1996 Olympic games, and lands controlled by the Toronto Harbour Commission were deemed appropriate for game facilities.

12 Principles of ecosystem planning management had been worked out in two prior institutional reports. However, before these reports, a citizens group Save the Oak Ridges Moraine (STORM) had called for a complete moratorium on moraine development, a preservation project also based on an ecosystem approach.

13 The social-democratic New Democratic Party government under Premier Bob Rae (1990–5) had put in place a regulatory framework which stipulated that proposed communities were subject to provincial approval and had "imposed strict criteria for protecting prime farmland and natural features" (Saunders 1996: A6). This enlarged

form of regulation was disliked intensely by developers and was gutted by the Progressive Conservative government of Mike Harris. A newspaper article quoted John Bousfield, "the father of some of Canada's best-known suburbs" as saying: "It had gone too far. The Red Guard was in charge" (ibid.).

14 In 1988 various cogs in Vaughan's local growth machine were implicated in series of scandals involving land speculators, civic officials and politicians.

15 Till Plain areas of the ORMC form a perimeter around the moraine. Most mappings of the area (including in this chapter) leave these recharge areas off the map.

16 Storm water management policies include provisions for multiple ponds, wet ponds, stormwater wetlands, infiltration trenches or basins, grassed or vegetated swales, filter strips, and special purpose stormwater management practices.

17 These range from encouraging "alternative development standards, providing opportunities for pedestrians and bicyclists, and ensuring [ominously] that neighboring developments are physically compatible and respect existing development conditions," to "ensuring that development complements the natural landscape, and protects and conserves the natural landform of areas having prominent physical features" (OPA 400: 30).

18 Edward Relph (1990: 15) who compared Toronto's growth to "water flowing into a series of ever larger containers," from the core by the Lake outward, observed recently that "in the fringe areas around Metro Toronto the new urban form is a sprawling, disconnected, multi-centered one of high-tech industrial parks, wide arterial roads, historicized and gentrified fragments of old towns, corn fields waiting to be developed, apartment buildings in the middle of nowhere, empty corridors for expressways to be built sometime" (p. 74).

19 This tends to occur predominantly in the suburbs. But inner city spaces are also subject to similar rhetorics. A recent ad for a downtown Toronto ravine development read: "A gated enclave of custom homesites nestled in this secluded ravined hillside" (The Toronto Star 12 April 1997: G1).

20 We are using ecological here not in a normative sense but in order to define a horizon of possibilities for more sustainable urban practices. While we sympathize with the emphatic use of ecology in much of the European discussion (see Lipietz 1995), we are aware of and agree with the critical discussion of ecology as an insufficient concept for the liberation of societal relationships with nature (see Trepl 1996).

REFERENCES

Aglietta, M. (1979) A Theory of Capitalist Regulation: The US Experience, London: Verso.

Amin, A. (ed.) (1994) Post-Fordism: A Reader, Oxford: Blackwell.

Bloch R. (1994) "Metropolis inverted: the rise of and shift to the periphery and the remaking of the contempory city," unpublished PhD dissertation, University of California, Los Angeles.

Caulfield, J. (1994) City Form and Everyday Life: Toronto's Gentrification and Critical Social Practice, Toronto: University of Toronto Press.

City of Vaughan (1993) Community Planning and Design Study, OPA 400 background report, Toronto: Berridge, Lewinberg, Greenberg Ltd.

—— (1993) Subwatershed Study – Background Report on Existing Environmental Conditions, Toronto: Berridge, Lewinberg, Greenberg Ltd.

—— (1995) Amendment Number 400 to the Official Plan of the Vaughan Planning Area, Toronto: Berridge, Lewinberg, Greenberg Ltd.

—— (1995) *Environmental Management Guideline: A Companion Document to Official Plan Amendment 400*, Toronto: Berridge, Lewinberg, Greenberg Ltd.

Cronon, W. (1991) *Nature's Metropolis: Chicago and the Great West*, New York: W. W. Norton.

Davis, M. (1990) *City of Quartz*, London: Verso.

Desfor, G., Goldrick, M. and Merrens, R. (1989) "A political economy of the water-frontier: planning and development in Toronto," *Geoforum* 20: 487–501.

Filion, P. (1995) "Fordism, post-Fordism and urban policy making: urban renewal in a medium sized Canadian city," *Canadian Journal of Urban Research* 4: 43–71.

Fowler, E. (1992) *Building Cities That Work*, Montreal and Kingston: McGill-Queen's University Press.

Garreau, J. (1991) *Edge City: Life on the New Frontier*, New York: Doubleday.

Girardet, H. (1992) *The Gaia Atlas of Cities: New Directions for Sustainable Urban Living*, New York: Anchor Books.

Goodwin, M., Duncan, S. and Halford, S. (1993) "Regulation theory, the local state, and the transition of urban politics," Environment and Planning D: *Society and Space* 11: 67–88.

Grady, W. *Toronto the Wild*, Toronto: Macfarlane, Walter and Ross.

Gregory, D. (1994) *Geographical Imaginations*, Oxford: Blackwell.

Hajer, M. (1995) *The Politics of Environmental Discourse: Ecological Modernization and the Policy Process*, Oxford: Clarendon Press.

Harvey, D. (1989) *The Condition of Postmodernity: An Inquiry into the Origins of Cultural Change*, Oxford: Blackwell.

—— (1993) "The nature of environment: dialectics of social and environmental change," in R. Miliband and L. Panitch (eds) *Socialist Register: Real Problems, False Solutions*, London: Merlin Press: 1–15.

—— (1996) "Cities or urbanization," *City* 1/2: 38–61.

Hassenpflug, D. (1993) *Sozialökologie: Ein Paradigma*, Opladen: Westdeutscher Verlag.

Howard, K., Boyce, J., Livingstone, S. and Salvatori, S. (1993) "Road salt impacts on groundwater quality – the worst is still to come," *GSA Today* 3: 1–19.

Howard, K., Eyles, N., Smart, P., Boyce, J., Gerber, R., Salvatori, S. and Doughty, M. (1997) "The Oak Ridges Moraine of Southern Ontario: a natural groundwater resource threatened by urbanization," in N. Eyles *et al.* (eds) *Environmental Geology of Urban Areas*, Scarborough: Geological Association of Canada Special Publication.

Jahn, T. (1996) "Urban ecology – perspectives of social-ecological research," *Capitalism, Nature and Socialism* 7: 95–101.

Keil, R. and Ronneberger, K. (1991) "Konflikt-und Möglichkeitsraum," *Kommune* 9(6), June: 46.

—— (1994) "Going up the country: internationalization and urbanization on Frankfurt's northern fringe," Environment and Planning D: *Society and Space* 12: 137–66.

Kipfer, S., Hartmann, F. and Marino, S. (1996) "Cities, nature and socialism: towards an urban agenda for research and action," *Capitalism, Nature, Socialism* 7: 5–19.

Lefebvre, H. (1972) *Das Alltagsleben in der modernen Welt*, Frankfurt: Suhrkamp.

—— (1975) *Die Stadt im marxistischen Denken*, Ravensburg: Otto Mailer.

—— (1976) *The Survival of Capitalism: Reproduction of the Relations of Production*, London: Allison and Busby.

—— (1991) *The Production of Space*, Oxford: Blackwell.

Lehrer, U. (1994) "Images of the periphery: the architecture of flexspace in Switzerland," Environment and Planning D: *Society and Space* 12: 187–205.

Lehrer, U. and Milgrom, R. (1996) "New (sub)urbanism: countersprawl or repackaging the product," *Capitalism, Nature, Socialism* 7: 49–64.

Lipietz, A. (1992) "A regulationist approach to the future of urban ecology," *Capitalism, Nature, Socialism* 3: 101–10.

—— (1995) *Green Hopes. The Future of Political Ecology*, Cambridge: Polity Press.

Marx, K. (1887) *Capital*, vol. 1, Moscow: Progress Publications.

Menninghaus, W. (1986) *Schwellenkunde: Walter Benjamins Passage des Mythos*, Frankfurt.

Painter, J. (1995) "Regulation theory, post-Fordism and urban politics," in D. Judge, G. Stoker and H. Wolman (eds) *Theories of Urban Politics*, London: Sage Publications.

Prigge, W. (1991) "Übergäenge," in T. Koenigs (ed.) *Vision offener Grünräeume*, Frankfurt: Campus: 1173–178.

Regional Municipality of York (1995) *York Region Profile*, Regional Municipality of York.

Relph, E. (1981) *Rational Landscapes and Humanist Geography*, London: Croom Helm.

—— (1990) *"The Toronto Guide, The City, Metro, the Region,"* Toronto: University of Toronto, Department of Geography.

Robinson, P. (1995) "Protecting the environment in a rapidly urbanizing community," *Plan Canada* 35: 22–5.

Royal Commission on the Future on the Toronto Waterfront (RCFTW) (1991a) *Pathways Toward an Ecosystem Approach: A Report on Phases I and II of an Environmental Audit of Toronto's East BayFront and Port Industrial Area*, Ottawa: Ministry of Supply and Services.

—— (1991b) *Planning for Sustainability: Towards Integrating Environmental Protection into Land Use Planning*, Ottawa: Ministry of Supply and Services.

—— (1992) *Regeneration: Toronto's Waterfront and the Sustainable City*, Toronto: Queen's Printer for Ontario.

Saunders, D. (1996) "Urban sprawl growing on a building spree," *Toronto Globe and Mail* 29 June: A1.

Sewell, J. (1993) *The Shape of the City: Toronto Struggles with Modern Planning*, Toronto: University of Toronto Press.

Todd, G. (1995) "Going global in the semi-periphery," in P. Knox and P. Taylor (eds) *World Cities in a World System*, Cambridge: Cambridge University Press: 192–212.

Tomalty, R., Gibson, R., Alexander, D. and Fisher, J. (1994) *Ecosystem Planning for Canadian Urban Regions*, Toronto: Intergovernmental Committee on Urban and Regional Research Publication.

Trepl, L. (1996) "City and ecology," *Capitalism, Nature, Socialism* 7: 785–95.

Walker, R. (1981) "A theory of suburbanization: capitalism and the construction of urban space in the United States," in M. Dear and A. Scott (eds) *Urbanization and Urban Planning in Capitalist Society*, London: Methuen.

Whatmore, S. and Boucher, S. (1993) "Bargaining with nature: the discourse and practice of environmental planning gain," *Transactions of the Institute of British Geographers* 18: 166–78.

Williams, R. (1973) *The Country and the City*, New York: Oxford University Press.

Wolch, J. (1996) "Zoöpolis," *Capitalism, Nature, Socialism* 7: 21–49.

Zukin, S. (1991) *Landscapes of Power: From Detroit to Disney World*, Berkeley: University of California Press.

6

ENVIRONMENTALISM, WISE USE AND THE NATURE OF ACCUMULATION IN THE RURAL WEST

James McCarthy

INTRODUCTION

On July 4, 1994, Nevada rancher Dick Carver, carrying a copy of the US Constitution, took a bulldozer to force open a closed road on US Forest Service land without waiting for federal approval. The incident was planned and publicized at the local level, and Forest Service employees were sent to stop Carver. He ended up forcing one of them out of the way as he plowed a short section of roadbed. Approximately 200 onlookers, many of them armed, encouraged Carver to continue, and some seemed eager for confrontation. Although the incident ended peacefully, Carver later claimed that the crowd would have shot the armed Forest Service employee on the spot had he reached for his gun (Larson 1995).

Carver's dramatic action has become the most widely publicized eruption of the Wise Use movement, a complex new social movement centered on the uses of rural environments – and the strongest anti-environmental backlash in the twentieth-century USA.[1] The movement seeks to "destroy environmentalism" and promote the "wise use" of natural resources (an intentionally ambiguous phrase strategically appropriated from Gifford Pinchot and the early conservation movement). More concretely, it attempts to increase private access to public resources and reduce state regulation of private land and resources. Wise Use has engendered considerable debate in the USA in recent years: is it the genuine populist grassroots social movement it claims to be, or the mere front group for industry environmentalists say it is? What does it mean for contemporary environmental politics? At stake are changing uses of rural environments in the USA: what economic activities will be permitted? Which will dominate? Where will "nature" be preserved? Who will benefit or lose?

Although I suspect he would hate the idea, I think that Dick Carver and many

126

of his Wise Use compatriots have a lot in common with long-dead French and British peasants. I see striking parallels between Wise Use and the landed aristocracy's defenses of the "traditional" rights of peasants in nineteenth-century France and eighteenth-century Britain, as explored in Marx's *The 18th Brumaire of Louise Bonaparte* (1963) and Williams' *The Country and the City* (1973) respectively. In each case, privileged property rights and labour arrangements in rural areas are threatened by a period of accelerated economic integration into an evolving capitalist economy. In each case, a group of capitalists deeply invested in rural production responds by mobilizing and using a larger class of rural workers and property owners to lend numbers and the ideological legitimacy of populism to its defense of existing property rights and relations of production. In each case, rural relations and meanings that are "supposed" to be overridden by capitalist development show surprising durability. In short, appeals to the "traditional" rights of rural land users, especially property rights, have long had a special potency in attempts to resist ongoing capitalist transformations.

WISE USE AND THE PRODUCTION OF NATURE

The existence of such strong parallels between Wise Use's conservative, reactionary aspects and other land-based class projects, centuries and continents away, suggests that Wise Use engages some perennial issues surrounding the production of nature in capitalist societies. At the same time, of course, much about the movement is specific to its immediate historical–geographical context, while the ways in which perennial social dynamics work are always locally mediated and contested.

This chapter will examine Wise Use in the light of contemporary theory regarding the production of nature. This approach serves two broad purposes. First, it provides tools to help us think about Wise Use's struggles over the production of nature: how and where they occur, by and for whom they operate, and how the dialectics between nature's produced and "natural" aspects and its discursive and material productions work in specific cases. By emphasizing the multiple arenas and complex ways in which Wise Use contests the production of nature, such an approach points beyond the predominant explanation of the movement, which is that it is a simple, functional "response" to recent economic restructuring and that rural communities and workers who support Wise Use are essentially dupes providing a cloak of legitimacy for intensified corporate exploitation of public resources.[2] What becomes apparent is the complexity of Wise Use's struggles, the unstable, porous boundaries of "the movement," and conflicts within it. Second, such sustained engagement with a complicated contemporary case tests theoretical approaches and suggests directions for future inquiry.

Debates about the "production of nature" under capitalism – including the degree to which it is produced rather than given, the functions of the ideological separation of nature and society, and what is at stake in competing concepts of "nature" – are fundamental to Marxist thought (Marx 1963; Williams 1980; Smith 1984). But they have gained new prominence on many theoretical agendas as the twentieth century draws to a close and the urgency and legitimacy of "environmental" issues has surged. Recent critical ferment has thus identified both important new sites and mechanisms of nature's production, and new questions regarding it.[3] Many of these have been tied to increased recognition that nature is profoundly political, not reducible to but inseparable from its discursive construction and from questions of power (Haraway 1992; Castree 1995) – a claim that challenges some basic tenets of environmentalism (Cronon 1995; White 1995).

Yet, much about the production of nature is not new at all. Social and ecological relationships are inextricable in any struggle over nature: to challenge or defend one is to do the same to the other (cf. Harvey 1996), and the pursuit of social goals from the environmental side of the dialectic is a classic strategy (Marx 1963; Glacken 1967; Williams 1973, 1980). Property rights remain a crucial arena in struggles over the production of nature in capitalist societies. Geography continues to matter: established, geographically specific patterns of land and resource ownership, use, and regulation often give rise to and powerfully shape such struggles. "Environmental" social movements may be new as a category, but many of the issues they form around are not. The commodification of new areas of nature, the displacement of social conflicts and agendas into "nature," and fears of the loss of an "authentic" nature, are all long-standing dynamics (Williams 1976, 1980). Above all, many struggles over nature remain fundamentally polit-ical and economic struggles over who will gain or lose depending upon what kinds of natures are produced where (Willems-Braun 1997).

Both the old and the new are embodied in the Wise Use movement, a coalition of over a thousand national, state, and local groups. Its existence by this name dates from a "Multiple-Use Strategy Conference" in 1988, attended by nearly 200 organizations, mainly in the western USA, including natural resource industry corporations and trade associations, law firms specializing in combating environmental regulations, and recreational groups. Wise Use groups have employed tactics ranging from grassroots activism to sophisticated media campaigns, and from legal and regulatory strategies to outright physical intimidation and violence. Their struggles thus engage at some point nearly every aspect of the state, capital, and civil society. Estimates of the movement's size range from a few tens of thousands of participants to several millions, but its rapid growth is beyond doubt (Brick 1995a). Large natural resource industry corporations remain its major funding source, and its national organizers admit to having set out to create a national "grassroots" movement, leading many to label it a mere front for corporations hiding behind populist rhetoric. Yet its strong, undeniable grassroots components make its class composition unclear.

128

The movement's size and diversity, and the ambiguity of "wise use," make defining a core Wise Use agenda difficult. Wise Use has been shaped – in some-times contradictory ways – by political-economic factors, by the mistakes of mainstream environmentalism, by Western political history, and by the effective-ness of populism and appeals to nature. Wise Use in the western USA focuses on defending continued commodity production on federal lands, and in the eastern USA on resisting regulation of private lands and development. But many of the movement's issues and constituencies cross these boundaries: priorities vary by region, by industry, and by organization. Moreover, Wise Use is full of internal tensions and contradictions. It includes complex alliances of corporations, labour, small businesses, and individual property owners in almost every conceivable configuration, while relevant governmental agencies intersect these and each other in myriad ways. Yet despite this heterogeneity, several issues and themes unite Wise Use, as will become clear (see Gottlieb 1989; Arnold 1993).

The structure of this chapter is as follows. I first argue that Wise Use should be understood primarily as a site of political-economic contradictions and struggle, both between classes and between capitalists. I then examine Wise Use's attempts to influence the production of nature in ways consonant with its political-economic goals in several key arenas: in state spheres, including regulation and property rights; in discourse; and in grassroots politics, as a social movement.[4] I conclude with an exploration of the conflict between Wise Use and the environmental movement and its lessons for current environmental politics.

WISE USE, COMPETITION AND CLASS CONFLICT

The Wise Use movement is first and foremost a vehicle and arena of political-economic struggle with particular class orientations. In contrast to the complexity of its means, its ends are often brutally clear: it functions mainly to defend privileged elite and corporate access to resources, and it may also be a wedge in a larger neoliberal project.

Struggles over rural environments in the USA are entwined with processes of economic and regulatory structuring at multiple spatial scales. Most accounts of Wise Use locate its origins in the decline of many natural resource industries in the rural West in the past two decades. These shifts in the region's economic base are best interpreted as part of post-1970 "global economic restructuring,"[5] which has included substantial changes in the international geography of production and consumption. Restructuring has been particularly pronounced in rural areas in the USA dominated by primary production, which enjoyed a massive boom during the 1970s followed by overproduction, excessive debt, and a severe crash in the early 1980s that virtually devastated many of them. The historic dominance and relative economic significance of natural resource industries in the rural West has declined and they appear unlikely to regain their dominance due to increased competition, resource exhaustion, declining federal subsidies, and increasing environmental regulations. Consequently, many rural localities have aggressively sought to diversify their local economies, accelerating their own restructuring by luring new users of rural environments, who often prize them for consumption-related values.[6] Such changes have led many contemporary observers to make millennial claims heralding the birth of a "New West" that is finally leaving behind its nineteenth-century extractive origins and whose twenty-first century economy will be based upon high-tech industries, tourism, and environmental amenities (e.g. Marston 1989; Power 1996).[7]

The standard analysis that Wise Use is a response to economic restructuring is not wrong, but it portrays the movement as a far simpler "response" than it is. I would like to develop this perspective by examining Wise Use within the framework of what Karl Polanyi (1944) identified as the "double movement" characteristic of modern capitalist societies: the ongoing commodification and subjugation to market logic of both humans and nature on the one hand, and efforts to mitigate the destructive consequences of this via various forms of social resistance and regulation on the other. The Wise Use agenda contains conflicting elements that correspond to the two sides of Polanyi's framework: it simultaneously advocates the marketization and deregulation of rural environments, and the conservative defense of established rural land uses and social relations against changing markets and regulatory structures. Both are historically common responses to heightened international competition – on the one hand, attempts to increase accumulation by removing any obstacles; on the other, attempts to strengthen barriers to competition – but which is chosen by a given actor or group depends on the specifics of its situation. Wise Use has, for now, brilliantly surmounted the tension between these two impulses within itself by defining the goals of each as "property rights": of private owners objecting to restrictions, or of users of the public lands defending their entitlements. Such contradictory tendencies reveal that Wise Use is in large part an alliance between private interests with somewhat conflicting goals that are best understood in terms of their class positions and historic access to rural land and resources. Let me lay out what interests are on each side of this contradictory dynamic.

Wise Use and radical marketization

> Privatisation is the highest priority for the environment . . . By denying ourselves material wealth today, by slowing the accumulation of wealth, we are denying our children. You deny the future by not using resources now.
>
> John Shanahan, of the Heritage Foundation
> (quoted in Rowell 1996: 51)

At first glance, the Wise Use movement would appear to belong squarely to the first side of Polanyi's dialectic: the reduction of all inputs to accumulation to commodity form. Its central goal is the continued development of natural resources in rural environments as the postwar economic and regulatory order erodes and subsidies to primary production decline. It vehemently opposes attempts to take these resources entirely out of the marketplace (such as wilderness reserves or endangered species protections). It also resists most regulation, preferring to let fights over environmental degradation take place between atomized individuals without government intervention. Despite its rhetorical flourishes, invocations of history, and important cultural facets, Wise Use is, in the end, most concerned with that most capitalist of objectives with regard to the production of nature: the creation or defense of private property rights in opposition to either public ownership or the state's right to regulate private property. Like Locke's, the Wise Use property rights agenda is specifically bourgeois. If enacted, it would amount to a significant redistribution of wealth and of environmental risks and benefits in favor of those who already own property (Echeverria and Eby 1995).

Wise Use also appears to be part of a larger, conscious program of economic liberalism designed to roll back much of the twentieth century's social protection legislation (notably the New Deal's) such as the National Labour Relations Act, minimum wage laws, civil rights legislation, and most entitlement programs (Ramos 1995). Its property arguments have their intellectual roots in the libertarian legal thought of Epstein (1985), who contends that virtually any reduction in the use or value of private property due to regulatory action constitutes a taking that must be compensated by the government. What this amounts to is de facto deregulation: by making regulation far more expensive and then attacking government deficits, deregulation is achieved without ever being debated as an overt policy goal (Bromley 1991). This argument thus supports a much broader deregulatory agenda intended to remove perceived constraints on capital accumulation. Wise Use's property rights campaign, centered on land-based property rights central to American ideology, may be the most politically palatable cutting edge for this broader assault on social protection legislation. In short, it can be viewed as a wedge for class warfare: Wise Use manages to make a truly radical movement in favor of capital look conservative and legitimate by tying it to popular ideologies and selectively constructed histories.

Struggles over property rights can also be viewed as an arena in which capitals compete to survive economic crisis. Downturns in natural resource industries and the more general decreased competitiveness of the USA have led to mechanization, struggles over wage rates, decreased subsidies, the reconfiguration of access to public resources, deregulation, and attempts at privatizing some infrastructure and resources. Wise Use's unremitting focus on property rights suggests that the movement can be read largely as a struggle over whose assets – whether real property or claims on property – will be eliminated or devalued as ongoing economic crisis leads to the restructuring of production. In this analysis, historical investments in particular produced natures and socio-ecological relationships are akin to fixed capital (O'Connor 1988; Harvey 1982, 1996). In sum, it seems clear that Wise Use is squarely on the side of capitalist modernization.

Wise Use and social protection

Yet, other elements of the Wise Use agenda correspond directly to the second half of Polanyi's dialectic: efforts to limit the ravages of the market and to preserve, at least to some degree, existing social relations and the environment. Such impulses arise from both economic self-interest and, as Polanyi emphasizes, non-economic motivations that complicate class analysis.

Straight market logic in recent years would often dictate turning rural environments over to other forms of commodification (the US Forest Service, for instance, announced in 1996 that revenue from recreational and ecological uses of the national forests now vastly outstrips that from timber harvesting (McHugh 1996)), or acknowledging that market forces now favor the relocation of much primary production elsewhere, as new supplies and cheaper producers have entered the international market (cf. Power 1996). Wise Use has responded with demands for economic and national protectionism. This conservative, reactionary side of the movement is dominated by a small number of land-based interests: primarily ranching, timber, and mining industries. It seeks not deregulation, but the freezing of certain environmental and regulatory relations, the retention of value-adding activities within the USA, and the maintenance of massive subsidies, such as below-cost timber sales or the mandatory sale of public lands to mining corporations. It is, in fact, a coordinated plea for entitlements – i.e. the exact opposite of any kind of "free market." The vital point here is that pressure to continue established rural land uses, to protect claimed property "rights," even when aimed towards increased commodity production, may run counter to the imperatives of maximizing capital accumulation in general.

Given that this side of the Wise Use agenda departs from free market dogma, it is hardly surprising that it also invokes cultural arguments. Much of its discourse centers on historic rights and attachment to the land and contains elements of a "moral economy" interpretation of Western history: i.e. how resources are used and sold should not be left entirely to the market. It often casts its demands as defenses of workers in these industries, or of small producers considered essential to US national identity. Wise Use thus can also be interpreted as a form of cultural resistance to capitalist modernization, a defense of tradition and established meanings and livelihoods in the face of accelerated economic restructuring and economic integration. Interpretations of it as a straightforward functionalist "response" are least useful in explaining this side of Wise Use, which is deeply shaped by its roots in Western history and culture.

WISE USE AND THE STATE

The state is a central arena in struggles over the production of nature. Wise Use demonstrates this clearly: given the great diversity of its tactics, it is striking how quickly most of them converge on the state, seeking specific legislative or regulatory changes. The main legacy of Dick Carver's bulldozer ride has been a tangle of slow-moving lawsuits. Yet, a powerful antipathy towards the federal government is central to the political culture of the western USA and to Wise Use, and shapes all of the movement's efforts. Both the centrality of state arenas and anti-federal sentiment are reminders of how much geography matters in the production of nature: the federal government is such a key site for Wise Use struggles in the western USA because its role is so great there (for numerous historical and geographical reasons). Here I will explore three state arenas crucial to the Wise Use agenda: the management of federal lands in the West, property rights, and environmental regulations.

Federal lands

Wise Use was first organized as a movement to contest the uses of the vast federal lands that make up half of the western USA, and these issues remain central to its agenda. It advocates continued or increased logging, mining, ranching, and other natural resource production on these lands; the recognition of traditional entitlements on the public lands, such as grazing allotments, as private property rights; the transfer of these lands to state, local, or private ownership and management; and changes or reinterpretations of some federal regulations governing the uses of these lands. These goals are seen as correctives to a trend towards environmental preservation and certain kinds of recreation becoming the dominant uses of the federal lands. At a deeper level, they are ways of protecting established social relations by defending particular ecological relationships and land uses on the federal lands, and of contesting who gets to be the "public" on the public lands. It is in these debates that Wise Use connects most directly to longstanding issues in western politics – particularly the pattern of resentful dependence on federal largesse.

One of the major tactics of this wing of the Wise Use movement has been the passage by many rural western counties, highly dependent on commodity production on federal lands within their borders, of a coordinated series of county ordinances asserting local control over those lands. They define the dominance of particular natural resource industries as central to "local custom and culture" and therefore legally protected under federal legislation. County officials insist that federal officials must consult them regarding changes in the management, use, and regulation of lands within their counties. Over 70 counties had passed such ordinances as of 1996, with more considering similar measures. County supremacy proponents go so far as to compare themselves to Native Americans as groups being forced off of western lands by a distant state (ignoring the ironies therein). They have been quite militant in some cases, engaging in direct physical confrontations with federal employees attempting to enforce environmental regulations on public lands. In Catron County, New Mexico, the first county to pass these ordinances, Wise Use activists have continued to elaborate the "local culture" argument: they now claim that their resentment of a distant federal government's control over local resources stems from their English, Scottish, and Irish descent, with its memories of oppression and enclosures (Dowie 1995).

The ordinances' legal foundation is dubious at best, but they do raise substantive issues regarding the state's role in the production of nature: how centralized should power and decision-making be regarding issues of resource use and access demanding knowledge of local conditions over time? As federal bureaucratic priorities change, local commodity producers are trying to shift decision-making processes towards local levels, where they have more power and influence. In this resurgent localism, many rural Westerners clearly regard out-of-state organizations – whether they be corporations, environmental groups, or the federal government – as the enemy (Erickson 1995).

Property rights

The Wise Use movement frames many of its struggles over resource control and access in terms of defending private property rights against their alleged "taking," applying slightly different versions of this argument to public and private lands. It claims first, that the usufructuary rights of traditional commodity producers on the federal lands have the status of private property rights, and second, that private property encompasses the right to use the property in any way the owner chooses, unaffected by governmental regulations. The movement thus argues that any governmental action that reduces or takes away these rights – e.g. the elimination of a grazing permit on federal lands, or regulations prohibiting the draining of wetlands, even on private lands – constitutes a taking of private property for public use, for which the government must pay monetary compensation. The implicit claim is that the production of nature should be a private affair, not a matter of democratic public debate (cf. Haraway 1997).

Property rights are currently the largest and most rapidly growing component of the Wise Use agenda. They have a much broader geographic and class appeal than the original focus on western federal lands, and so enabled Wise Use to grow into a broad-based national movement. They also resonate powerfully with American thought regarding the state's relationship to property and nature, which was largely forged around the experience of appropriating, transferring to private ownership, and developing the vast natural resources of North America.

The Wise Use property arguments are not entirely spurious: there are many legal ambiguities concerning the regulation of private property,[8] and, while the assertion of private property rights on the public lands sounds oxymoronic, it has some basis in both institutional history and local patterns of customary resource use and access.[9] The attempt to attach a monetary value to the differences between various permitted uses of nature, by defining them as takings of property, can also be defended as a valid tactic in a capitalist society (cf. Harvey 1996). But overall, the takings argument, although compelling in its simplicity, is deeply flawed. It relies on a slippage from real, tangible property to estimated values as "property." It ignores the fact that regulations also maintain and enhance property values. It relies on the historically untenable assumption that private property rights are absolute (Echeverria, in Echeverria and Eby 1995).[10] The fundamental problem is that the property rights and values that Wise Use asserts are inseparable from its overall Lockean theory of property, with its inherent class bias.[11] For example, Wise Use defends grazing rights on the federal lands with populist images of family ranchers, but fully half of the public grazing lands are controlled by the top 10 per cent of permit holders, including many large corporations (Helvarg 1994).

Environmental regulations

Environmental regulations are another major target of Wise Use's anti-federal sentiment. The movement contends that they restrict private liberty unequally and raise costs, putting industry (particularly smaller producers) at a competitive disadvantage and thereby contributing to US economic decline and the re-location of production elsewhere.[12] It therefore seeks to overturn or neutralize much of the major domestic environmental legislation of the past twenty-five years and advocates the increased use of cost/benefit analysis, revised approaches to risk assessment, and the use of "economic impact statements" and new appeals procedures that would effectively neutralize nearly all new regulations.

Wise Use also resists US participation in international agreements and institutions that, in its view, will dictate and limit access to land and resources in the USA (e.g. the World Heritage Treaty, the UN Commission on Population and Development, the global Biodiversity Treaty, and the Convention on Climate Change). It sees them as mechanisms to further erode US power and autonomy; to oppose them is thus to combat economic integration. Such stances form an important part of Wise Use's appeal to rural white workers who feel threatened by a rapidly changing international economy and regulatory structure virtually beyond their power to affect. While such fears are easily ridiculed when they take the form of warnings against black helicopters and "one world order" (as they have recently in the rural West) they have more than a germ of legitimacy: the fear of environmental protection and conservation becoming an institutional form through which oppressive state control of rural land uses and people is exercised is hardly a fanciful one, either now on a global scale (Neumann and Schroeder 1995; Peluso 1996) or in relatively recent US history (DuPuis 1996).

THE DISCOURSE OF WISE USE

We are the true environmentalists. We are the stewards of the earth. We're the farmers and ranchers, the hunters and trappers, the fishermen and watermen – people who have cared for the land and the waters for

generations. We're the miners and the loggers and the energy producers
who have provided this nation with the building blocks of a modern
civilization. . . . we hard working Americans built this country.

(Pendley 1994: vii)

Discourses are practices: what the Wise Use movement says is no small part of
what it does. Dick Carver's bulldozer ride was largely symbolic: the fact that
it consciously played on myths of the Old West and American patriotism
and received extensive media coverage was at least as significant as the physical
challenge to federal land managers. Discursive analysis has important analytical
limitations, but there is no question that ideas matter in shaping the material
production of nature.[13] Ideas of nature in particular have an enduring, nearly
archetypal, power in social conflicts. Discourses and representations concerning
nature and the West have been among the central and most effective means by
which Wise Use has attempted to affect material productions of nature. They help
to create the conditions for – and cannot be separated from – its success in
other arenas (cf. Harvey 1996; Willems-Braun 1997). Wise Use's discourses all
attempt to establish or contest legitimacy in struggles over the production of
nature in the rural USA. They fall into two broad categories: those relating
directly to the environment, and those centered on the intersection of class and
geography.

In the first category, Wise Use has strategically appropriated the language of
the early conservation movement in the USA – "wise use" was Pinchot's mandate
for the uses of the public lands – and thereby positioned its agenda as one of
restoring an historic balance that has been thrown awry in the direction of
preservationism in recent years. Wise Use, like its cousins "conservation" and
"sustainable development," is intentionally vague, is difficult to be against, and
assumes that economic use of resources will continue. It is no coincidence that
critiques of sustainable development's appealing ambiguity and lack of a
fundamental critique of development (Sachs 1992; Escobar 1995) are directly
applicable to "wise use." Likewise, the discourse is constantly evolving: many
Wise Use participants have abandoned that now controversial term, saying
instead that they advocate "multiple use" or are engaged in "resource advocacy"
(Switzer 1996). Wise Use participants also reject the "anti-environmental" label
commonly applied to them, insisting that they represent a truer environmentalism than does the environmental movement, because they relate to nature
through work rather than through consumption (Echeverria and Eby 1995;
White 1995).

In the second category, Wise Use's privileging of direct producers is part of a
larger agrarian, productivist populism espoused by the movement, which is
ultimately rooted in a physiocratic theory of value. Wise Use's version centers on
small property owners and producers (the petit bourgeosie). This class plays a
pivotal role in the political mythology of both the USA and the West, and is
a position with which both corporations and labour are often happy to identify.

136

This self-representation allows the movement to draw on Jeffersonian ideals and de-emphasizes its corporate aspects – in fact, some Wise Use rhetoric is explicitly anti-corporate, in a classic American trust-busting vein (e.g. Greve 1993; see Snow 1994; Maughan and Nilson 1995). Overall, it facilitates Wise Use's attempts to position itself as a more legitimate representative of the "public" than the environmental movement.

Wise Use activists also invoke geography. They assert that their claims are more legitimate because they are asserted by local rather than distant actors, and based on local rather than "expert" knowledge (this valorization of the "local" will be explored further below). They appeal to selective versions of history and to a pervasive sense of victimization among many inhabitants of the rural West. Wise Use draws much of its rhetoric from previous western resource-based protest movements, including claims that the region is treated as a colony of the East and that the extent of federal ownership and bureaucratic control in the West is fundamentally undemocratic, and many Wise Use activists situate themselves in this lineage of regional protest.[14] While dated at best, these claims remain surprisingly widespread and potent rallying cries in the region. Current changes in the regulations and subsidies relating to the public lands industries are thus being interpreted in much of the rural West as a "war on the West."

WISE USE AS AN ENVIRONMENTAL NEW SOCIAL MOVEMENT

Given this breadth of popular appeal, it is not surprising that Wise Use has had considerable success at the grassroots level. It can thus be viewed as a "new social movement" focused on the environment. The recent emergence of such movements has received substantial critical attention, often being hailed as an important new phenomenon in civil society centrally concerned with contesting the production of nature at the turn of the millennium (see e.g. Shiva 1989; Alvarez 1992; Escobar 1995). Although this work has largely focused on movements in the "South," Wise Use fulfills nearly all of the criteria advanced for such movements, supporting an interpretation of it as the legitimate populist environmental movement it claims to be.[15] Analyzing it thus provides an excellent opportunity to scrutinize this evolving theoretical category, including the dangers of its populist aspects, and to explore how it matters when struggles over the production of nature take the form of a "social movement," and whether they are labeled as such.

To label a given set of struggles over the production of nature an "environmental new social movement" matters in several respects. In the case of Wise Use, it highlights important commonalties between it and similar movements outside of advanced capitalist nations rarely considered within the same analytical frameworks, helping to break down the "North–South" dichotomy in categorizing struggles over resources and perhaps helping Wise Use critics to see it in a new light. More broadly, calling any organized political activity a "social movement" increases its legitimacy and capacity to gain political and financial support: we do not call political action committees (PACs) "social movements." Notably, environmentalists have resisted calling Wise Use a social movement, referring to it instead by labels such as "industry campaign." Last, to categorize these political activities as social movements implicitly recognizes the importance of civil society as an arena of contestation over nature.

There are however serious problems with "environmental new social movement" as an analytical category: each of the terms is highly ambiguous. Social movement theory offers, in fact, little guidance on how to define a social movement, measure one's effects, or understand when or whether one arises (Rochon and Mazmanian 1993; Peet and Watts 1996). It has also been plagued by methodological individualism, voluntarism, and liberal models of public action. Many recent authors (see Buttel 1992) have posited the existence of "new social movements" (NSMs), whose issues of concern and modes of operations in the political sphere differ from widely accepted models. Buttel characterizes environmental movements as NSMs, structurally linking their rise over the past two decades to a general drift towards neoliberalism, the collapse of postwar social democratic parties, and significant changes in how class politics work. Similarly, some authors have interpreted Wise Use as a NSM (Snow 1994; Maughan and Nilson 1995). But while the emphasis of NSM theory on recent structural changes is valuable, it would be a serious mistake to read Wise Use as being in any way "beyond" class politics. There are also problems with the NSM theory's linkages between economic change and social struggles (cf. Harvey 1989).

Compounding the ambiguity of "new social movement" as a category is disagreement over what makes some "environmental." Yearley (1994) claims that they are characterized by their internationalism, the importance of science in them, and their visions of a radically alternative society. Others argue that their distinctiveness lies in the fact that they often utilize conceptions of time and space that differ sharply from those dominant in capitalist society (Harvey 1990). Still others contend that they are part of a larger "greening" of capitalism, in which greater attention to environmental concerns and constraints has become essential to the ongoing expansion of capital (O'Connor 1993; Altvater 1994; Escobar 1996). One could also argue that they are distinctive insofar as many environmental concerns lie in the realm of use values produced outside of circuits of capital.

In contrast, I would suggest that what is perhaps most curious about many of these movements is that they are not distinct, that many of the issues at stake are neither new nor specifically environmental. Many of Wise Use's core struggles concern the terms of increasing economic integration and the accompanying loss of local control, challenges to identity created by these processes, the advantages and dangers of "free markets," the extent of a social safety net, and the role of the state. These issues are neither specially "environmental" nor remotely new in the history of capitalism. It is also very easy to overstate and romanticize the "authentic" aspects of Wise Use and other environmental social movements (in what DuPuis and Vandergeest 1996 call "green orientalism": i.e. fetishizing the authenticity of the environmental knowledge and managerial capacities of "local" people). The "local" is often cynically manipulated, is largely the outcome of interactions with extra-local processes, and is by no means a necessarily progressive construction (Watts and McCarthy 1997). Nor is there any necessary correlation between "localness" and environmental knowledge or responsibility (Harvey 1996). For example, despite its rhetorical championing of "local" control, Wise Use has rapidly grown into a coordinated national movement using faxes and e-mail, with transnational corporations being major beneficiaries.

Why, then, have "environmental new social movements" achieved such prominence in struggles over the production of nature at the end of the millennium? Possible answers include: that they are strategic appeals for funds and legitimacy given that environmental concerns appear to have largely edged out social justice issues on many development and aid agendas; that they are chosen as political vehicles because they are issue specific and able to cut across class boundaries (Buttel 1992); and that they are now the most acceptable framework within which to critique capitalist modernization in the wake of the apparent discrediting of socialism. Or perhaps the movements are fundamentally like earlier ones, but we are simply labeling them differently.

I would like to suggest one other possible answer. If capitalist modernity is encapsulated in Marx's famous phrase "All that is solid melts into air," then nature becomes that which does not melt into air, the most solid possible anchor for threatened social relations, quite literally the ground upon we take our stands. Social relations and meanings relating to "nature" show surprising durability in the face of capitalist transformations, in part because people invest so much in them as the repository or site of specifically non-economic values. Environmental movements, particularly those defending established patterns of resource use and livelihood, are thus particularly effective ways to contest the constantly de-stabilizing effects of capitalism. Both environmentalists and Wise Use supporters employ nature in this way, and modern history is full of invocations of the ideal of an environmentally rooted "moral community" as a defense against capitalism's demystification of the world and constant destruction of existing relations.[16] But moral community relating to nature is of course mainly about the community, making it essential to scrutinize the material and political-economic relationships being defended or pursued under the rubric of any given "environmental" movement.

CONCLUSION: WISE USE AND ENVIRONMENTALISM

The environmental extremists' vision of the West is of a land nearly devoid of people and economic activity . . . everything from the 100th meridian to the Cascade Range becomes a vast park through which they drive, drinking Perrier and munching on organic chips, staying occasionally in the bed-and-breakfast operations into which the homes of Westerners have been turned, with those Westerners who remain fluffing duvets and pouring cappuccino.

(Pendley 1995: 7)

This brings us at last to the confrontation between two of the most powerful visions of environmentally rooted moral communities: those of Wise Use and mainstream environmentalism. Like a photographic negative, the Wise Use movement highlights the very areas the environmental movement has left dark. In this sense, it may be the most important development in environmental politics in the USA in decades. Wise Use's existence and rapid growth, its particular configurations and issues, its stance towards environmentalists, and environmentalists' responses to it all point to larger themes in contemporary environmental politics. Wise Use is part of a larger international "green backlash" (Rowell 1996) which, while predominantly neoliberal in orientation, does pinpoint important flaws in how environmentalists have thought about the production of nature. There are major constituencies in Wise Use that are not corporate, including many rural communities, wage labourers, and small property owners. Wise Use clearly speaks to them in ways that mainstream environmentalism does not, and it is important to understand why rather than to dismiss them as victims of false consciousness.

Wise Use mounts a powerful critique of mainstream environmentalism. It charges – with considerable evidence – that environmental protection has been in large part an elite project representing not a generic "public" but a group segregated by race, class, and place. It further characterizes environmentalists as anti-human, arrogant, ignorant of local conditions and issues, extreme, violent, and religious (pagan) in outlook. In this light, the struggle between the two movements is in large part a struggle to stake out the perceived "center" of civil society, and the legitimacy that comes with it, via contrasting definitions of the "public good." Wise Use has succeeded to a surprising degree in displacing environmentalism from the moral high ground of an altruistic champion of various public goods that it occupied for much of the postwar period, questioning its definition of the public good as an aggregate national construct and casting environmental protection as, rather, an elite interest. Wise Use simultaneously casts itself as an insurgent populist movement that represents "the people" rather than special interests and multiple local "publics" rather than one national one. It presents itself not as "anti-environmental," but as a viable alternative to the mainstream environmental movement. Wise Use has thus been shaped largely in conversation with environmentalism, and the cornerstone of its appeal is that it places livelihoods and the necessity of the social transformation of nature at the center of its ideologies in a way that environmentalism never has (Snow 1994; White 1995).

Wise Use's critique is not entirely unfounded. Mainstream environmentalism's historic refusal to acknowledge nature as produced has led it to embrace an inaccurate schism between nature as a valued "wilderness" and a dismissed "society" encompassing everything else (Smith 1984; Cronon 1996). This failure to engage with the necessity of the transformation of nature or the many shades of produced natures has too often amounted to environmentalists neglecting issues of work, livelihood, and the production and distribution of wealth in society. In short, environmentalism has failed to theorize the labour process and effectively abandoned this terrain to the opposition (White 1995).

This fundamental theoretical misstep has led to many omissions and failures on the part of the mainstream environmental movement.[17] First and foremost, it has been slow to recognize that environmental protection must be reconciled with production and social justice, or to trace many environmental problems back to their roots in capitalist dynamics (Harvey 1996). Second, historically it has displayed too little concern for local economies, communities, working-class issues, and urban areas (Wilderness Society 1993; Snow 1994; Erickson 1995). Third, it has focused on achieving policy goals via access, legislation, and litigation at the federal level, neglecting the grassroots and specific local issues (Baca 1995; Dowie 1996). Fourth, it has tended to advocate ecological projects while ignoring both the desires of those most directly affected and the economic consequences, except as an afterthought (Brick 1995b; DuPuis and Vandergeest 1996) (although, ironically given their overall advocacy of taking many resources out of the market, environmentalists sometimes implicitly seem to champion capitalist modernization when it comes to natural resource industries).[18] Fifth, it has often not distinguished between resource use and resource depletion, collapsing the former entirely into the latter. Encompassing all of the above, it has failed to see ecological and political economic projects as inextricably linked: it has attempted to change relations to "nature" without recognizing or admitting that to do so was simultaneously to change social relations (Snow 1994; White 1995; Harvey 1996). These areas of neglect lend credence to Wise Use's critique of environmentalism.

Sadly, environmentalists do not appear to be rushing to correct their mistakes. While some have seen Wise Use as a call to redefine their goals and methods, most have been slow to recognize the legitimate issues it raises and have restricted themselves to trying to discredit it (in part on the strikingly disingenuous basis that it receives corporate money and corporate officials sit on the boards of its organizations). As a result, most of their accounts of Wise Use are flawed and partial in numerous respects. Typical responses have included plans for systematic smear campaigns against Wise Use groups, stronger appeals to their existing supporters, attempts to "hold the line" on existing environmental legislation, and calls for ideological warfare to "capture the hearts and minds of the local people" (Helvarg 1994) – in sum, retrenching rather than rethinking.

To conclude, it seems to me that Wise Use espouses a more accurate view of the production of nature than many environmentalists. I certainly do not endorse its view entirely: it tends to recognize only the enabling aspects of nature's produced nature and to ignore the continued presence of natural constraints. More important, its net effect is largely to support and obfuscate corporate capital's continued privileged role in shaping contemporary productions of nature. Despite its populist rhetoric, Wise Use has had nothing to say about, for instance, the relationships between urban pollution and exploitation of the poor. Yet in its basic recognition of the interactive and constitutive relationships between work and "nature," Wise Use – the allegedly reactionary, corporate movement – may in fact champion a far more dialectical view of human environment relations than the environmental movement.

Environmentalists are thus making a strategic and theoretical mistake by trying to counter Wise Use with continued defenses of pure "nature." They are fighting a rearguard action, as even scholars who describe themselves as committed environmentalists mount attacks on "nature" or "wilderness" as ontological categories. The most recent flare-up has centered on William Cronon's (1996) essay, "The trouble with wilderness," in which he ably deconstructs the concept of wilderness and suggests, rightly I think, that an excessive focus on it has been detrimental to American environmentalism (see also White 1997). The reaction of environmentalists has been dismaying. Rather than recognizing positive critique, they have dug in their heels and defended wilderness. While some authors (Foreman 1997; Waller 1997) have raised valid criticisms of Cronon's argument (e.g. its lack of ecological criteria and its sweeping generalizations regarding a complex movement), most of the responses have missed his point entirely, insisting that efforts to incorporate social issues into environmentalism is a coopatation of the latter's agenda, and in some cases resorting to ad hominem attacks (e.g. Snyder 1996/7; Sessions 1997). Most revealing are those that concede Cronon's point but caution him against publishing it lest it be used by anti-environmental forces. The latter of course implicitly admit that "wilderness" is a political and discursive category, always being actively redefined, but they cannot bring themselves to make this admission explicit. Environmentalism would be better served if they would.

ACKNOWLEDGEMENTS

I would like to thank Michael Johns, Richard Walker, Louise Fortmann, Donald Moore, Julie Guthman, Scott Prudham, Anne Walker, Jonathan London and the editors for detailed and challenging comments on an earlier version, and Nancy Peluso and Fred Buttel for related conversations.

NOTES

1 Opposition to the environmental movement and regulations in the USA is of course hardly new; what is new is the exceptional breadth and organization of the Wise Use coalitions.

2 While containing much that is accurate, this analysis is functionalist, biased, and overly simplistic. It lacks historical depth and attention to geographic variation and social complexity, and minimizes the roles played by environmental regulations and growing environmental and recreational claims on resources. Central and representative examples include: Wilderness Society (1993); Helvarg (1994); Echeverria and Eby (1995); Tokar (1995); Rowell (1996).

3 See Smith (1984); Fitzsimmons (1986); Haraway (1992); Altvater (1993); Latour (1993); O'Connor (1993); Sachs (1993); Castree (1995); Escobar (1995, 1996); Peet and Watts (1996); Smith (1996); Watts and McCarthy (1997); Willems-Braun (1997).

4 Contestation over scientific definitions of 'environmental' problems is also important in Wise Use's efforts to affect contemporary productions of nature, but I will not address this here.

5 This is a collective term for a complex set of related processes. See Dicken (1986); Harvey (1989); Storper and Walker (1989); Flora (1990); Pred and Watts (1992); Sayer and Walker (1992).

6 See Flora (1990) and Buttel (1992) for broad overviews of recent trends in rural restructuring, and Albert et al. (1989), White (1991), Hess (1992), Snow (1994), Maughan and Nilson (1995), and Power (1996), for accounts of effects on the rural West in particular.

7 These changes are broadly congruent with an analysis of "rural restructuring" in the advanced capitalist countries put forward in recent work by a group of (predominantly) British academics attempting to extend the analysis of the French regulationist school to rural areas (e.g. Lowe et al. 1990, 1993; Marsden et al. 1990, 1993). Nearly all the research underlying this work has been specific to the British countryside, however. The USA presents enormous geographical and historical differences, particularly in the West, including: greater state and corporate land ownership, an extraordinary military presence, extensive forestry and ranching industries, Native American claims to land and resources, and considerable climatic constraints (cf. Davis 1993; Goodman and Watts 1994).

8 Historically, 'takings' have applied to condemnation of tangible property under the government's power of eminent domain, such as taking property for a road, not to regulatory changes that reduced value, or even eliminated it entirely. However, the Supreme Court has also said at times that when enough value is removed it is a "regulatory taking." But it has never provided any kind of formula or guidance regarding how much is "enough." There are also other legal ambiguities and interpretations that favor the Wise Use position. Therefore, Wise Use efforts to push the issue are by no means unfounded (see Echeverria and Eby 1995 and Yandle 1996 for reviews of and commentaries on precedents and contemporary cases).

9 Systems of extralegal property rights existed on these lands prior to intensive federal administration; the federal land agencies regulating these uses were for decades classic examples of agencies "captured" by the interests they supposedly regulated, meaning that policies were largely dictated by users with relatively little outside interference, leading to low fees and extensive local control; commodity production has long had both customary and legal status. Many families, particularly ranchers, have, as they claim, been working the same land for generations, in some cases since before the institution of federal bureaucratic controls (Hess 1992; CQ Researcher 1994; Dowie 1996; Klyza 1996; cf. DuPuis and Vandergeest 1996).

10 Property rights do not give property owners permission to do things that will harm others, such as pollute. Wise Use activists respond that others who feel that they have been harmed are free to sue, but this is hardly a tenable solution to problems such as dispersed air pollution or degraded water quality.

11 The Wise Use view of property is basically Lockean: the role of the state should be to facilitate individuals' appropriation and transformation of nature, which is best done by converting the commons into private property and protecting those rights; security of property provides an essential precondition for productive investment and a check on the power of the state (Bentham 1978; Locke 1978; MacPherson 1978; Roush 1995). Glendon (1995) argues that, despite the fact that private property rights have always been subject to regulation, the assertion of their absoluteness has been especially central in the USA because of Locke's strong influence on American thought about the state.

12 The net effect of environmental regulations on economic competitiveness is unclear, but much evidence suggests minimal or positive effects (Brick 1995a; Institute for Southern Studies 1995; Trocki 1995; World Resources Institute 1995). It should be noted that most Wise Use attacks on environmental regulations more broadly have failed thus far (White 1997), with the notable exception of the March 1997 Supreme Court ruling that anyone economically harmed by the Engangered Species Act could bring suit under the act.

13 The precise pathways by which they come to matter, and the range of their causal powers, have been the subject of fruitful explorations in recent engagements between Marxist and poststructuralist thought (Castree 1995; Escobar 1995, 1996; Peet and Watts 1996). Marx (1986) explicitly recognized that it is often within the realm of ideology (including, for him, law and politics) that we become conscious of material conflicts and fight them out, while more recent authors have argued that such struggles have both a high degree of autonomy from, and may be decisive in the transformation of, material changes (Eagleton 1991; Peet and Watts 1996), and millennia of historical evidence suggest that reading social conflicts into nature is highly effective (Glacken 1967; Williams 1976, 1980). Nonetheless, there are problems with relying too heavily on discourse analysis. First, while discourses are of course at some level material practices, much analytical power is lost by collapsing of those categories. Second, discursive analysis often provides no way of evaluating competing discourses, or of understanding how they compete – i.e. the material and power relations that allow one discourse to become hegemonic while another is marginalized (Watts 1995). Third, the basic poststructural turn in social science has been founded on an analytical analogy between language and political-economic systems that Anderson (1984) has cogently argued is fundamentally untenable. See Watts and McCarthy (1997) for a detailed discussion of the problems in applying poststructuralism to political ecology.

14 Wise Use's continuities with major themes in Western history, politics, and culture are beyond the scope of this chapter. See De Voto (1934); Hays (1959); Nash (1985); Robbins (1986); Limerick (1987, 1995); Graf (1990); Truettner (1991); Cronon et al. (1992); Steen (1992); CQ Researcher (1994, 1995); Snow (1994); Maughan and Wilson (1995); Klyza (1996).

15 For example, it is has a strong local component and is outside of the state sphere; it relies on constructions of specific "indigenous" or local cultures; it asserts the superiority of local environmental knowledge over that of professional scientists; it refashions local communities and traditions as hybrids; and it defends the local in the face of the global (see Watts and McCarthy 1997 for a synopsis of these criteria).

16 Such an interpretation parallels Habermas' characterization of environmental movements as efforts to resist the colonization of the lifeworld. This follows in the Frankfurt school tradition, which moved away from a strict focus on class struggle,

examining more broadly instrumental reason and the domination of nature as the sources of domination and repression in society (Eckersley 1992; Castree 1995; Harvey 1996).

17 Speaking of environmentalists as a unitary whole is a necessary shorthand here, but of course unfair in some respects. Some environmental organizations have begun to work more cooperatively with natural resource dependent workers and communities and to broaden their agendas, while other grassroots groups have been doing so all along (Gottlieb 1993). I refer here primarily to the large national environmental organizations, the "Big Ten" being the most emblematic, and to most propopents of "deep ecology."

18 Environmentalists often appear eager to see the remaining representatives of extractive industries in the West fade away into the past where they belong, as "market forces" support a more thorough separation of production and consumption in which much primary production moves out of the USA entirely. Environmentalists thus often eagerly participate in the construction of rural primary production as marginal, backward, and inefficient (see DuPuis and Vandergeest 1996). Many reports and recommendations by environmental organizations on the future of rural western areas have this tone: they suggest that a transition away from a resource-based economy is inevitable, progressive, and well underway (cf. DuPuis 1996), and promote tourism as an alternative.

REFERENCES

Albert, K., Hull, W. and Sprague, D. (1989) *The Dynamic West: A Region in Transition*, San Francisco, CA: Council of State Governments, Western Office.

Alvarez, C. (1992) *Science, Development and Violence*, Delhi: Oxford University Press.

Alvater, E. (1993) *The Future of the Market*, London: Verso.

—— (1994) "Ecological and economic modalities of time and space," in M. O'Connor (ed.) *Is Capitalism Sustainable? Political Economy and the Politics of Ecology*, New York: Guilford.

Anderson, P. (1984) *In the Tracks of Historical Materialism*, Chicago: University of Chicago Press.

Arnold, R. (1993) *Ecology Wars: Environmentalism as if People Mattered*, Bellevue, WA: Free Enterprise Press.

Baca, J. (1995) "People for the West!," in J. Echeverria and R. Eby (eds) *Let the People Judge: Wise Use and the Private Property Rights Movement*, Washington, DC: Island Press.

Bentham, J. (1978) "Security and equality of property," in C. B. Macpherson (ed.) *Property*, Toronto: University of Toronto Press.

Brick, P. (1995a) "Determined opposition: the wise use movement challenges environmentalism," *Environment* 37: 16–20, 36–42.

—— (1995b) "Taking back the rural West," in J. Echeverria and R. Eby (eds) *Let the People Judge: Wise Use and the Private Property Rights Movement*, Washington, DC: Island Press.

Bromley, D. (1991) *Environment and Economy: Property Rights and Public Policy*, Oxford: Blackwell.

Buttel, F. (1992) "Environmentalization – origins, processes, and implications for rural social change," *Rural Sociology* 57: 1–27.

Castree, N. (1995) "The nature of produced nature: materiality and knowledge construction in Marxism," *Antipode* 27: 12–48.

CQ Researcher (1994) "Public land policy," *CQ Researcher* 4, Washington, DC: Congressional Quarterly.

—— (1995) "Property rights," *CQ Researcher* 5(22), Washington, DC: Congressional Quarterly.

Cronon, W. (1996) "The trouble with wilderness," in W. Cronon (ed.) *Uncommon Ground*, New York: W. W. Norton.

Cronon, W., Miles, G. and Gitlin, J. (eds) (1992) *Under an Open Sky: Rethinking America's Western Past*, New York: W. W. Norton.

Davis, M. (1993) "Dead West: ecocide in Marlboro Country," *New Left Review* 200: 49–73.

De Voto, B. (1934) "The West: a plundered province," *Harper's Magazine* 169: 355–64.

Dicken, P. (1986) *Global Shift: Industrial Change in a Turbulent World*, New York: Harper & Row.

Dowie, M. (1995) "With liberty and firepower for all," *Outside* November.

—— (1996) *Losing Ground: American Environmentalism at the Close of the Twentieth Century*, Cambridge: MIT Press.

DuPuis, M. (1996) "In the name of nature: ecology, marginality, and rural land use planning during the New Deal," in M. Dupuis and P. Vandergeest (eds) *Creating the Countryside: The Politics of Rural and Environmental Discourse*, Philadelphia: Temple University Press.

DuPuis, M. and Vandergeest, P. (1996) *Creating the Countryside: The Politics of Rural and Environmental Discourse*, Philadelphia: Temple University Press.

Eagleton, T. (1991) *Ideology*, New York: Verso.

Echeverria, J. and Eby, R. (eds) (1995) *Let the People Judge: Wise Use and the Private Property Rights Movement*, Washington, DC: Island Press.

Eckersley, R. (1992) *Environmentalism and Political Theory*, New York: SUNY Press.

Epstein, R. (1985) *Takings: Private Property and the Power of Eminent Domain*, Cambridge: Harvard University Press.

Erickson, T. (1995) "Finding the ties that bind," in J. Echeverria and R. Eby (eds) *Let the People Judge: Wise Use and the Private Property Right Movement*, Washington, DC: Island Press.

Escobar, A. (1995) *Encountering Development*, Princeton, NJ: Princeton University Press.

—— (1996) "Constituting nature: elements for a postructuralist political ecology," in D. Peet and M. Watts (eds) *Liberation Ecologies*, London: Routledge.

Fitzimmons, M. (1986) "The matter of nature," *Antipode* 21: 106–20.

Flora, C. (1990) "Rural peoples in a global economy," *Rural Sociology* 55: 157–77.

Foreman, D. (1997) "Around the campfire," *Wild Earth* Winter 1996/7: 1–4.

Glacken, C. (1967) *Traces on the Rhodian Shore*, Berkeley: University of California Press.

Glendon, M. (1995) "'Absolute' rights," in J. Echeverria and R. Eby (eds) *Let the People Judge: Wise Use and the Private Property Rights Movement*, Washington, DC: Island Press.

Goodman, D. and Watts, M. (1994) "Reconfiguring the rural or fording the divide?," *Journal of Peasant Studies* 22(1): 1–49.

Gottlieb, A. (ed.) (1989) *The Wise Use Agenda: The Citizen's Policy Guide to Environmental Issues*, Bellevue, WA: Free Enterprise Press.

Gottlieb, R. (1993) *Forcing the Spring*, Washington, DC: Island Press.

Graf, W. (1990) *Wilderness Preservation and the Sagebrush Rebellions*, Savage, MD: Rowan and Littlefield.

Greve, M. (1993) "The importance of property rights," paper delivered at first annual ECO conference, Reno, NV, 19 February.

Haraway, D. (1992) "The promise of monsters," in L. Grossberg, C. Nelson and P. Treichler (eds) *Cultural Studies*, New York: Routledge.

—— (1997) *Modest_Witness@Second_Millennium*, New York: Routledge.

Harvey, D. (1982) *The Limits to Capital*, Oxford: Blackwell.

—— (1989) *The Condition of Postmodernity: An Enquiry into the Origins of Cultural Change*, Oxford: Blackwell.

—— (1990) "Between space and time: reflections on the geographical imagination," *Annals of the Association of American Geographers* 80(3): 418–34.

—— (1996) *Justice, Nature and the Geography of Difference*, Oxford: Blackwell.

Hays, S. (1959) *Conservation and the Gospel of Efficiency*, Cambridge, MA: Harvard University Press.

Helvarg, D. (1994) *The War Against the Greens: The "Wise Use" Movement, the New Right, and Anti-Environmental Violence*, San Francisco: Sierra Club Books.

Hess, K. (1992) *Visions Upon the Land: Man and Nature on the Western Range*, Washington, DC: Island Press.

Institute for Southern Studies (1995) *Gold & Green*, Durham, NC: Institute for Southern Studies.

Klyza, C. (1996) *Who Controls Public Lands?*, Chapel Hill: University of North Carolina Press.

Larson, E. (1995) "Unrest in the West," *Time* 23 October.

Latour, B. (1993) *We Have Never Been Modern*, Cambridge, MA: Harvard University Press.

Limerick, P. (1987) *The Legacy of Conquest: The Unbroken Past of the American West*, New York: W. W. Norton.

—— (1995) "A history of the public lands debate," paper presented at the conference Challenging Federal Ownership and Management: Public Lands and Public Benefits, 11–13 October, Boulder, CO: Natural Resources Law Center, University of Colorado.

Locke, J. (1978) "Of property," in C. B. Macpherson (ed.) *Property*, Toronto: University of Toronto Press.

Lowe, P., Marsden, T. and Whatmore, S. (1990) *Technological Change and the Rural Environment*, London: David Fulton.

Lowe, P., Murdoch, J., Marsden, T., Munton, R. and Flynn, A. (1993) "Regulating the new rural spaces: the uneven development of land," *Journal of Rural Studies* 9: 205–22.

McHugh, P. (1996) "Forest recreation's growing impact," *San Francisco Chronicle* 19 September.

Macpherson, C. (1978) "The meaning of property," in C. B. Macpherson (ed.) *Property*, Toronto: University of Toronto Press.

Marsden, T., Lowe, P. and Whatmore, S. (eds) (1990) *Rural Restructuring: Global Processes and Their Responses*, London: David Fulton.

Marsden, T., Murdoch, J., Lowe, P., Munton, R. and Flynn, A. (1993) *Constructing the Countryside*, London: UCL Press.

Marston, E. (ed.) (1989) *Reopening the Western Frontier*, Washington, DC: Island Press.

Marx, K. (1963) *The 18th Brumaire of Louise Bonaparte*, New York: International Publishers.

—— (1986) [1859] "Preface to the *Critique of Political Economy*," in J. Elster (ed.) *Karl Marx: A Reader*, Cambridge: Cambridge University Press.

Maughan, R. and Nilson, D. (1995) "What's old and what's new about the Wise Use movement," *Green Disk 3*.

Nash, G. (1985) *The American West Transformed: The Impact of the Second World War*, Lincoln: University of Nebraska Press.

Neumann, R. and Schroeder, R. (1995) "Manifest ecological destinies: local rights and global environmental agendas," *Antipode* 28(4): 321–4.

O'Connor, J. (1988) "Capitalism, nature, socialism: a theoretical introduction," *Capitalism Nature Socialism* 1: 11–38.

O'Connor, M. (1993) "On the misadventures of capitalist nature," *Capitalism Nature Socialism* 4/3: 7–40.

Peet, D. and Watts, M. (1996) "Liberation ecology: development, sustainability, and environment in an age of market triumphalism," in D. Peet and M. Watts (eds) *Liberation Ecologies*, London: Routledge.

Peluso, N. (1996) "'Reserving' value: conservation ideology and state protection of resources," in M. Dupuis and P. Vandergeest (eds) *Creating the Countryside: The Politics of Rural and Environmental Discourse*, Philadelphia: Temple University Press.

Pendley, W. (1994) *It Takes A Hero: The Grassroots Battle Against Environmental Oppression*, Bellevue, WA: Free Enterprise Press.

—— (1995) *War on the West: Government Tyranny on America's Great Frontier*, Washington, DC: Regnery Publishing.

Polanyi, K. (1944) *The Great Transformation*, New York: Rinehart.

Power, T. (1996) *Lost Landscapes and Failed Economies*, Washington, DC: Island Press.

Pred, A. and Watts, M. (1992) *Reworking Modernity: Capitalism and Symbolic Discontent*, New Brunswick, NJ: Rutgers University Press.

Ramos, T. (1995) "Regulatory takings and private property rights," unpublished manuscript, Portland, OR: Western States Center.

Robbins, W. (1986) "'The Plundered Province' thesis and the recent historiography of the American West," *Pacific Historical Review* 55: 577–97.

Rochon, T. and Mazmanian, D. (1993) "Social movements and the policy process," *The Annals of the American Academy of Political and Social Science* 528: 75–87.

Roush, J. (1995) "Introduction," in J. Echeverria and R. Eby (eds) *Let the People Judge: Wise Use and the Private Property Rights Movement*, Washington, DC: Island Press.

Rowell, A. (1996) *Green Backlash*, London: Routledge.

Sachs, W. (ed.) (1992) *The Development Dictionary*, London: Zed Books.

—— (1993) "Global ecology and the shadow of development," in W. Sachs (ed.) *Global Ecology: A New Arena of Political Conflict*. London: Zed Books.

Sayer, A. and Walker, R. (1992) *The New Social Economy: Reworking the Division of Labor*, Cambridge, MA: Blackwell.

Sessions, G. (1997) "Reinventing nature? The end of wilderness?," *Wild Earth* Winter.

Shiva, V. (1989) *Staying Alive*, London: Zed Books.

Smith, N. (1984) *Uneven Development*, Oxford: Blackwell.

—— (1996) "The production of nature," in G. Robertson *et al.* (eds) *Future Natural*, New York: Routledge.

Snow, D. (1994) "Wise Use and the public lands in the West," *Northern Lights*, Missoula, MT: Northern Lights Institute.

Snyder, G. (1996/7) "Nature as seen from Kitkitdizze," *Wild Earth* Winter.

Steen, H. (ed.) (1992) *The Origins of the National Forests: A Centennial Symposium*, Durham, NC: Forest History Society.

Storper, M. and Walker, R. (1989) *The Capitalist Imperative: Territory, Technology, and Industrial Growth*, New York: Blackwell.

Switzer, J. (1996) "Women and Wise Use: the other side of environmental activism," paper delivered at the annual meeting of the Western Political Science Association, San Francisco, 14–16 March.

Tokar, B. (1995) "The 'Wise-Use' backlash: responding to militant anti-environmentalism," *The Ecologist* 25(4): 150.

Trocki, L. (1995) "Science, technology, environment, and competitiveness in a North American context," in J. Echeverria and R. Eby (eds) *Let the People Judge: Wise Use and the Private Property Rights Movement*, Washington, DC: Island Press.

Truettner, W. (ed.) (1991) *The West as America: Reinterpreting Images of the Frontier 1820–1920*, Washington, DC: Smithsonian Institution Press.

Waller, D. (1997) "Wilderness redux," *Wild Earth* Winter.

Watts, M. (1995) "A new deal of the emotions," in J. Crush (ed.) *Power of Development*, London: Routledge.

Watts, M. and McCarthy, J. (1997) "Nature as Artifice, Nature as Artefact: Development, Environment and Modernity in the Late Twentieth century," in R. Lee and J. Wills (eds) *Geographies of Economies*, London: Edward Arnold.

White, R. (1991) *"It's Your Misfortune and None of My Own": A New History of the American West*, Norman, OK: University of Oklahoma Press.

—— (1995) "Are you an environmentalist or do you work for a living?," in W. Cronon *et al.* (eds) *Uncommon Ground*, New York: W.W. Norton.

—— (1997) "The current weirdness in the West," *Western Historical Quarterly* 28, Spring: 5–16.

Wilderness Society (1993) *The Wise Use Movement: Strategic Analysis and Fifty State Review*, Washington, DC: Wilderness Society.

Willems-Braun, B. (1997) "Buried epistemologies: the politics of nature in (post)colonial British Columbia," *Annals of the Association of American Geographers* 87: 3–31.

Williams, R. (1973) *The Country and the City*, New York: Oxford University Press.

—— (1976) *Keywords: A Dictionary of Culture and Society*, Oxford: Oxford University Press.

—— (1980) *Problems in Materialism and Culture*, London: New Left Books.

World Resources Institute (1995) *Jobs, Competitiveness, and Environmental Regulation*, Washington, DC, World Resources Institute.

Yandle, B. (ed.) (1996) *Land Rights: The 1990s Property Rebellion*, Lanham, MD: Rowman and Littlefield.

Yearley, S. (1994) "Social movements and environmental change," in T. Benton and M. Redclift (eds) *Social Theory and the Global Environment*, London: Routledge.

7

THE NATURE OF DENATURALIZED CONSUMPTION AND EVERYDAY LIFE

Allan Pred

> *No historical category without natural substance; no natural substance without its historical filter.*
>
> (Walter Benjamin[1])

> *The links between nature and everyday life have been battered by modernity.*
>
> (Cindi Katz and Andrew Kirby[2])

****

> *The erosion of the nation-state, national economies and national cultural identities is a very complex and dangerous moment.*
>
> (Stuart Hall[3])

> *The disintegration of cultural forms is endemic to modernity. Its temporality is that of fashion, the relentless production of the new – and therefore, just as relentlessly, the production of the outmoded.*
>
> (Susan Buck-Morss[4])

****

In geographically variant ways, the present moment is a moment of hyper-modernity. An extended moment of danger in which the practices, relations and experiences encountered in everyday life are best characterized as modernity magnified, as an accentuation and speeding up of circumstances associated with industrial modernity and the socially engineered, or "high," modernity which followed in its wake.[5] Local manifestations of the fleeting, volatile, turbulent and fragmented conditions of hypermodernity are almost always enmeshed with a series of interrelated processes which transcend national boundaries and which,

150

for a lack of a more appropriate term, are together usually referred to either as "globalization," "global restructuring" or "the globalization of capital." Hyper-commodification is one of the major elements synonymous with the ephemeral and repeatedly reconstituted, again and again deconstituted, conditions of hyper-modern everyday life. Within hypermodernity virtually every nook and cranny of everyday, everynight life is subject to colonization by the commodity form.[6]

Throughout European and North American commodity societies there now reigns a politics of resentment and profound discontent, if not rage. Because of a prolonged decline in real income, and actual or threatened unemployment, the majority of people live with a sense of insecurity, a sense of uncertainty and, not least of all, a sense of dissatisfaction with their levels of consumption. In popular and mainstream political discourses these sentiments are rarely translated into an apt critique of capitalism's current workings. Instead, in nationally and locally variant ways, they are translated into multiple scapegoatings: the scapegoating of immigrants and refugees; the scapegoating of long-resident minorities and the perennially poor; and the scapegoating of government, especially for its assessment of taxes and, thereby – explicitly or implicitly – especially for its intrusions upon the "freedom" to consume.

It is in this context that my preliminary thoughts on some all too neglected attributes of the nature of consumption and everyday life are to be heard – in part. For there is one other framing argument within which what follows should be heard. Given the long march of environmental degradation that is interfused with the relentless internationalization of capital, that is ensnarled with hyper-commodification and other hypermodern processes, the nature of consumption at this precarious moment needs to be re-cognized – seen again and thought anew – in such a manner that its inseparability from nature becomes every bit as explicit as its deep entanglement with politics, the economy and culture.

★★

It is, in practice, hard to see where "society" begins and "nature" ends.

In a fundamental sense, there is in the final analysis nothing unnatural about New York City.

We are all (no matter whether we are ecologically minded or not) implicated in putting monetary valuations on "nature" by virtue of our daily practices.

(David Harvey[7])

The integration of human and physical systems, I suggest, is not so much an epistemological problem as an ontological one. In these terms it is resolved everyday that men [and women] appropriate their material universe in order to survive. The two worlds are necessarily connected by social practice, and there is nothing in this which requires them to be connected through a formal system of common properties and universal constructs.

(Derek Gregory[8])

Urbanization as a process has constituted the city and the countryside, society and nature, a "unity of opposites" constructed from the integrated, lived world of human social experience. At the same time, the "urbanization of consciousness" constitutes Nature as well as Space.

(Margaret Fitzsimmons[9])

****

The conception of the progress of civilization as the unlimited increase of objects produced for sale was a defining moment of modernity.

(Susan Buck-Morss[10])

****

In key ways which we are normally unaccustomed to thinking of, consumption is a process intricately entangled in the situated practices, power relations and meanings of hypermodern everyday life, in the commonplace spatialities of individual and collective existence. It is so completely and complexly entangled with those phenomena that it may be regarded onto itself only through acts of academic legerdemain.[11] Moreover, insofar as consumption also encompasses the act-ual usage of products of reworked material nature, it is in a double sense central to the nature of everyday life.

As a process, consumption neither begins nor ends with the acts of shopping and purchasing, of buying objects produced for sale. The purchase of a good, however common or extraordinary, however inexpensive or dear, demands foreknowledge and the awakening of a want or desire, the emergence of a need or requirement. However, advertising is rarely the sole or even the most impor-tant source of pre-purchase knowledge regarding the existence and qualities of a particular good, seldom the single stimulator of want and desire, only exceptionally the primary means through which awareness of need arises in everyday life.[12]

It is through situated practice,
    through social interaction at sites of work, education,
           and other institutionally embedded activities,
    through formal and informal conversations
           participated in during the conduct of daily life,
    through everyday discourses and representations
           encountered in public spaces, private spaces and the mass media,
    through visual and aural observations
           made in the course of site-to-site movements,
that consumer knowledge is accumulated,
that the desire to possess is aroused,
that needs and wants are constructed,
that requirements and usage possibilities become apparent,
that tastes take shape.
    If the acquisition of consumption knowledge – like the acquisition of any

other type of knowledge – cannot be separted from doing; if the formation of wants and needs, desires and requirements, are interfused with numerous forms of daily doing; then it follows that these elements of the consumption process cannot be severed from the power relations associated with the practices of every-day life in a particular place or area. They cannot be extricated from those social relations which define who – individually or collectively – may or may not do what, when and where, under what conditions of control or surveillance, if any.

Power relations – including those existing between men and women, parents and children, citizens and regulating governmental authorities – are furthermore pervasively associated with the usage of the already purchased, with the who, when, where and how rules which govern practical utilization. In addition, they permeate everyday nonconsumption as well as everyday consumption. Class relations and other power relations that produce differences in disposable income or money availability by extension produce circumstances where the desired and within sight may remain unacquired because unacquirable, where the wanted or needed and within reach may remain unobtained because unaffordable, where the inability to purchase the much wished for may yield social tensions and the cultural and political reworking thereof. Individual and group consumption differences may thus be as much a matter of power relations as a matter of taste, preference and the quest for distinction. This is doubly the case, for tastes, preferences and notions of distinction are never the product of autonomous mind, but always directly or indirectly constructed through participation in quotidian practices and their associated power relations.[13]

Everyday urban life in contemporary commodity societies is suffused with consumption in the sense of act-ual usage of purchased goods and services. There are few, if any, moments of the day during which women and men are not in corporeal contact with commodities purchased by themselves or others: with the clothing borne on their bodies; with the food which they prepare and eat; with chairs, tables, beds and other items of furniture; with the floor or pave-ment beneath their feet; with the countless utensils, pieces of equipment and machinery, and other artifacts repeatedly employed at home, work and elsewhere. This virtually incessant contact with useful exchange-value items is for the most part unreflected upon, is second natural, is usually synonymous not only with a failure to recognize the social relations which have made item purchase possible, but also with a failure to recognize that the items are nature once removed, are a consequence of the appropriation and transformation of material nature.[14] (The contemporary decoupling of consumption from material nature is under-scored by those who – despite their critical bent, or even their sensitivity to the workings of capitalism – have reduced the study of consumption to a "reading of culture," to an interpretation of either the symbolic expression and resistance of consumers, or the semiotic content of commodities and shopping sites).[15]

How – despite a widespread concern for environmental and ecological issues – has such a selective nonconciousness been constructed and reproduced in Western commodity societies? How is it possible that material nature has become divorced from culture and society in this particular way,[16] that acts of consumption have become so denaturalized (except where proximity to nature is marketed by real estate developers and in the tourism and resort industries where the experience of nature – as physical setting – is explicitly commodified)?[17] Why is it generally unrecognized that every instance of localized everyday life, every moment of situated practice, is a conjuncture of here-and-now humanly embodied nature and products whose ultimate natur(e)al resource origins are more or less geographically dispersed? The selectively nonconscious, unrecognized, denaturalized qualities of everyday consumption are over-determined in historically and geographically variant ways, as well as socially differentiated for any given time and place. All the same there are some familiar and not so familiar general answers to these questions – answers that are inescapably partial and mutually entangled, answers that are here offered in the form of a Benjamin-inspired montage of the present, of a generalized historical geography of the present whose most extended details involve a reworking of Benjaminian themes.

****

Industrial capitalism, armed with the factory system, organized the work process in a manner that transformed the relation between the worker and nature into a travesty of even its former very limited self. Because the worker was reduced to a "thing," the worker became alienated from his or her product, from the manner of producing it, and, ultimately from nature itself.

(David Harvey[18])

Nature as we know it was invented in the differentiation of city and countryside, in the differentiation of mental and manual labor, and in the abstraction of contemporary culture and consciousness from the necessary productive social work of material life.

(Margaret Fitzsimmons[19])

****

The denaturalization of consumption was and remains consistent with the transformation of everyday practices, social relations and forms of consciousness that accompanied the appearance of industrial capitalism in its various locally, regionally and nationally specific forms. As Harvey, Fitzsimmons and others have pointed out, the spread of industrial capitalism, of everyday practices based on wage labor and class relations, usually coincided with a corporeal removal from work processess directly involving the transformation of nature's raw materials. Or, industrial capitalism widely resulted in a "reconstitution of the relationship between humans and the material world"[20] – between human labor and human knowledge of natural processes; in an alienation of working people "from any sensuous and immediate contact" with "nature."[21] While this divorcing of production practices from any palpable engagement with "raw" natural phenomena is arguably the most fundamental, the most obvious and deeply sedimented contributor to denaturalized consumption, its impact has been compounded, reinforced and embellished by a number of other key developments.

With the urbanization of industrial practices and the industrialization of urban practices during the nineteenth century there eventually came a romanticization or sentimentalization of a removed-from-the-scene, more or less distant nonurban nature – a nature to be encountered and "consumed" during leisure or vacation, if at all. On the part of the bourgeoisie this in some measure involved a fanciful flight from the social tensions and class antagonisms of city life, the creation of an imaginary breathing space far from an environment contaminated and made threatening by the frequently encountered "dangerous classes."[22] On the part of factory, construction and transportation workers and their poorly housed families this often operated as a mental antidote to the daily experience of physical hardship, drudgery and humiliation, to the daily experience of living in an alienating, frequently inhuman environment. (Ironically, for those among the bourgeosie who regarded the working classes as "primitive" and threatening, the planned urban park – an exceptional enclave of (re)de-signed nature – was "seen as providing 'mental refreshment' for the working masses,"[23] as a means for pacifying the crude and dangerous).

Sooner or later the discourses of nationalism served to reinforce the sentimentalization of primordial nature and nonurban landscapes, to further the social construction of Nature as something abstractly beyond the city.[24] Almost invariably such discourses included an appeal to landscape images that equated nature and the national soul, to images that blurred identification with undefiled nature and identification with the nation. Alternatively, such discourses produced images that were meant to evoke a harmonious, socially frictionless, unfragmented, authentic rural past as confirmation of the True Nature of national community – and thereby meant to obscure the conflict-ridden, discordant nature of modern everyday urban existence.

★★★★★★★★★★★★★★★

The clock stands for mechanical unnature.

(Christoph Asendorf[25])

★★★★★★★

The distancing of urbanized consciousness from nature was further enabled by the imposition of rigid time discipline on factory work and other types of employment and the attendant colonization of educational, family and "freetime" activities by similar synchronization demands, time pressures and forms of clock awareness.[26] Whereas the sense of time held by people had rested on a direct knowledge of natural rhythms, the time consciousness now permeating every recess of daily life was founded on an arbitrary mechanical ordering little connected to the workings of nature.

155

****

The advent of a money economy, Marx argues, dissolves the bonds and relations that make up "traditional" communities so that "money becomes the real community." We move from a social condition, in which we depend directly on those we know personally, to one in which we depend on impersonal and objective relations with others.

(David Harvey[27])

In exchange, things crystallize into commodity values, independently of their material form. For it is precisely the stripping away of their "bodily shape" that makes it possible for commodities to carry out the operation of an exchange value.

(Christoph Asendorf[28])

****

The proliferation of exchange relations based on money was as important as anything to the denaturalization of consumption. The thorough monetarization of everyday urban life which accompanied industrial capitalism was not only a dissolver of traditional social bonds and a masker of the social relations of production, but also a redefiner of the taken-for-granted nature of things. As the natural world, including bodily labor, and products of transformed nature became reduced to a single universe of comparable values, and as more and more things became commodified, property and goods at the moment of sale or purchase became stripped of their material qualities, metamorphosed, charged with unnatural meanings. This universe of unnatural meanings became everyday nature, or second natural, whatever the particular antecedent meanings of desire, whatever the singular meanings subsequently attached to goods through the contexts, politics and experiences of act-ual usage.[29]

****

For the first time, the most recent past becomes distant.

(Walter Benjamin[30])

There have been [for Benjamin], then, two epochs of nature. The first evolved slowly over millions of years; the second, our own, began with the industrial revolution and changes its face daily.

(Susan Buck-Morss[31])

He was standing at one of the windows, looking . . . into the brownish street, and for the last ten minutes, watch in hand, he had been counting the cars, carriages, and trams, and the pedestrians' faces, blurred by distance, all of which filled the network of his gaze with a whirl of hurrying forms . . .

"It doesn't matter what one does," the Man Without Qualities said to himself, shrugging his shoulders. "In a tangle of forces like this it doesn't make a scrap of difference."

(Robert Musil[32])

****

Seen through Walter Benjamin's distinctive glasses, the forms that industrial capitalism began assuming in the burgeoning cities of nineteenth-century Europe had a "natural history" of their own; one in which consumption and its associated technologies – "the entire world of matter" – took on a "new nature." "The short half-life of technologies and commodities, the rapid turnover in style and fashion," were, for him, one with a "new nature" – at once alluring and threatening – whose essence was "even more transient, more fleeting than the old."[33] The "new nature" of consumption, of one transitory fad or "necessity" repeatedly replacing another transitory fad or "necessity," lent itself to images of decay, ruin, accelerated fossilization. And, the "new nature" of commodities was such, the ephemeral quality of capitalist culture was such, that the consumption process took on an "always-the-same-again" quality.[34] In the beginning there are meanings associated with promise and enchantment, with the possibilities of the desired/acquired fetish providing dream fulfillment; in the end – whether or not the object has been put to creative or rewarding use, whether or not pleasure or rich satisfaction actually has been achieved – there are meanings associated with "the discarded fetish . . . hollowed out of life,"[35] with the betrayal of promise, with disenchantment, with the crushed dream – which is quickly supplanted by other dreams, by newness repeating itself yet again. This phantasmagoria of commodity circulation, this shocking lack of permanence and permanence of new shocking impressions, this incessant human-like flow of inanimate objects and ceaseless flow of object-like humans, this constant motion of things made possible by the symbolic and practical fluidity of money, was experienced, read and represented by *fin-de-siècle* Viennese modernists, among others, as a crisis of perception, as an obstruction to "any sort of empathic contemplation" of the nature of urban life.[36]

****

[T]he arcades are the most important architecture of the nineteenth century.

"We have," so says the illustrated guide to Paris from the year 1852, . . . "repeatedly thought of the arcades as interior boulevards, like those they open onto. These passages, a new discovery of industrial luxury, are glass-covered, marble-walled walkways through entire blocks of buildings, the owners of which have joined together to engage in such a venture. Lining both sides of these walkways which receive their light from above are the most elegant of commodity shops, so that such an arcade is a city, a world in miniature, where the willing buyer can find everything he wants.

157

During unexpected rain showers they serve as a place of refuge for all who have been taken by surprise. . . . "

[T]he arcades, in contrast to [railway stations, market halls, and other prototypical] *iron-frame structures are not actually about illuminating interior space but about reducing* [awareness of] *the outside.*

[The arcades were] *the original temple of commodity capitalism . . . the mold from which the image of "the modern" is cast . . . the precursors of the department stores* [as well as of today's malls and gallerias] *. . . a center for the luxury-goods trade. . . . People of the times did not tire of admiring them.*

[There] *commodities* [were] *suspended and shoved together in such boundless confusion, that* [they appeared] *like images out of the most incoherent dreams.*

Dada was the father of Surrealism; its mother was an arcade.

(Walter Benjamin[37])

[T]he Passage de l'Opéra is a big glass[-lidded] coffin.

(Louis Aragon[38])

****

The arcades which became widespread in Paris during the 1820s and 1830s, and then flourished there and in numerous other European cities for most of the remainder of the nineteenth century,[39] were regarded by Benjamin as the "primeval landscape of consumption,"[40] as the ur-space of modern consumption and commodity fetishism. As such, they may be seen and heard anew, may be recognized, as the ur-site, or ur-place form, of denaturalized consumption.[41] As a "world in miniature," each glass-roofed arcade was literally a world unto itself, a shopping world cut off from the elements, a shelter from wind and cold, a rain-free space, an out-of-this-world display world which provided an escape from nature. There was a (narrow) world of "suffocated perspective," a "space without [open] air."[42] There, unlike the streets, was a world free of horsedrawn carriages, wagons and omnibuses; free of frequent encounters with fetid horse manure. There the combination of "countless lanterns maintained by shopowners" and the pioneering usage of multi-armed hydrogen-gas "streetlamps" resulted in "daylit evenings,"[43] in light after dark, in a blurring of the natural distinction between day and night, in a preview of the nightworld invasion by "daytime" consumer-oriented activities that eventually became commonplace with the spread of street lighting and the introduction of electricity.[44] There was an other-worldly world of commodity availability; a beyond-nature "fairyland grotto" or "fairyland castle" of consumption alternatives; an eroticized surrealm of purchasing possibilities – of goods issuing seductive invitations from behind their "mirror walls;"[45] a dreamworld of desire and pleasure – of elegant restaurants, theater offerings, gambling rooms and countless luxury establishments offering specialty goods of the moment as well as the latest in fleetingly fashionable apparel; a phantasmagoria where everything was on the move, ephemeral – not only the array of items for sale and the throngs en passage, but even the names on the fashion-wear shops;[46] a *zeit-(t)raum*, or time-(dream)space, where at one and the same time "collective consciousness sinks into an increasingly deeper sleep" and individual consciousness becomes "self-engrossed" with the (denatured) images and stimulations it there encounters.[47] There was a space of dream-like ambiguity, a "perfectly ambiguous space where the external and the internal were unnaturally (con)fused with one another; an artificially enclosed house-like space whose interiorized thoroughfares were at once outdoor street and indoor parlor,[48] and where, consequently, every passing person was confronted by shopdoors that were simultaneously building entrances and (street) room exits, by shopwindows which faced out toward them and yet through which they could look out – but only into an interior; a severed-from-nature space which dis-oriented one's thoughts as well as one's footsteps with its "profusion of mirrors" that distorted distance and dimension, that made thing's attractively appear and disappointingly disappear in the blink of an eye, that in rapid succes-sion suggested now Nirvana, now Nothingness.[49] There, in that thoroughly commodified, unnatural dreamworld nature could logically be given a visible presence only as a commodity; only as a not-here, sentimentalized Nature, available for a price; only as a panorama, as a mimesis of "True Nature" to be detachedly viewed on a cylindrical wall from a raised centrally situated platform, as an imitation of a Swiss mountainside or some other supposedly pristine site made "real" by tree shadows, the sound of rushing water, and by "theatrical tricks" which compressed the diurnal cycle of light changes into a span of 15 to 30 minutes – as, in short, an illusion for sale.[50] Likewise, nature there could be given a visible presence only as a diorama, "cosmorama," "stereorama," "georama," or some other "rama" that suggested illusions superior to those of a panorama, that suggested the ultimate in purchasable verisimilitude; only – later in the century – as a photograph or as a post card, as an image-laden magazine, as a "magic lantern" show or a stereoscopic picture, as, in each case, a mechanically reproduced "exact representation" of some removed and uncontaminated landscape, as a representation marketed in unlimited numbers so as to pay off in (auraless) pleasure.[51] There, in that male-constructed dreamworld, in that ur-space of commodity fetishism and the gazing *flâneur*, it was also fully logical that embodied nature be ontologically transformed with profit in mind, that subjects be sold as objects, that subjects be enticed to make themselves over as objects, that made objects be offered for sale as subjects. There, where shopwindow displays and posters repeatedly fused female sex appeal with commodity appeal, where teasingly attractive goods abounded, were prostitutes in profusion, were a multitude of "sirens" and "tempting odalisques" beckoning from the gaslight, were "angels nesting" in the heavens one flight up, were "dream(worldly)girls" in the flesh, were women who were salesperson and commodity in one, were thinking, feeling women who sold themselves as use-value things.[52] There, in Benjamin's view, the woman who succumbed to

with discourses aimed at defusing the possibility of cultural and political resistance,

with discourses which attempt to stabilize or "rearticulate the complex relations among the state, the economy and culture."[59]

Nowhere has the melting together of consumption and would-be hegemonic discourses been more evident than in conjunction with the discursive rhetoric surrounding the development and operation of spectacular consumption spaces. At international expositions from the Crystal Palace onward, at state and county fairs, at annual trade shows, at mega-malls and theme parks, at sports arenas, at concert halls and other sites of what Werckmeister terms "citadel culture,"[60] it usually has been – and remains – difficult fully to distinguish those discourses through which business enterprises market their products from those discourses through which the state or ruling elites attempt to market their legitimacy.

The interfusion of consumption and would-be hegemonic discourses at nineteenth-century and early twentieth-century international expositions and world fairs contributed to the denaturalization of consumption in pivotal ways that ought not be overlooked. As the most spectacular of all spectacles, as the most distractive of all distractions, in each instance these events made a profound and lasting impression on the popular imagination via the out-of-the-ordinary experiences taken home by millions of visitors, via the intense, months-long vivid coverage of the daily press and widely read periodicals, via the cascading circulation of souvenirs, posters and word-of-mouth accounts. (The World's Columbian Exhibition held in Chicago during 1893 drew more than 27.5 million attendees at a time when the US population stood at approximately 67 million. Seven years later, the *Exposition Universelle* in Paris attracted over 48.1 million visitors, or about ten million more people than the then population of France.)[61]

At these spectacular spaces, discourses pertaining to the selling of the new were repeatedly interwrapped with discourses pertaining to progress, to the (re)construction of national identity, to national and racial superiority, to a just-around-the-corner paradise where the now better would get even better interwrapped with discourses which, regardless of specific content, were meant to paper over social conflict and discontent, to make some things visible and others invisible, to inculcate particular ways of seeing and not seeing, to facilitate social control.[62] Not least of all, women and men attendees were confronted with discursive and visual practices whereby the new goods and industrial technology to which they were exposed were to be recognized as conclusive proof of "man's" triumph over nature, of his mastery of nature.[63] The new commodity wonders, and the machinery which made them possible, were to be re-cognized as telling proof that the urban population of the host nation was civilized, that it was cultivated and cultured, because it had risen above the wild and disorderly, had come to dominate nature and thereby was superior not only to nature, but to all those colonized, non-Western or class others who were uncivilized, who were of a savage nature.[64] The conquering/harnassing/exploitation of nature by industrial capitalists was to be re-cognized in such a way that it appeared totally apart from the exploitation/harnessing/disciplining of human labor which made it possible. The sights and messages of international exhibitions and world's fairs in some sense thus served as a culminating confirmation of the separateness of "external nature" and "human nature"; as an ultimate illustration of the success of the Enlightenment scientific tradition and its claims that control over nature and its forces could be achieved through attaining knowledge of its mechanistic laws; as a crowning justification for a bourgeois intellectual optimism so widespread that it contributed to an "eco-blindness" in the later writings of Marx.[65]

Unavoidably, the experiences and would-be hegemonic messages encountered at international exhibitions were reworked differently, were ascribed somewhat different meanings by women and men, by people who differed as to class, age, occupation, regional origins, and past-path history. While some of these meanings may even have been oppositional on some counts,[66] I wish to argue that, in the context of other practice-based elements of urban consciousness, the connections drawn between new forms of consumption and a supposedly complete domination of nature made it all the more easy for the vast majority of large city residents further to distance themselves from nonurban nature, further to ignore it as anything other than a nonpresent otherness.[67] Over time, with intergenerational transmission, such a socially produced ignoring could be transformed into virtually blind ignorance, or selective nonconsciousness, could help objects of transformed nature signify anything but nature, could help make denaturalized consumption part of the nature of everyday life.

✷✷✷✷✷✷✷✷✷✷✷✷✷✷✷

It isn't that the past casts light on what is present or that what is present casts light on what is past; rather, an image is that in which the Then [and There] and the [Here and] Now come together into a constellation like a flash of lightning.

The nineteenth century is, as the Surrealists would have it, the noise which intervenes in our dreams and is interpreted by us during the process of awakening.

(Walter Benjamin[68])

✷✷✷

The intertwinings of social and ecological projects in daily practices as well as in the realms of ideology, representation, aesthetics and the like are such as to make every social ... project a project about nature, environment and ecosystem, and vice versa.

(David Harvey[69])

✷✷✷✷✷✷✷✷✷✷✷✷✷✷✷

All too much has been presented in a summary fashion.

All too little has been shown.

All too many questions have been unasked.

Given what is re-cognized here about the nature of denaturalized consumption and everyday life, how should we reframe our critical questions about the mutual entanglements of politics, economy and culture under the conditions of hypermodernity?

How are the geographic specifics of denaturalized consumption, hyper-commodification, other elements of global restructuring, and the politics of resentment to be brought into constellation with one another?

How are we to move from a historical geography of the present – of the extended moment of danger which characterizes hypermodern commodity societies – to an engagement with the social and ecological future? All too much remains to be shown. All too much remains to be done. Naturally!

****

Within the area we are dealing with there is only lightning-like knowledge. The text is the thunder which rumbles on long afterward.

(Walter Benjamin[70])

When trying to reconstruct what the arcades, [international] *expositions, urbanism and technological dreams were for Benjamin, we cannot close our eyes to what they have become for us* [to – among other things – how they have become relayered within our own denaturalized consumption]. *It follows that in the service of truth, Benjamin's own text must be "ripped out of context," sometimes, indeed, with a seemingly brutal grasp.*

(Susan Buck-Morss[71])

[For Benjanim] *the past and the present are not illuminated by one another but group themselves into a constellation "charged with tensions" around a dialectical image that functions as a catalyst. It is only proper to ask oneself at the proper time – soon – whether the historian's task* [the task of the historical geographer of the present, of the heretical empiricist] *can become the principal of concrete action.*

(Pierre Missac[72])

ACKNOWLEDGEMENT

Special thanks to Dick Walker for his usual constructive critical remarks.

NOTES

1 Walter Benjamin (1982) *Gesammete Schriften*, vol. 5, *Das Passagen-Werk*, edited by Rolf Tiedemann (Frankfurt am Main, Suhrkamp Verlag), 1034, as translated in Susan Buck-Morss (1989) *The Dialectics of Seeing: Walter Benjamin and the Arcades Project* (Cambridge, MA: MIT Press) p. 59.
2 Cindi Katz and Andrew Kirby (1991) "In the nature of things: the environment and everyday life," *Transactions of the Institute of British Geographers* 16: 261.
3 Stuart Hall (1991) "The local and the global: globalization and ethnicity," in Anthony D. King (ed.) *Culture, Globalization and the World-System: Contemporary Conditions for the Representation of Identity* (Binghamton, NY: Department of Art and Art History), quote from p. 25.
4 Susan Buck-Morss (1975) "The city as dreamworld and catastrophe," *October 73*, quote from p. 4.
5 For a spelling out of this point, and related observations on the conflation of "postmodern" academic discourses with a lived and experienced "postmodern" world see Michael J. Watts and Allan Pred (1994) "Heretical empiricism: the modern and the hypermodern," *Nordisk Samhällsgeografisk Tidskrift* 19: 3–26; Allan Pred (1995) *Recognizing European Modernities: A Montage of the Present* (London and New York: Routledge); and *idem*, "Out of bounds and undisciplined: social inquiry and the current moment of danger," *Social Research* 1995, Winter: 1065–91.
6 In various of his writings Henri Lefebvre emphasized the colonization of everyday life by the commodity form as a defining characteristic of contemporary life. This "colonization" is not merely a figure of speech, but a term consciously employed so as to draw upon "the implications of occupation, dispossession and territorialization with which it is freighted" (Derek Gregory 1994, *Geographical Imaginations*, Oxford: Blackwell, p. 403).
7 David Harvey (1993) "The nature of environment: the dialectics of social and environmental change," in Ralph Miliband and Leo Panitch (eds) *Real Problems, False Solutions* [*The Socialist Register*] (London: The Merlin Press) pp. 31, 28, 4.
8 Derek Gregory (1978) *Ideology, Science and Human Geography* (New York: St. Martins) p. 75, emphasis added.
9 Margaret FitzSimmons (1989) "The matter of nature," *Antipode* 21: 108.
10 Susan Buck-Morss (1995) "Envisioning capital: political economy on display," *Critical Inquiry* 21, Winter: 456.
11 Although not pursued here, *The World of Consumption* by Ben Fine and Ellen Leopold (London and New York: Routledge, 1993) buttresses this assertion. They argue that the consumption of specific goods is united in distinctive ways with "various material and cultural practices" – as well as the economic and social processes – comprising the production and "system of provision" associated with those goods.
12 Cf. Paul D. Glennie and Nigel J. Thrift (1992) "Modernity, urbanism, and modern consumption," *Society and Space* 10: 423–43; W. Leiss, S. Kline and S. Jhally (1986) *Social Commodities in Advertising: Person, Products and Images of Well-Being* (Andover: Methuen); M. Schudson (1984) *Advertising, the Uneasy Persuasion: Its Dubious Impact on American Society* (New York: Basic Books); and J. Sinclair (1987) *Images Incorporated: Advertising as Industry and Ideology* (London: Croom Helm).
13 Cf. Pierre Bourdieu (1984) *Distinction: A Social Critique of the Judgement of Taste* (Cambridge, MA: Harvard University Press).
14 While material nature here refers to mineral, vegetable and animal matter, it is neither to suggest that humans – as biological beings – are not a part of material nature, not to suscribe to the ideological dualism of external and universal (human) nature. Cf. Neil Smith (1984) *Uneven Development: Nature, Capital and the Production of Space*

(Oxford: Basil Blackwell); and Noel Castree (1995) "The nature of produced nature: materiality and knowledge construction in Marxism," *Antipode* 27: 12–48, and the literature cited therein).

15 This is not to deny the significance of the work of Willis, Hebdige and others concerned with the counterhegemonic dimension of consumption. On the contrary, key items in this genre have had a considerable impact on my own work (Pred, *Recognizing European Modernities*). Cf. the constructive critical commentary contained in Orvar Löfgren, "Creativity and consumption: some reflections on the pairing of two concepts," forthcoming.

16 Because of the particular questions posed, as well as length limits, the subsequent contents of this chapter do not come close fully to engaging the range of discourses and practices through which "nature" became uncoupled from culture and society in conjunction with "the urbanization of consciousness."

17 Cf. S. Britton (1991) "Tourism, capital, and place: towards a critical geography of tourism," *Society and Space* 9: 451–78; and Katz and Kirby, "Nature of things," 266–8. Stores specializing in nature-inspired products are another, although secondary, exception (see Sharon Corwin 1994, "Consuming nature: the commodification of nature imagery and the 'natural,'" *Critical Sense* 2: 9–24). Nature has become so "squeezed out" of everyday life in contemporary cities that, arguably, even food consumption has become largely denaturalized. Cf. Michael Hough (1995) *Cities and Natural Processes* (London and New York: Routledge).

18 David Harvey (1985) *Consciousness and the Urban Experience* (Oxford: Basil Blackwell, 1985) p. 53, citing Karl Marx (1964) *The Economic and Philosophic Manuscripts of 1844* (New York: International Publishers).

19 FitzSimmons, "Matter of nature," p. 110.

20 FitzSimmons, "Matter of nature," p. 108.

21 Harvey, "Nature of environment," p. 10.

22 For those members of the bourgeoisie who could afford it, romanticized nature was made concrete by the acquisition of a country mansion or a more modest rural summer residence. In some instances the bourgeoisie seriously questioned the ability of lower-class women and men to appreciate or love nonurban landscapes and nature.

23 Katz and Kirby, "Nature of things," p. 267, citing Galen Cranz (1982) *The Politics of Park Design: A History of Urban Parks in America* (Cambridge, MA: MIT Press).

24 Cf. Orvar Löfgren (1987) "The nature lovers," in Jonas Frykman and Orvar Löfgren *Culture Builders: A Historical Anthropology of Middle-Class Life* (New Brunswick: Rutgers University Press); Orvar Löfgren (1989) "Landscapes and mindscapes," *Folk* 31: 183–208; and Kenneth Olwig (1984) *Nature's Ideological Landscape* (London: George Allen and Unwin).

25 Christoph Asendorf (1993) *Batteries of Life: On the History of Things and Their Perception in Modernity* (Berkeley: University of California Press) p. 140.

26 Cf. E. P. Thompson's (1967) classical essay "Time, work-discipline and industrial capitalism," *Past and Present* 38: 56–97. Also note Allan Pred (1981) "Production, family, and free-time projects: a time-geographic perspective on individual and societal change in nineteenth-century U.S. cities," *Journal of Historical Geography* 7: 3–36; Asendorf, *Batteries of Life*, pp. 140–52; and David Harvey, "Money, time, space, and the city," chapter 1 of *Consciousness and the Urban Experience*, pp. 1–35. While measured time and clock consciousness made their first European inroads in conjunction with fourteenth-century forms of market capitalism, they did not come to dominate urban life until the spread of factory production and the availability of affordable timepieces. See, for example, Nigel Thrift (1981) "Owners' time and own

time: the making of capitalist time-consciousness, 1300–1880," in Allan Pred (ed.) *Space and Time in Geography: Essays Dedicated to Torsten Hägerstrand* (Lund: CWK Gleerup) pp. 56–84; and Jacques LeGoff (1980) *Time, Work, and Culture in the Middle Ages* (Chicago: University of Chicago Press).

27 David Harvey (1989) *The Condition of Postmodernity: An Enquiry into the Origins of Cultural Change* (Oxford: Basil Blackwell).

28 Asendorf, *Batteries of Life*, p. 32, reworking Marx's *Capital*.

29 Cf. Arjun Appadurai (1986) "Introduction: commodities and the politics of value," and Igor Kopytoff, "The cultural biography of things: commoditization as process," in Arjun Appadurai (ed.) *The Social Life of Things: Commodities in Cultural Perspective* (Cambridge: Cambridge University Press), pp. 3–63 and 64–91 respectively.

30 Benjamin, *Passagen-Werk*, p. 1250, as translated in Buck-Morss, *Dialectics of Seeing*, p. 65.

31 Buck-Morss, *Dialectics of Seeing*, p. 20.

32 Robert Musil (1930) *The Man Without Qualities*, vol. 1 (London, Picador 1979) pp. 7, 8.

33 Buck-Morss, *Dialectics of Seeing*, p. 64. Although Benjamin did not use the expression "new nature," Buck-Morss employs it, arguing that it most clearly captures the particular set of meanings he attached to "productive forces" and related terms.

34 Cf. Buck-Morss, *Dialectics of Seeing*, especially pp. 159–64, 293, regarding ruin; Pierre Missac (1995) *Walter Benjamin's Passages* (Cambridge, MA: MIT Press) pp. 171–3; and Richard Wolin (1994) *Walter Benjamin: An Aesthetic of Redemption* (Berkeley: University of California Press).

35 Buck-Morss, *Dialectics of Seeing*, p. 160.

36 Asendorf, *Batteries of Life*, pp. 178–211; quote from p. 210.

37 All of the above quotes are passages from Benjamin's *Passagen-Werk*, either as they have been translated (from pp. 45, 83, 86, 993) by Buck-Morss (*Dialectics of Seeing*, pp. 383, 254, 293), or as they have been retranslated by myself from Ulf Peter Hallberg's Swedish translation [*Paris 1800-talets huvudstad- Passagearbetet*] (Stockholm: Symposium Bokförlag, 1990) pp. 7, 24, 37, 61, 87, 115, 452, 691).

38 Louis Aragon (1926) *Paris Peasant* (London: Pan Books, 1980) p. 47.

39 The first Parisian arcade actually dated back to 1800. On the history of arcades in Europe and elsewhere see Johann Friedrich Geist (1982) *Arcades: The History of a Building Type* (Cambridge, MA: MIT Press).

40 Benjamin *Passagen-Werk* [Swedish translation, hereafter simply *Passagearbetet* p. 678].

41 As "the first international style of architecture" the arcades themselves were, in effect, a commodity sold by architectural firms (Buck-Morss, *Dialectics of Seeing*, p. 39).

42 Benjamin, *Passagen-Werk* [*Passagearbetet*, p. 685].

43 J. A. Dulaure (1835) *Histoire de Paris . . . depuis 1821 jusqu' à nos jours, II* (Paris) p. 29 and Eduard Devrient (1840) *Briefe aus Paris* (Berlin) 34, as quoted in Benjamin, *Passagen-Werk* [*Passagearbetet*, pp. 164, 93].

44 Even during mornings and afternoons there was a blurring of the natural day/night distinction in the arcades, owing to the diffused half-light which filtered through their opaque glass roofs. Regarding the spread of street lighting and its consequences see Wolfgang Schivelbusch (1988) *Disenchanted Night: The Industrialization of Light in the Nineteenth Century* (Berkeley: University of California Press).

45 Benjamin, *Passagen-Werk* [*Passagearbetet*, pp. 484, 690, 455]. Although he later chose to reject it, the original subtitle for Benjamin's never completed arcades project was "*A Dialectical Fairy-Tale*."

46 It was not uncommon for a *Magasin de Nouveautés* to be renamed after a currently popular theatrical or vaudeville show (Benjamin, *Passagen-Werk* [*Passagearbetet*, p. 692]).

47 Except for the parenthetical "denatured," Benjamin spoke in these terms with regard to both the nineteenth century in general and the arcades in particular *Passagen-Werk* [*Passagearbetet*, p. 315].

48 For some the arcades were at once outdoor street and indoor bedroom: "And those who can't pay for a night's rest? They sleep where they can find a place in the arcades, the passageways" (Benjamin, *Passagen-Werk*, as translated in Missac, *Benjamin's Passages*, p. 195).

49 Benjamin, *Passagen-Werk* [*Passagearbetet*, pp. 753, 15–16] spoke in these terms with regard to both the nineteenth century in general and the arcades in particular (*Passagen-Werk* [*Passagearbetet*, p. 315]).

50 Panoramas were highly popular in Paris and elsewhere from the 1820s onward. And, the most popular of all Parisian arcades, especially on Sundays, was the "*Passage des Panoramas*, named for the two panoramas built on each side of the entrance" (Paul D'Artiste [1930] *La vie et le monde du boulevard – 1830–1870* [Paris], as quoted in Benjamin, *Passagen-Werk* [*Passagearbetet*, p. 97]).

51 Cf. Benjamin on art and aura in "The work of art in the age of mechanical reproduction," in Hannah Arendt (ed.) (1969) *Illuminations* (New York: Schocken), pp. 217–51. Factory produced representations of nature made from woodcuts were also sold in the arcades, especially between 1835 and 1845.

52 Prostitution was so central to Benjamin's image of the arcades that he regarded the banning of the former as one of the key reasons for the decline of the latter – or, as he would express it, for their death, mortification, decay, fall into ruin and hollowed out meaning (*Passagen-Werk* [*Passagearbetet*, p. 122]).

53 Buck-Morss, *Dialectics of Seeing*, p. 101, reworking Benjamin, *Passagen-Werk*, pp. 111, 118, 126, 139.

54 Benjamin, *Passagen-Werk* [*Passagearbetet*, 8, 571].

55 Anders Ekström (1994) *Den utställda världen: Stockholmsutställningen 1897 och 1800-talets världsutställningar* (Stockholm: Nordiska Museets Förlag) p. 93.

56 Benjamin, *Passagen-Werk* [*Passagearbetet*, pp. 44–5, 814].

57 Timothy Mitchell (1991) *Colonising Egypt* (Berkeley: University of California Press) p. 9.

58 As Glennie and Thrift have elsewhere observed ("Modernity, urbanism, and modern consumption," p. 441), advertising may be regarded as just one source among many; not so much a locus of authority as a point on a continuum of discourses influencing consumers (also see sources in note 12, above).

59 Lawrence Grossberg (1992) *We Gotta Get Out of This Place: Popular Conservatism and Postmodern Culture* (New York and London: Routledge) p. 247.

60 "Citadel culture," or "the dominant artistic and intellectual culture of the democratic industrial societies during the years 1980–87," was "funded generously by variable combinations of state subsidies, corporate investments, advertising accounts and returns from large audiences"; O. K. Werckmeister (1991) *Citadel Culture* (Chicago: University of Chicago Press) pp. 4, 15.

61 Attendance statistics from John Allwood (1977) *The Great Exhibitions* (London: Studio Vista) p. 182. The fact that international exhibition attendance figures included tourists from abroad, as well as a repeated counting of those making more than one visit, in no way diminishes the truly enormous impact those expositions had on the popular imagination.

62 Cf. Tony Bennett (1988) "The Crystal Palace as icon," in Gerry Kearns and Chris Philo (eds) *Selling Places: The City as Cultural Capital, Past and Present* (Oxford: Pergamon Press) pp. 103–31; Graeme Davison (1982) "Exhibitions," *Australian Cultural History* 2: 5–21; Ekström, *Utställda världen*; Paul Greenhalgh (1988) *Ephemeral Vistas: The Expositions Universelles, Great Exhibitions and World's Fairs,*

1851–1939 (Manchester: Manchester University Press); David Ley and Kris Olds (1988) "Landscape as spectacle: world's fairs and the culture of heroic consumption," *Society and Space* 6: 191–212; Mitchell, *Colonizing Egypt*; Pred, *Recognizing European Modernities*, and Robert W. Rydell (1984) *All the World's a Fair: Visions of Empire at American International Expositions, 1876–1916* (Chicago: University of Chicago Press).

63 Inasmuch as discourses of the period persistently denoted culture as masculine and nature as feminine, all rhetorical claims as to the mastering of nature helped legitimate the practical subordination and hierarchical downgrading of women – a circumstance radical ecofeminists and others have repeatedly pointed to.

64 This discursive imagery, visually magnified in unprecedented ways at these spectacular spaces, had a history dating as far back as Adam Smith's (1776) *An Inquiry into the Nature and Causes of the Wealth of Nations*. There he argued: "Having an abundance of 'objects of comfort' is the litmus test that distinguishes 'civilizing and thriving nations' from 'savage' ones" (Buck-Morss, "Envisioning capital," p. 456, restating and quoting Smith).

65 Ted Benton (1989) "Marx and natural limits," *New Left Review* 178: 51–86; and Castree, "Nature of produced nature," p. 22. To the extent Marx conceived of "animals as natural and thus without culture and human beings as transcending nature and thus having culture" he subscribed to the bourgeois dualism of external and human nature; quote from Hilary Rose (1994) *Love, Power and Knowledge: Towards a Feminist Transformation of the Sciences* (Bloomington: Indiana University Press) p. 42; cf. Smith (1984) *Uneven Development*.

66 Cf. Pred, Recognizing *European Modernities*, pp.78–84.

67 In a manner that space does not permit spelling out, this line of reasoning may be linked to Timothy Mitchell's arguments regarding the process of enframing, or the mode of object display and distanced viewing via which the entire world might be regarded as an exhibition (*Colonizing Egypt*; also see Gregory, *Geographical Imaginations*, pp. 34–42).

68 Benjamin, *Passagen-Werk*, p. 578, as translated in Margaret Cohen (1993) *Profane Ilumination: Walter Benjamin and the Paris of Surrealist Revolution* (Berkeley, University of California Press, 1993: 10); *ibid.* [*Passagearbetet*, p. 685].

69 Harvey, "Nature of environment," p. 41.

70 Benjamin, *Passagen-Werk* [*Passagearbetet*, p. 735].

71 Buck-Morss, *Dialectics of Seeing*, p. 340 (quoted phrases from *Passagen-Werk*, pp. 595, 592; parenthetical sense my own).

72 Missac, *Benjamin's Passages*, p. 118 (parenthetical phrase my own reworking rather than Missac's sense). Cf. Mike Savage (1995) "Walter Benjamin's urban thought: a critical analysis," *Society and Space* 13: 201–16.

Part 3

ACTORS, NETWORKS AND THE POLITICS OF HYBRIDITY

INTRODUCTION

What happens to analysis and politics when the familiar identities of modernism – Nature and Society – are no longer taken as givens? What if in their place exist only imbroglios in which science, politics, organisms, religion, law, economy, technology and so on are mixed together in a skein of relations so dense, so entangled, that it is no longer possible to assign objects to either pole – Nature or Society? Such imbroglios challenge the very foundations of modern epistemology as well as social and political theory. This is a central argument in the following chapters, which suggest that our modern dualisms leave us blind to what is really going on. Knowledge is never just about a "reality" that is pre-given, but about mixtures of machines, institutions and social relations as well as scientists and those wily actors we have imagined to be passive nature. Likewise, society is never just about humans, but about the many things – technologies, organisms, texts – that make possible the association of humans and non-humans. In the words of Donna Haraway (1997), the time has come to rethink "kinship" – to recognize that social relationships include nonhumans as well as humans as socially active partners. Analytical and political hope lies precisely in tracing "networks" and "mediations" where previously we saw only "pure" entities and "interactions".

The four chapters in this section represent a concerted attempt to think the networks and politics of late-modern nature–culture complexes. **David Demeritt** (Chapter 8) begins by addressing one of the most contentious aspects of recent attempts to trace imbroglios of nature and culture: the notion that scientific knowledges are "constructed". With Science playing a central role in social life today, it is no wonder that there is fierce debate over what it tells us, and what authority it should be granted. Much current debate has centerd on "social

169

constructivism" and Demeritt provides a lucid exposition of the multiple – and often contradictory – approaches to science, knowledge and nature that today travel under the constructivist banner. In its strongest forms constructivist accounts reverse the causal relation between representation and things that modernism assumed, such that science "constructs", in an ontological sense, the world it represents. This, Demeritt suggests, has had unfortunate results, since it allows the self-styled "defenders" of Science to dismiss "constructivism" as either absurd or polemic, and thus carry on with a conventional understanding of Science as the progressive unfolding of more accurate representations of a real, independent, and pre-existing natural world. Demeritt seeks to shift debate away from these caricatures, and points to "artifactual constructivism" as a useful approach which does not deny the ontological existence of the world but instead emphasises the Heideggerian point that the world's apparent "reality" depends on a configuration of social practices within which it becomes manifest. As Demeritt explains, this allows for ways of talking about science, knowledge and nature without recourse to either the object or an ontologically given Nature while also refusing the conceit of neo-Kantian constructivism where representation is all there is. Demeritt illustrates his arguments through examples drawn from forest ecology and global warming; in both instances, knowledge about nature, and the way that environmental problems are framed, emerge from specific historical practices in which "nature" is certainly an active agent, but which are also shaped by particular social, institutional and discursive contexts. Moreover, from these conceptual frames follow particular "programs of action" which have vital ecological and social implications. Demeritt's conclusion is thus critical for any project of building future natures: to the extent that scientific knowledges are today woven into every level of the social fabric (and political life) we need no nonsense accounts of the powerful and productive practices by which the truth of representations of nature are realised and produced.

In Chapter 9 the analysis of "networks" and practices of "mediation" (by which hybrids of nature and culture are formed) is taken further by **Margaret Fitzsimmons** and **David Goodman**, who trace the entanglement of "society" and "nature" in the political economy of modern agro-food complexes. They begin by developing an important critique of two prominent recent attempts to bring "nature" back in to social theory – environmental history and Marxist political economy. Environmental history has insisted on the autonomous agency of nature (as ecological limits). Marxist political economy, at least in its "green" form, has insisted on a dialectical relation between "nature" and "society". Yet, for all their obvious advantages, Fitzsimmons and Goodman find that they fall into the trap of enlightenment ontology. In both, nature and society are retained as separate categories that "interact", and thus they are unable to explain the "conjoined materiality" of nature and society in late modernity. Fitzsimmons and Goodman suggest that actor-network theory provides new and promising avenues of analysis and political engagement where agency is seen as an effect of networks and practices of mediation, rather than residing in pre-existing categories of "nature" and "society". Through a variety of sites – corn production, bovine spongiform encephalophy (BSE) and anorexia – the authors illustrate how actor-network theory allows new insights into agro-food systems as well as cultural practices related to food consumption.

In Chapter 10 **Bruno Latour** tackles the declining fortunes of the ecology movement in France, and begins to develop an alternate political philosophy consistent with his insistence that we unlearn our modern nature–society dualism. Latour suggests a number of reasons why the ecology movement has failed to develop an original form of politics – from its absorption into existing political philosophies (often conservative) to the problems that follow from its claim to represent a separate ontological realm of "nature". Indeed, one of the chief flaws of the ecology movement, Latour argues, is that it has invested too much in the absolute separation of nature and culture and he suggests that a reconfigured "political ecology" must begin by abandoning the false conceit that it is concerned with "nature" as such. Indeed, in a provocative section, Latour shows how ecology never has been about "nature". Although it claims to talk about nature, it actually only ever talks about endless imbroglios which always involve some level of human participation. But rather than viewing this as a shortcoming, he insists that this is precisely what is most useful about ecology, since in this "failure" we can begin to see the outlines of a reconfigured "political ecology". According to Latour, what political ecology does do – and does extremely well! – is to bring attention to complicated forms of association between beings: regulations, equipment, consumers, institutions, habits, animals, and so on. At the same time, it opens up the question of humanity: what would a human be without elephants, plants, lions, cereals, ozone or plankton? Latour explains that "political ecology" cannot be about shifting our allegiance from the "human" realm to the "natural", a project that is dubious philosophically and futile politically, but rather about thinking possible associations between things and people without any of these entities – rivers, animals, workers, and so on – being used as a simple means by the others. A politics of "ecologizing", by this definition, means attending to networks of quasi-objects while at the same time leaving open the question of interrelatedness. Not about prudence, precaution or scepticism, this is a politics aligned against the violent (en)closures of modernist dualisms and notions of epistemological certainty. It is about making uncertainty a guiding principle, about foregrounding associations, about developing procedures that continuously introduce into political life the question of means and ends. Only by reconfiguring "political ecology" along these "amodern" lines, Latour suggests, will it be possible to drag it from its present state of stagnation, and permit it to occupy the position that the Left has left open for too long.

Finally, in Chapter 11, **Michael Watts** brings together many of the book's themes in his exploration of the knotting together of nature (oil), capital, cultural identity and state rationality in Ogoniland, Nigeria. In the West, the Ogoni struggle is best known through the events surrounding the execution of novelist Ken Saro-Wiwa and eight other Ogoni activists in 1995. But, as Watts explains, these events must be understood in terms of how social identities like "oil", the "Ogoni" and the "state" have been constructed and contested in relation to each other. In short, the very structure of the post-colonial Nigerian state, the multiple and shifting identities of groups like the Ogoni and contests over "development" and "modernization" occur as part of a continuum with the social production of nature in global circuits of capital and modern technoscientific practices. None of the "actants" in this story have an "essence" which pre-exists their constitution within these complex networks. From this Watts develops three important points. First, the Ogoni struggle can never be contained as only "environmental" but is simultaneously "cultural" as well as "socio-economic". Second, these imbroglios are infused with politics. Far from using the language of "networks", "mediation" and "translation", in a detached, analytical manner, Watts insists on the ways in which these networks, mediations and hybrid identities matter. As he explains, at a particular historical moment, the construction of these identities and relations is made to pass through the body of Saro-Wiwa in an act of ferocious state violence. Here the politics and the costs of some sorts of hybrid identities become brutally clear. Third, Watts provides a cautionary tale for critiques of development discourse. Pointing to the case of Ogoni resistance – with its mixing of modernity, hybrid cultural identities, and local community organization – Watts suggests that anti-development critiques too readily appeal to the "local", to the "popular" and to the "cultural" (or "identity") as resources through which to resist "modernity" and "development" without recognizing the degree to which these are themselves constitute within rather than outside dynamics of modernization and its political and social imaginary. Thus, like the other chapters in this section, Watts situates agency – and politics – firmly within the complex local and global nature–culture networks that characterize the *fin de siècle*.

REFERENCE

Haraway, D. (1997) *Modest_Witness@Second_Millennium.FemaleMan©Meets_Onco-Mouse™*, London: Routledge.

8

SCIENCE, SOCIAL CONSTRUCTIVISM AND NATURE

David Demeritt

INTRODUCTION

Science, it seems, is in the news these days. Reports of further cancer risks, pollution problems, and environmental nightmares compete for space in the headlines with the technical triumphs of modern science: miracle medical cures, computer wizardry, and awe-inspiring discoveries about the origins of the human species and the universe itself. But science in this day and age is as commonplace as it is extraordinary. From the scientifically engineered food we eat to the space age materials like Goretex™ and Kevlar™ that clothe and shelter us, modern science and its technical creations have become ubiquitous, indeed indispensible, if also largely taken for granted, aspects of everyday life – at least in the industrialized world. Yet despite this success, because of it in fact, the sciences are met with increasing public unease and skepticism. Assurances from the grave men in white lab coats are no longer sufficient to ease public concern about toxic chemicals, nuclear contamination, and the other environmental "side effects" of industrial society.

This loss of public faith in the sciences is a characteristic of what Ulrich Beck (1992) calls the emergent "risk society." Whereas previously industrial society was organized around the application of scientific knowledge for the production and distribution of wealth, now, according to Beck (1992: 19–20), the defining feature of contemporary society is the distribution and management of hazards such as global warming that result "from techno-economic development itself." As the chief cause of these modern environmental problems as well as "the medium of [their] definition, and the source of solutions" (p. 155), the sciences occupy a controversial and contradictory position in the risk society. In the face of global environmental changes that seem to make them "more and more necessary," the sciences are "at the same time, less and less sufficient for the socially binding definition of truth" (p. 156).[1]

Beck's notion of the "risk society" provides a useful starting point from which to begin making sense of the recent controversies about science, social constructivism, and nature. Against this backdrop of uncertainty about the risks associated with scientific and technological progress, the status of scientific knowledge has become the object of fierce, academic dispute. The controversy pits a variety of cultural critics who emphasize the socially contingent manner in which scientific knowledge is constructed against self-styled defenders of science, many of them practising scientists themselves, who uphold a conventional understanding of science as the progressively more accurate explanation of a real, independent, and pre-existing natural world. This commonsense explanation of science is epistemologically realist. It posits that scientific knowledge is true if it represents the world as it in fact really is. While a variety of philosophers have questioned the logical underpinnings for such a correspondence theory of truth (Rorty 1979; van Fraasen 1980; Rouse 1987), historians, sociologists, and anthropologists who study practising scientists have emphasized the ways in which the validation of scientific theories is determined socially rather than by correspondence to an independent reality.[2] These critiques of scientific and epistemological realism have become so widespread in the academy that the New York Academy of Sciences and the British Royal Society both sponsored high-profile symposia to defend science against social constructivism. The media, ever fixated by controversy, have given the debate considerable play, both as a news item and as grist for the op-ed page.

The academic debate over social constructivism is more complex than either the sensationalist coverage or the often glib dismissals would suggest. In its various forms, social constructivism poses fundamental questions about public trust and scientific credibility. What makes some knowledge scientific? How does science work? Why should we believe it? What accounts for its apparent success in explaining the world? Why are the sciences to be preferred over other ways of knowing and relating to nature? Who should decide?

There are a number of compelling, if not necessarily compatible, answers to these questions, but it is important to recognize that they concern more than just the foundations of scientific knowledge. The debate about social constructivism is also about social power and legitimacy, which is one reason why the furore has become so heated of late. My concern in this chapter is less to adjudicate the epistemological status of science (to my mind, largely a pointless exercise), than it is to use the debate to clarify what is at stake in the practice of science and the social construction of nature.

My discussion will be divided into three parts. First, I will review the controversy surrounding science and social constructivism. Although the debate is commonly staged in simplistic either/or terms such as science/anti-science and realist/constructivist, it is considerably more diverse and multi-faceted. A number of very different approaches to science, knowledge, and nature travel under the deceptively simple banner "social construction" (Sismondo 1993). Having described what it might mean to talk about the social construction of science and nature, I shall discuss in turn the examples of forest conservation and global warming. Science has been crucial in constructing these environmental problems. In each case, the nature of the problem and the sort of techniques applied to address it have depended fundamentally upon the particular metaphors and scientific practices by which it has been constructed and represented. Indeed, it is difficult to imagine an environmental phenomenon less directly observable, more remote from everyday experience, and more dependent on the technical apparatus of science for constructing its apparent "reality" than the so-called greenhouse effect (Cronon 1994: 41). For climate change skeptics, the fact that atmospheric scientists must endlessly tune, correct, and parameterize their global circulation computer models (GCMs) in order to represent the facts of future climate change provides a reason to dismiss the entire problem as a phantasmic social construction, born of paranoid "hype" by "environmental pressure groups" and unproven by any "solid fact" or independently verifiable scientific observation of actual anthropogenic climate change (Singer 1992: 34). Such charges are vigorously denied by the scientists involved as well as by environmentalists, many of whom fear that social constructivism, by focusing on the human interests and agencies involved in constructing and promoting particular environmental problems, denies them any objective reality, thereby sapping political will for protecting the environment (cf. Dunlap and Catton 1994; Soule and Lease 1995). This analysis of social constructivism is incomplete and unhelpful. It rests on a problematic distinction between nature and society that confounds our understanding of the practice of science and the representation of nature. I try to address this difficulty in this chapter by outlining a theory of artifactual social constructivism that reconciles a recognition of the productive activity of scientists in constructing and representing the facts of science with an appreciation of the role of other, heterogeneous agencies in realizing the nature of the world.

SOCIAL CONSTRUCTIVISM

Social constructivism has become a popular, catch-all term to describe a variety of very different approaches to science, knowledge, and nature. Sergio Sismondo (1993) identifies four distinct uses of the construction metaphor, each describing a different object of construction (see Table 1).[3] First, there is what might be called social object constructivism. This refers to the construction, through the interplay of actors, institutions, habits, and other social practices, of subjective belief about reality that over time "congeals for the man on the street" into a "taken-for-granted 'reality'" (Berger and Luckman 1966: 3; see also Searle 1995).

Feminists have been among the most enthusiastic proponents of social object constructivism. Many, though by no means all, distinguish sharply between gender, the subjective and socially constructed beliefs about sexual difference that constitute a changeable, but no less real, "social reality," and sex itself, the

Table 1 A Typology of Social Constructivisms

	Common-sense realism	Social object constructivism	Social institutional constructivism	Artefactual constructivism	Neo-Kantian constructivism
Chief tenets	Observational statements refer directly to a pre-existing, independent, and, in this sense, objective reality	Taken-for-granted beliefs about reality, e.g. gender, constitute a social reality no less "real" in its causal effects than reality itself	Science is a social construction in the sense that its institutions and the social contexts of its discoveries are socially conditioned and constructed	The reality of the objects of scientific knowledge is the contingent outcome of social negotiation among heterogenous human and non-human actors	The objects of scientific thought are given their reality by human actors alone
Key proponents	Gross and Levitt (1994)	Berger and Luckman (1966); Searle 1995	Merton ([1938] 1970)	Latour (1987); Haraway (1992)	Woolgar (1988); Collins and Pinch (1993)
Ontology	Nature/ society, subject/ object, mind/ matter are ontologically distinct realms	Socially constructed reality distinct from objective facts given by nature, e.g. sex	Objective reality distinct and independent from belief about it	No absolute ontological distinction between representation and reality, nature and society	Nature is whatever society makes of it
Epistemology	Truth value determined by correspondence between representation and reality	Scientific truth explained by nature; socially constructed belief is the cause of scientific falsehood	Ignorance and socially constructed bias explain belief in scientific falsehood	Ultimate truth is undecidable	Truth is what the powerful believe it to be

biologically given, immutable material reality of those differences. This distinction provides the basis for a well-established tradition of liberal feminism that Sandra Harding (1986; 1991) has dubbed feminist empiricism. Feminist empiricists criticize practising scientists for failing to live up to their own high standards of objectivity and allowing gendered and socially constructed beliefs to bias their representations of nature. Though under assault as part of the general right-wing counter-attack on social constructivism (Gross and Levitt 1994), this feminist use of the construction metaphor actually supports a conventional understanding of science and objectivity. Social object constructivism preserves the ontological distinction between a social reality of human making (gender) and an underlying material reality not of human construction (sex) that provides the epistemic basis for distinguishing true and objective scientific knowledge from subjective and socially constructed belief. In the hands of feminist critics, social object contructivism provides a way to expose sexist bias in science without giving up on the ideals of science as a means of exposing the objective reality of women's oppression.

A second variety of social constructivism is social institutional constructivism. This describes the development of the institutions of science and the social processes of theorizing, experimenting, and arguing by which scientists establish their knowledge of an objective reality. Much of the work of this type has been historical, tracing the social pressures influencing the conduct and direction of scientific research (cf. Merton [1938] 1970; Rudwick 1985). Even professed opponents of social constructivism acknowledge that these are legitimate subjects for social science research, for they speak to the ever-present problem of bias, which a rigorous scientific method is designed to weed out of science. Like social object constructivism, social institutional constructivism is what David Bloor (1976: 4–5) calls asymmetrical. It distinguishes sharply between, on the one hand, the properly sociological explanation of the social context for particular scientific discoveries or incorrect beliefs and, on the other hand, the explanation of scientifically valid knowledge, which is largely unquestioned. Indeed, in the classical tradition of sociology of knowledge, science was explicitly declared off-limits for sociological explanation (Mannheim [1929] 1954). As a result, social institutional constructivism, like social object constructivism, is not at all inconsistent with epistemological realism and the claim that scientific knowledge is true and objective because it describes the world as it in fact actually is, quite independent of any human volition or activity.

By contrast, a third variety of social constructivism, artifactual constructivism, poses more of a challenge to realism. Artifactual constructivism refers to the construction, through material interventions and interactions, of the artifacts and other phenomena of the laboratory. The purified samples and carefully calibrated apparatus, as well as the theories and technically mediated observations that scientists build up and work with, constitute a "highly preconstructed artificial reality" (Knorr-Cetina 1983: 119). As such, artifactual constructivists maintain that the objects of scientific knowledge are the outcome of carefully contrived practice, not pre-existing objects waiting to be discovered and correctly represented by science (Hacking 1983; Latour 1987; Haraway 1992). This poses several challenges to epistemological realism. In common with empiricist arguments against epistemological realism (van Fraasen 1980), artifactual constructivism deflates the sense of metaphysical truth on which realism depends. For artifactual constructivists, questions of abstract truth are undecidable, if not altogether meaningless. The criterion of success for scientific theory is empirical adequacy and pragmatic achievement, not ultimate truth or falsity.

By emphasizing the productivity of scientific knowledge and practice, artifactual constructivism also denies the sharp break postulated by realism between reality and scientific descriptions of it. Latour and Woolgar (1979: 64) articulate the criticism this way: "It is not simply that phenomena depend on certain material instrumentation; rather the phenomena are thoroughly constituted by the material setting of the laboratory. The artificial reality, which participants describe in terms of an objective [i.e. existing independent of human agency] entity has in fact been constructed . . . through material techniques." The so-called real world against which the truth of a particular scientific representation might be tested can only be grasped through other representations, because reality appears as such only as a condition and result of the specific, productive activities of its representation (Demeritt 1997). Such artifactual constructivism, it is important to emphasize, does not deny the ontological existence of the world, only that its apparent reality is never pre-given; it is an emergent property that "depends upon the configuration of practices within which [it] becomes manifest" (Rouse 1987: 160–1). This Heideggerian insight is a difficult one. It is easy to slip from artifactual constructivism that is ontologically realist about entities but epistemologically anti-realist about theories (the things we call electrons are real objects, but our ideas about them are constructed) into a much stronger use of the construction metaphor that is anti-realist about both theories and entities (electrons have no objective existence; our belief in them as social objects is what gives them their apparent "reality").

This much stronger, neo-Kantian sense of the metaphor is the fourth variety of social constructivism. Neo-Kantians like Steve Woolgar (1988) use social construction in the very strongest and most literal sense: the social construction of the objects of scientific thought and representation. They reverse "the presumed relationship between representation and object, [claiming] that representation gives rise to the object" (Woolgar 1988: 65). Other sociologists have been more circumscribed in their approach, adopting neo-Kantian constructivism as a methodological principle for explaining the production of scientific knowledge symmetrically: without reference to its ultimate "truth and falsity, rationality and irrationality, success or failure" (Bloor 1976: 7). From this perspective, the actual nature of reality plays no role in determining our beliefs about it. Methodological relativism allows the "apparent independence of the natural world" to be described as something "granted by human beings in social negotiation" (Collins and Yearley 1992: 320). It leads to the polemical conclusion of Collins and Pinch (1993) (quoted in Mermin 1996: 11) that "the truth about the natural world [is] what the powerful believe to be the truth about the natural world."

This neo-Kantian variety of social constructivism, not surprisingly, has drawn fierce criticism. While the merits of this strong programme have been the subject of intense discussion within the field of science studies itself (Scott *et al.* 1990; Pickering 1992; Pels 1996; Wynne 1996), the loudest, or at least the best publicized protests have come from a number of self-appointed defenders of science, many of them practising scientists. They complain that social constructivism is relativist, irrational, and patently absurd. Its refusal to acknowledge any objective criteria for scientific verification makes it impossible to distinguish "reliable knowledge from superstition" (Gross and Levitt 1994: 45), thereby opening the door to our most "irrational tendencies" (Weinberg 1996). These realist defenders of science invite their constructivist critics to test the absurdity of the claim that "the laws of physics are mere social conventions . . . [by] transgressing those conventions from the windows of my [twenty-first floor] apartment" (Sokal 1996: 62). Gross and Levitt give a more elaborate homily on the foolishness of social constructivism:

> Imagine that a few of us are cooped up in a windowless office, wondering whether or not it's raining. Opinions vary. We decide to settle the issue by stepping outside, where we note the streets are beginning to fill up with puddles, that cars are kicking up rooster-tails of spray, that thunder and lightning fill the air, and, most significantly, that we are being pelted incessantly by drops of water falling from the sky. We retreat into the office and say to each other, "Wow, it's really coming down!" We all now agree it's raining. Insofar as we are disciples of Latour, we can never explain our agreement on this point by the simple fact that it is raining. Rain, remember is the outcome of our "settlement," not its cause! Badly put, this seems ridiculous. Nevertheless, if we accept the validity of Latour's putative insight, we are ineluctably obliged to accept this analysis of a rainy day.
>
> (Gross and Levitt 1994: 58)

This appeal to commonsense is compelling but ultimately deceptive. Commonsense would indicate that the sun revolves around the earth and that heavy objects fall more quickly than light ones, but the laws of physics say otherwise. Many objects of scientific knowledge do not lend themselves to verification as easily as rain, though the observation of even this everyday phenomena involves prior theoretical commitments about what constitutes "rain" (Hesse 1980: 65–83). The observation of molecular structures like DNA, if this is even the right word to describe the process of visualization involved in an electron microscope, is ever so much more complex a social and technical achievement. Other unobservable entities, like quarks and neutrinoes, can only be known indirectly, by observing the effects of their manipulation in multi-million dollar particle accelerators. It is difficult, therefore, to argue that the truth of our representations of these phenomena is in any way self-evident, as the most vocal opponents of social constructivism seem to be contending.

But the neo-Kantian account of scientific knowledge is no better. By denying the natural world any role in constraining scientific knowledge of it, neo-Kantian constructivism seems to suggest that nature is whatever science makes it out to be. This makes it difficult to understand how science could ever fail or a scientific theory be invalidated. The neo-Kantian case is much easier to sustain in the case of the unobservable entities of particle physics, a favored object of neo-Kantian explanation (cf. Collins and Pinch 1993), than it is for the more familiar objects of applied science, whose obdurate reality seems much harder to deny. The claim of Andrew Ross (1991: 217) that "the only difference" between modern atmospheric scientists and rain dancers "is that they appeal to differently organized systems of rationality" discounts the much great predictive success enjoyed by contemporary weather forecasters, with their satellite images and computer simulation models. Surely, weather forecasters know something about the weather; this is why we pay attention to them.

Realists take this practical and technical success as proof of the objective truth of scientific theory (Boyd 1984), but there are problems with this abductive argument for epistemological realism. The standards of empirical adequacy that define successful "working" and prediction are themselves socially determined norms and not given self-evidently as data by the nature of reality itself. Scientific standards of proof are prime examples of social object constructivism. No one would deny that airplanes and the other complex artifactual constructions of modern science do indeed work and that this success depends crucially (but not entirely) upon the successful predictions of scientific theory. The issue, for realists and anti-realists, is whether this explanatory and predictive success is in any way indicative of the truth of scientific representation. At one time, both phlogiston theory and Ptolemaic planetary models "worked": they provided reliable frameworks to predict and explain the available evidence, but both have now been discredited. While realists take this as evidence that scientific knowledge is converging on truth, it is not entirely clear what converging toward truth might mean when the historical development of scientific theory is taken as evidence of the very thing it is supposed to be explained by: convergence on truth (Laudan 1984).

Ultimately, the issue of scientific truth is not a very interesting one. It tells us nothing about whether a particular scientific theory works or why. And yet, the debate about science and social constructivism has been fixated by the objective truth of representation. Realists uphold truth as correspondence, and neo-Kantians deny it, by collapsing the realists' dualism into a single, socially constructed monism, thereby conflating anti-realism about scientific theories and the epistemological claim that the grounds for representations are arbitrary and socially constructed with anti-realism about scientific entities and the ontological claim that reality itself is made up. Realists dismiss this out of hand as absurd, which it is, but their dualism is no less problematic. Focused exclusively on the correspondence of scientific representation to reality, they ignore the fact that this reality is only ever realized as an artifact of scientific representations.

Scientific knowledge depends crucially upon the human relationships described so insightfully by the work of social object and social institutional constructivists, but it also depends upon a variety of nonhuman actors. Scientists are, after all, struggling to understand the natural world. While their knowledges are figured in culturally specific (and materially significant) ways, they are "about" something more than just culture. The difficulty comes in acknowledging the active role played by the objects of scientific knowledge in shaping or constraining this knowledge without falling back into some kind of epistemological realism in which true knowledge is said to reflect the world as it is ontologically (pre-) given.

Artifactual constructivism provides a way out of this dead end. It refigures the actors in the construction of what is made for us as nature and society. The social in these social constructions is not just "us": it includes other humans, non-humans, and even machines and other, non-organic actors. Artifactual constructivism provides a way of acknowledging that these agencies "matter" without taking the particular configuration of their matter or the process by which it is realized for granted (Butler 1993). This makes it possible to talk about science, knowledge, and nature without recourse either to the objective and ontologically given Nature of epistemological realism or to the omnipotent and all-knowing Society of neo-Kantian constructivism. Instead, artifactual constructivism focuses on the powerful and productive practices of science by which the reality of nature and our socially constructed knowledge of it are produced and articulated, thereby dispelling the modern dualism on which the debate about science and social constructivism has turned.

Science appears rather differently once we abandon the illusion that it must either be a purely objective reflection of the world or an entirely subjective construction of it. Questions about scientific representation and correspondence to an external and ontologically given natural world give way to questions about scientific practice and the mediated relationships among humans and their ever-active, non-human partners in the social production of knowledge and nature. Artifactual constructivism makes these interactions visible. It makes it possible to interrogate the culturally specific knowledges and ways of being that scientific interventions in and reconfigurations of the natural world realize and produce. "Biology," as Donna Haraway (1992: 298) insists, "is a discourse, not the living world itself. But humans are not the only actors in the construction of the entities of any scientific discourse . . . So while the late twentieth century immune system, for example, is a construct of an elaborate system of bodily production, neither the immune system nor any other of biology's world-changing bodies – like a virus or an ecosystem – is a ghostly fantasy." These objects of scientific knowledge are co-constructions. This makes them no less real or materially significant. It simply highlights the complex and negotiated process of scientific practice and representation by which they are materialized and produced for us as natural–technical objects of human knowledge.

For too long we have been debilitated by the notion of disembodied, Olympian truth and the correspondence theory of knowledge. This has made it difficult to appreciate the diversity of the sciences and the differences in the ways they render the world. Silviculturalists, for example, represent the forest very differently than ecologists whose theoretical concern with ecological com- munities discloses interspecific aspects of the forest discounted by silviculture, for which the single-species age class was long the fundamental unit of analysis. These differences are consequential, but they cannot be explained in terms of the (un)truth of silvicultural representation of the forest. They are the products of practice, not representation. The issue is not whether one better reflects the underlying nature of the forest but how this nature is figured and realized and with what effects. In the next two sections, I explore these general issues around the artifactual construction of nature through a discussion of sciences of forest conservation and global warming.

SAVING THE FOREST

The science of forestry is founded on a series of metaphors representing the nature of the forest and dictating the practices that should be applied to conserve it. Concerned by the rapid depletion of the American forest, conservationists and professional foresters of a century ago, like US Forest Service Chief Gifford Pinchot (1905), argued that the only way to save the forest was to manage it scientifically: by which they meant treating the forest as a kind of natural capital to be conserved, rather than exploiting it shortsightedly, as had previously been the case, as if it were a mine and a non-renewable resource with no future beyond its immediate stumpage value. Since the turn of the century, this scientific construction of the forest as capital and the practice of sustained yield forestry that flowed from it have been the model, if perhaps not quite the norm, for forest management in North America. Recently, however, scientific forestry has come under fire from those who complain that the old conception of the forest as "working capital whose purpose is to produce successive crops" (Pinchot 1905: 41) is dangerously narrow. In place of the old ideal of sustained yield, proponents of so-called "new forestry," such as Jerry Franklin (1989), speak of sustaining forest ecosystems and forest health. The struggle between industrial advocates of sustained yield forestry and promoters of new forestry and ecosystem management turns as much on representing the nature of the forest as on the question of what should be done to it.

Very different programs of actions flow from these competing constructions of the forest as a quantity of capital to be conserved and an ecosystem whose health is to be protected. Turn-of-the-century conservationists seized on the comparison of the forest to "a savings bank from which you could draw interest every year" as part of their campaign against the prevailing practice of cut-and-run logging (US Division of Forestry 1887: 9–10). The analogy provided a basis for protecting future forest supplies, but it was harder to justify the protection of non-market public goods derived from forests, such as flood abatement, in terms of natural capital. If flood abatement happens at all, it is only as an unanticipated byproduct of conservative timber harvesting. Strict profit-maximizers have no economic incentive to look after unpriced public goods, whose social value the market, and thus the notion of natural capital, does not account for. Mostly, the comparison of the forest to accumulating capital was didactic, made without much sense of the tensions inherent between the forest's value as a fixed source of socially necessary materials and as a fluid financial asset.

The idea of forest capital appealed to the interests of large forestland owners. It highlighted the difference between destructive so-called timber mining, which depleted the supply of natural capital, and conservative forestry in which "only the interest [was] taken . . . [and] the principal of the investment [was] retained" (Cary 1899: 161). In a very real sense, the forest was already being represented by the dollar sign; loggers sized it up in terms of its immediate monetary value. Turn-of-the-century conservationists simply took advantage of this fact to promote scientific, sustained yield forestry to a skeptical forest industry. In this way, then, the rise of scientific forestry might be understood as an example of social institutional constructivism in which the social power of the forest products industry explains the relatively rapid uptake and institutionalization of sustained yield forestry (cf. Hays 1959). But such an external influence explanation discounts the degree to which the new scientific understanding of the forest as an accumulation of natural capital affected the actual details of scientific practice.

Scientific representation of the forest as naturally accumulating capital illuminated a variety of forest properties and relationships that had long escaped notice. Since, as the forester C. A. Schenck (1911: 21) explained, "growth in conservative forestry is the making of revenue," foresters studied the growth rates of merchantable species as well as other physiological and ecological processes affecting their development and reproduction. Non-merchantable species, long ignored by loggers as little more than a nuisance, were suddenly reconstituted as competition. Foresters experimented with technical treatments, such as girdling, thinning, and herbicides, to favor the growth of merchantable species over their competitors. To this Darwinian view of the world, life in the forest is a zero-sum game, in which resources are limited and their use necessarily reduces the available supply. Scientific concern with competition dictated new harvesting practices. In the mountainous West, the US Forest Service mandated selective rather than the customary clear-cut logging to insure that sufficient trees were left to seed the next generation. While foresters tried to minimize inter-specific competition, intra-specific competition and dense thickets of young timber were ideal, because competition among seedlings was thought to insure that the hardiest and most vigorous individuals survived (Langston 1995: 31).

The idea of conserving the forest as if it were capital and the confidence in science that it reflected has led foresters to treat the forest as an assemblage of individual objects that can be managed more or less in isolation from one another. Foresters sought to maximize the yield without much thought to how larger quantities of a merchantable species would interact with the rest of the forest. In the spruce fir forest of the Northeast, selective logging for spruce has transformed uneven aged stands dominated by mature red spruce into much simpler stands with two age classes: a canopy dominated by balsam fir and a sprinkling of immature, red spruce left as "seed trees" and a dense understory of suppressed firs, which reproduces more prolifically than spruce. As this fir-dominant understory matured, it became vulnerable to infestation by spruce budworm, an insect whose preferred food, in fact, is balsam fir. At periodic intervals, growing more devastating with each occurrence, spruce budworm outbreaks have devastated the forest, killing most mature fir and many spruce trees as well (Seymour 1992). This high mortality, combined with massive salvage efforts to harvest vulnerable stands before they are attacked, has created the ideal habitat for spruce budworm to thrive: large areas of dense, even-aged, fir-dominant forest stressed from fierce competition for canopy space and thus less able to fend off attack (Lansky 1992). Similar problems plague the Blue Mountain forests of eastern Oregon, where the suppression of fire, thought necessary to protect the second growth of ponderosa pine, actually led to its elimination by fire-intolerant and low-value grand fir that grew in dense, tangled stands which were vulnerable to insect infestation and subsequent fires that consumed millions of acres of forest. By the early 1990s, the diseased and fire-plagued forests of the Blue Mountain were widely condemned in the press as a "Man-made blight" brought on by well-meaning but misguided forest management focused narrowly on lumber production without regard to the wider ecological effects of this management strategy (*Seattle Post Intelligencer* 1991, quoted in Langston 1995: 6).

Within the forestry profession, problems such as these have led to recent calls for the development of a "new forestry," focused "on the maintenance of complex ecosystems and not just the regeneration of trees" (Franklin 1989: 38). Ecosystem management, designed to sustain the "health" of forest ecosystems, has recently been adopted as the official policy objective of the US Forest Service (1992). This new conception of forest conservation is undoubtedly a response by the Forest Service and by the forestry profession in general to outside pressure from environmentalists, but it is also a product of some of the different ways in which the forest is now being framed as an object of scientific knowledge. The aims and objects of so-called new forestry depend fundamentally upon scientific ideas and techniques, first developed by ecologists but now widespread in other scientific disciplines as well, that set up the forest as a coherent eco-system whose interrelated parts are connected by flows of matter and energy (Hagen 1992). These practices make it possible to imagine the forest as an ecosystem whose health might be monitored and managed, rather than, as in more traditional silviculture, as a disparate collection of age classes (Costanza *et al.* 1992). For all its apparent simplicity, however, ecosystem "health" has proven frustratingly difficult to define and thus to manage for or sustain (Suter 1993; O'Laughlin *et al.* 1994).

In some sense the conceptual ambiguity of forest health and ecosystem management has proven to be its greatest appeal. Scientists, industry officials, and environmentalists can all heartily endorse these new constructions of the forest without necessarily agreeing about the specific technical details required to achieve them. There is fierce disagreement, for example, about whether clear-cutting mimics natural disturbance processes, and thus is consistent with new forestry principles of ecosystem management, or whether it is ecologically damaging. Nature and the naturalism of forest practices at issue here are certainly framed by science, but it is hardly the case, as advocates of neo-Kantian constructivism would contend, that scientists are the only actors of consequence when it comes to constructing the nature of the forest. Foresters struggling with insect infestations have learned from long experience that they are not free to make the forest in any way they choose. Other actors matter too. This realization was crucial in the recent development of new forestry and its concern, however instrumental, with forest characteristics and processes beyond the accumulation of merchantable fibre. It is important to acknowledge the ontological inde-pendence of this nature without losing sight of the ways in which its reality is only ever realized and produced for us as an artifact and object of scientific practice and representation.

GLOBAL WARMING

Global warming presents a rather different set of issues and actors, but like the struggle to conserve the forests, the nature of climate change is as much a politically charged production of science as it is a straightforward reflection of some independent biophysical reality. Although theories about anthropogenic climate change date as far back as ancient Greece (Glacken 1967), and scientific discussion of an enhanced greenhouse effect, caused by changes in the earth's radiation balance due to the accumulation in the atmosphere of carbon dioxide (CO_2) and other radiatively sensitive gases, began in the late nineteenth century (Rowlands 1995), global warming did not emerge as a serious environmental concern until the late 1980s, when, almost overnight, it burst onto the scene as "the most important problem facing mankind over the next fifty years" (Gribbin 1990, quoted in Buttel *et al.* 1990: 57).

This rapid take-off owed much to the ambitions of government bureaucracies and Western environmental lobby groups, who seized upon the issue as an organizing rationale for a wide range of environmental protection and pollution control policies, and to the promoters of so-called alternative energy sources such as nuclear and hydro-electricity that were struggling in the cheap energy markets of the 1980s (Boehmer-Christiansen 1994). Indeed, many of the participants in the Intergovernmental Panel on Climate Change (IPCC), the international organization of scientific experts created in 1988 to advise governments and parties to the UN Framework Convention on Climate Change, came to Working Group Three, which was devoted to policy responses to climate change, directly from energy modeling and the late 1970s debate about how best to respond to the imminent depletion of fossil fuels.[4] As important as these social institutional factors were to the acceptance of global warming as an environmental crisis caused by fossil fuel consumption, tropical deforestation, and other anthropogenic emissions of greenhouse gases (cf. Boehmer-Christiansen 1990; Rowlands 1995), I would like to focus here instead upon the internal scientific practices contributing to this construction and to the political implications of the representation of climate change as a global scale environmental problem demanding global environmental, rather than say local and regional or cultural and economic, solutions.

From the outset, global warming has been constructed in narrowly scientific terms as a problem of atmospheric emissions largely divorced from their social context. Whereas the Bruntland Commission of the United Nations and its notion of sustainable development addressed themselves to the whole range of cultural and economic imperatives contributing to environmental problems, atmospheric scientists leading the IPCC and other international scientific bodies have defined the problem of global warming in terms of flows of matter and energy, thereby excluding from the analysis the political economy responsible for producing greenhouse gas emissions in the first place. This reductionism makes climate change scientifically manageable, unlike sustainable development, which is an analytical abstraction difficult to define or work with in practice. By constructing global warming as a matter of simple physics, scientists are able to model it mathematically. Their computer visualizations represent the facts of future climate change in alarming hues of red and orange, making the problem "real" for policy-makers and the public at large. Indeed, global climate change is difficult to imagine apart from the massive general circulation models (GCMs), computer simulations of the global climate system so sophisticated and expensive to run that there are only a few worldwide. These models have provided the most authoritative evidence of future global warming (Shackley and Wynne 1995).

The reductionism of climate change science is aligned to both a moral–liberal and a rational–technocratic view of politics and science (Taylor and Buttel 1992). In either case, the reductionist conception of global warming as an exogenous environmental force affecting humanity as a whole appeals to the common and undifferentiated interests of a global citizenry. It bypasses the complex, locally specific problems of sustainable development, reducing them to the single question of controlling global greenhouse gas emissions. The only difference is how emission reductions are to be achieved. The moral–liberal formulation depends on communicating scientific knowledge of the objective risks of climate change to sway self-serving, naïve, or scientifically ignorant behavior contributing to global warming, while the rational–technocratic relies on science to identify the optimal policy to which individuals must then submit. Both assume that the proper role for science is to provide certain knowledge on which to found political decisions and that therefore the first obstacle to addressing climate change is scientific uncertainty, which impedes the formation of democratic consensus (moral–liberal) and the optimization of policy (rational–technocratic).

This emphasis on objective scientific knowledge as the basis for rational political action serves well both scientists and policy-makers. The authority of science provides legitimacy for controversial public policy decisions, while the promise of still greater certainty secures funding for more research. To this end, researchers are developing a whole new generation of integrated assessment models in which global climate and physical impact models are linked to land use, energy, and general equilibrium economic models to evaluate the impacts of different policy and climate change scenarios (Parson 1995; *Climatic Change* 1996) Although these integrated assessment models incorporate many more socio-economic dimensions of climate change than the GCMs, for which society was black boxed as a source of emissions and a sink for climate impacts, the general approach remains resolutely reductionist. As such they are tied to the same alliance of moral–liberal and rational–technocratic politics in which the solution to global warming depends on the resolution of scientific uncertainty and the successful communication of this objective knowledge to policy-makers and the public (Shackley and Darier 1997).

Concentrating upon the political use of climate change science, such an interest-based analysis of the social construction of global warming leaves unquestioned the status of scientific knowledge. It discounts the degree to which social commitments are built into the technical details of scientific practice as well as their subsequent use in the public sphere. For example, Shackley and Wynne (1995) have shown how in global climate modeling the practice of flux-correction, which is necessary to keep the separate ocean and atmosphere components of coupled GCMs in synch over long time scales (Kattenberg *et al.* 1996), has been constructed as the technique of choice, despite concerns about the massive "fudge-factor" involved (Kerr 1994), on the basis of a political desire for long-term prediction, which, until recently, was only possible with flux correction, as much as on its technical merits alone. Thus, Wynne (1996: 372) concludes, "the *intellectual* order of climate scientific prediction, and the *political* order of global management and universal policy control, based as it is on the promise of deterministic processes, smooth changes, long-term prediction and scientific control, mutually construct and reinforce one another."

Similar political commitments are built into the scientific calibration of the global warming potential (GWP) of the various greenhouse gases, which each have different atmospheric residence times and radiative properties depending in turn upon their relative concentrations (Smith 1993). These processes are too complicated to be integrated into a GCM, so it is necessary to calculate the GWP for each individual greenhouse gas, converting its relative radiative forcing per unit to a standard measure (typically a CO_2 equivalent) that can be aggregated to produce an overall radiative forcing function to drive GCM predictions of future global warming. Activists from developing countries complain that the luxury emissions of CO_2 from fossil fuel consumption in industrial countries cannot justly be compared to the survival emission of methane from agricultural production in developing countries (Agarwal and Narain 1991; Bodansky 1993). Furthermore, because the decay rates of these gases differ so greatly, the choice of time horizon strongly influences the calibration of their relative GWP, and thus any calculation of national emission profiles or the cost benefits of CO_2 versus methane emissions. These are questions of considerable import for the ratification and enforcement of the international emission reduction convention recently negotiated at Kyoto, Japan, but climate change science is not set up to answer them. The GCMs, which construct global warming as a global environmental problem, depend upon reducing the differences between various atmospheric emissions to a universal system of equivalence. They are indifferent to the social meanings of the entities that have been artificially unified through the GWP index and the globalizing ambitions of climate change science (Wynne 1996: 376–7).

Such a social constructivist critique of the way in which science has set up and framed the threat of global warming does not imply that the problem is unreal or that our socially constructed and contingent knowledge of it is simply false. It does question the authority for that knowledge and the legitimacy of what has been done in its name. For this reason many critics of social constructivism, such as Scott *et al.* (1990), have concluded that it is inherently conservative in its political orientation because it provides entrenched interests with the intellectual tools to refute any scientific criticism of their actions. It is certainly true that the fossil fuel industry has founded its opposition to greenhouse emission reductions upon the uncertainties inherent to flux correction and other scientific practices (Singer 1992; Global Climate Coalition 1994), but they have not been alone in their resistance to the political solutions proffered by the moral–liberal and rational–technocratic formulation of global climate change. In the initial blush of excitement about global warming in the late 1980s, a coalition of Western environmentalists, government regulators, and atmospheric scientists called for global carbon taxes, tropical rainforest protection, and other measures to reduce greenhouse gas emissions and avert certain climate catastrophe. Their appeal to the universal interests of a global citizenry was founded on scientific certainty, rather than the more difficult work of making global warming meaningful to a differentiated international public. This has proven to be neither a very democratic nor an especially effective way of constructing a political response to global warming. Social activists complain that the narrow focus on greenhouse gas emissions, whose effects will not be felt for a generation or more, divorces them from their social context and displaces attention from far more pressing and immediate concerns, such as poverty and hunger. Developing countries are resisting pressure to reduce their greenhouse emissions as a new form of environmental colonialism, designed to keep them poor and underdeveloped (Parikh 1992; Parikh and Painuly 1994). In industrialized countries, individuals hesitate to alter their lifestyles in response to climate change public education and outreach campaigns because they do not trust the corporations to do likewise or believe that individual action will make much of a difference (Hinchliffe 1996). What is worse, perhaps, is that continued scientific uncertainty has become the principal rationale for inaction in the face of climate change. To the extent that the narrowly scientific focus on global climate change addresses itself to an undifferentiated global "we" and relies exclusively on the authority of science to create this sense of global citizenship, "we" are likely to act more as spectators than participants in the shaping of our related, but different futures (Taylor and Buttel 1992: 406).

CONCLUSION

Through these brief discussions of forest conservation and global warming I have tried to articulate a theory of artifactual constructivism that is sensitive both to the cultural politics of scientific representations of nature and to the independent, if also ineluctably framed and socially mediated, reality of nature. If this makes the practice of science seem more problematic than it once did, it makes it no less essential for making our way in the world, as the example of global warming suggests. Here is an environmental problem that would be difficult even to imagine, let alone address, without the considerable technical abilities of atmospheric science and computer modeling. The image of a dangerously warmer global climate that comes out of the GCMs is unquestionably a social construction – after all, it would not exist, nor, arguably, would the present-day concern with global warming, without the intervention of scientists and their supercomputers. That its apparent reality is only ever realized as an artifact of scientific representation should not make the potential threat of climate change any less real for us. Such a reaction, born of age-old distinctions between nature and society, seems strangely misplaced in this day of artificial life and genetic engineering. It leaves us with an inflexible, take-it or leave-it understanding of scientific knowledge: either real, objective, and therefore true or artificial, subjective, and thus socially constructed. By dissolving these dualisms, artifactual constructivism tempers the tendency either to worship science for its God-like objectivity or to demonize it for failing to live up to our unrealistic expectations. It moves us away from the schoolboy philosophy squabbles of the social constructivism debate and its fixation with the truth of scientific representation. Instead, artifactual constructivism focuses upon the powerful and productive practices by which the truth of representation is realized and produced. This is long overdue. With science responsible for producing so many of our environmental problems and yet also indispensable to their solution, there can be no question of dispensing with science altogether. The challenge is how to live it better. This demands a more pragmatic and more critical understanding of science and the politics of its constructions of nature.

NOTES

1 I have removed the emphasis from the original.
2 It should be said that van Fraasen (1980), though fiercely critical of the correspondence theory of truth underwriting the conventional understanding of scientific knowledge, is himself an empiricist and as such is not at all supportive of the stronger forms of social constructivism.
3 In his book, Sismondo (1996) breaks out several more varieties of social constructivism, but for my purposes here, I have consolidated his (1993) third kind of constructivism (the construction of artifacts and other phenomena in the laboratory) with his (1996) "heterogeneous construction."
4 I owe this observation to John Robinson, one of the lead authors for Working Group Three.

REFERENCES

Agarwal, A. and Narain, S. (1991) *Global Warming in an Unequal World*, New Dehli: Centre for Science and Environment.
Beck, U. (1992) *Risk Society: Towards a New Modernity*, trans. M. Ritter, London: Sage Publications.
Berger, P. L. and Luckman, T. (1966) *The Social Construction of Reality: A Treatise in the Sociology of Knowledge*, Garden City, NY: Doubleday.
Bloor, D. (1976) *Knowledge and Social Imagery*, London: Routledge and Kegan Paul.
Bodansky, D. (1993) "The UN Framework Convention on Climate Change: a commentary," *Yale Journal of International Law* 18: 451–558.
Boehmer-Christiansen, S. (1990) "Energy policy and public opinion: manipulation of environmental threats by vested interests in the UK and West Germany," *Energy Policy* 18: 828–37.
—— (1994) "A scientific agenda for climate policy?," *Nature* 372: 400–2.
Boyd, R. (1984) "The current status of scientific realism," in J. Leplin (ed.) *Scientific Realism*, Berkeley: University of California Press.
Butler, J. (1993) *Bodies That Matter*, New York: Routledge.
Buttel, F. H., Hawkins, A. P. and Power, A. G. (1990) "From limits to growth to global change: constraints and contradictions in the evolution of environmental science and ideology," *Global Environmental Change* 1: 57–66.
Cary, A. (1899). "How to apply forestry to spruce lands," *Paper Trade Journal* 27 (19 February): 157–62.
Climatic Change (1996) Special issue on integrated assessment, 34: 315–95.
Collins, H. M. and Pinch T. (1993) *The Golem: What Everybody Should Know About Science*, Cambridge: Cambridge University Press.
Collins, H. M. and Yearley S. (1992) "Epistemological chicken," in A. Pickering (ed.) *Science as Culture and Practice*, Chicago: University of Chicago Press.
Costanza, R., Norton, B. G., Haskell, B. D. (eds) (1992) *Ecosystem Health: New Goals for Environmental Management*, Washington, DC: Island Press.

Cronon, W. (1994) "Cutting loose or running aground?," *Journal of Historical Geography* 20: 38–43.

Demeritt, D. (1997) "Representing the 'true' St. Croix: knowledge and power in the partition of the Northeast," *William and Mary Quarterly* 54: 515–48.

Dunlap, R. E. and Catton, W. R. Jr (1994) "Struggling with human exemptionalism: the rise, decline, and revitalization of environmental sociology," *The American Sociologist* 25: 5–30.

Franklin, J. (1989) "Toward a new forestry," *American Forests* 95 (November–December): 37–44.

Glacken, C. J. (1967) *Traces on the Rhodian Shore: Nature and Culture in Western Thought from Ancient Times to the End of the Eighteenth Century*, Berkeley: University of California Press.

Global Climate Coalition (1994) *Potential Global Climate Change*, Washington, DC: Global Climate Coalition.

Gribben, J. R. (1990) *Hothouse Earth: The Greenhouse Effect and Gaia*, London: Bantam Press.

Gross, P. R. and Levitt, N. (1994) *Higher Superstition: The Academic Left and Its Quarrels with Science*, Baltimore: Johns Hopkins University Press.

Hacking, I. (1983) *Representing and Intervening: Introductory Topics in the Philosophy of Natural Science*, Cambridge: Cambridge University Press.

Hagen, J. B. (1992) *An Entangled Bank: The Origins of Ecosystem Ecology*, New Brunswick: Rutgers University Press.

Haraway, D. J. (1992) "The promises of monsters: a regenerative politics for inappropriate/d others," in L. Grossberg, C. Nelson and P. A. Treichler (eds) *Cultural Studies*, New York: Routledge.

Harding, S. (1986) *The Science Question in Feminism*, Ithaca: Cornell University Press.

—— (1991) *Whose Science? Whose Knowledge?*, Ithaca: Cornell University Press.

Hays, S. P. (1959) *Conservation and the Gospel of Efficiency: The Progressive Conservation Movement, 1890–1920*, Cambridge: Harvard University Press.

Hesse, M. (1980) *Revolutions and Reconstructions in the Philosophy of Science*, Bloomington: University of Indiana Press.

Hinchliffe, S. (1996) "Helping the earth begins at home: the social construction of socio-environmental responsibilities," *Global Environmental Change* 6: 53–62.

Kattenberg, A., Giorgi, F., Grassl, H., Meehl, G. A., Mitchell, J. F. B., Stouffer, R. J., Tokioka, T., Weaver, A. J. and Wigley, T. M. L. (1996) "Climate models: projections of future climate," in J. T. Houghton, L. G. Meira Filho, B. A. Callander, N. Harris, A. Kattenberg and K. Maskell (eds) *Climate Change 1995: The Science of Climate Change*, Cambridge: Cambridge University Press.

Kerr, R. (1994) "Climate modeling's fudge factor comes under fire," *Science* 265 (9 September): 1528.

Knorr-Cetina, K. (1983) "Towards a constructivist interpretation of science," in K. Knorr-Cetina and M. Mulkay (eds) *Science Observed: Perspectives on the Social Study of Science*, Beverly Hills: Sage Publications.

Langston, N. (1995) *Forest Dreams, Forest Nightmares: The Paradox of Old Growth in the Inland West*, Seattle: University of Washington Press.

Lansky, M. (1992) *Beyond the Beauty Strip: Saving What's Left of Our Forests*, Gardiner, ME: Tilbury Publishers.

Latour, B. (1987) *Science in Action: How to Follow Scientists and Engineers through Society*, Cambridge: Harvard University Press.

Latour, B. and Woolgar, S. (1979) *Laboratory Life: The Social Construction of Scientific Facts*, London: Sage Publications.

Laudan, L. (1984) "A confutation of convergent realism," in J. Leplin (ed.) *Scientific Realism*, Berkeley: University of California Press.

Maine State Forest Commissioner (1891) *Annual Report*, Augusta, ME, Maine State Forest Commissioner.

Mannheim, K. ([1929] 1954) *Ideology and Utopia*, New York: Harcourt Brace.

Mermin, N. D. (1996) "What's wrong with sustaining this myth?," *Physics Today* 49 (April): 11–13.

Merton, R. K. ([1938] 1970) *Science, Technology, and Society in Seventeenth-Century England*, New York: Howard Fertig.

O'Laughlin, J., Livingston, R. L., Thier, R., Thornton, J., Toweill, D. E. and Morelan, L. (1994) "Defining and measuring forest health," *Journal of Sustainable Forestry* 2: 65–85.

Parikh, J. K. (1992) "IPCC strategies unfair to the South," *Nature* 360: 507–8.

Parikh, J. K. and Painuly, J. P. (1994) "Population, consumption patterns and climate change: a socioeconomic perspective from the South," *Ambio* 23: 434–7.

Parson, E. A. (1995) "Integrated assessment and environmental policy making: in pursuit of usefulness," *Energy Policy* 23: 463–75.

Pels, D. (1996) "The politics of symmetry," *Social Studies of Science* 26: 277–304.

Pickering, A. (ed.) (1992) *Science as Culture and Practice*, Chicago: University of Chicago Press.

Pinchot, G. (1905) *A Primer of Forestry: Practical Forestry*, Washington: US Department of Agriculture, Bureau of Forestry, Bulletin 24.

Rorty, R. (1979) *Philosophy and the Mirror of Nature*, Princeton: Princeton University Press.

Ross, A. (1991) *Strange Weather: Culture, Science and Technology in the Age of Limits*, New York: Verso.

Rouse, J. (1987) *Knowledge and Power: Toward a Political Philosophy of Science*, Ithaca: Cornell University Press.

Rowlands, I. H. (1995) *The Politics of Global Atmospheric Change*, Manchester: Manchester University Press.

Rudwick, M. J. S. (1985) *The Great Devonian Controversy: The Shaping of Scientific Knowledge Among Gentlemanly Specialists*, Chicago: University Chicago Press.

Schenck, C. A. (1911) *Forest Policy*, Darmstaadt: C. F. Winter.

Scott, P., Richards, E. and Martin, B. (1990) "Captives of controversy: the myth of the neutral social researcher in contemporary scientific controversies," *Science, Technology, & Human Values* 15: 474–94.

Searle, J. R. (1995) *The Construction of Social Reality*, London, Allen Lane.

Seymour, R. S. (1992) "The red spruce-balsam fir forest of Maine: evolution of silvicultural practice in response to stand development patterns and disturbances," in M. J. Kelty, B. C. Larson, and C. D. Oliver (eds) *The Ecology and Silviculture of Mixed-Species Forests*, Dordrecht: Kluwer.

Shackley, S. and Darier, E. (1997) "The seduction of 'Groping the Dark': a dialogue on global modelling," unpublished manuscript, Centre for the Study of Environmental Change, Lancaster University.

Shackley, S. and Wynne, B. (1995) "Global climate change: the mutual construction of an emergent science-policy domain," *Science and Public Policy* 22: 218–30.

Singer, F. (1992) "Warming theories need warning label," *Bulletin of the Atomic Scientists* (June): 34–9.

Sismondo, S. (1993) "Some social constructions," *Social Studies of Science* 23: 515–53.

—— (1996) *Science Without Myth: On Constructions, Reality, and Social Knowledge*, Albany: State University of New York Press.

Smith, K. (1993) "The basics of greenhouse gas indices," in P. Hayes and K. Smith (eds) *The Global Greenhouse Regime – Who Pays?: Science, Economics, and North–South Politics in the Climate Change Convention*, London: Earthscan.

Sokal, A. (1996) "A physicist experiments with cultural studies," *Lingua Franca* 6 (May/June): 62–4.

Soule, M. E. and Lease, G. (eds) (1995) *Reinventing Nature? Responses to Postmodern Deconstruction*, Washington, DC: Island Press.

Suter, G. W. II (1993) "A critique of ecosystem health concepts and indexes," *Environmental Toxicology and Chemistry* 12: 1533–9.

Taylor, P. J. and Buttel, F. H. (1992) "How do we know we have global environmental problems? Science and the globalization of environmental discourse," *Geoforum* 23: 405–16.

US Division of Forestry (1887) *Report*, Washington, DC: Government Printing Office.

US Forest Service (1992) *Ecosystem Management of the National Forests and Grasslands*, policy letter 1220–1, 4 June, Washington, DC: Government Printing Office.

Van Fraasen, B. (1980) *The Scientific Image*, New York: Oxford University Press.

Weinberg, S. (1996) "Sokal's hoax," *New York Review of Books* 48 (8 August): 11–15.

Whitford, H. N. and Craig, R. D. (1918) *Forests of British Columbia*, Ottawa: Commission of Conservation.

Woolgar, S. (1988) *Science, The Very Idea*, Chichester: Tavistock.

Wynne, B. (1996) "SSK's identity parade: signing-up, off-and-on" *Social Studies of Science* 26: 357–91.

9

INCORPORATING NATURE

Environmental narratives and the reproduction
of food

Margaret FitzSimmons and David Goodman

INTRODUCTION

Forebodings of environmental crisis have lent a sense of urgency and of political
relevance to the "greening" of social theory in recent years. The concerns of
this increasingly widely accepted intellectual project are discernible in a broad
range of scholarship in the social sciences and humanities. Contributors to
this endeavour share an agenda: to bring "nature" "back in" to social theory by
contesting its abstraction from "society." However, these reconstructive efforts
have remained largely confined to separate academic fields and often imply
different conceptions of nature–society intersections. With rare exceptions few
have attempted to find a common ground among these perspectives or explored
the implications of this scholarship for fields where dualist thought still holds
sway. In this chapter, we propose to remap part of this shared terrain. Until
recently, it has become a commonplace in social theory to ignore the specific
"agency" and "materiality" of nature or, where that agency has been admitted, to
conceive of it within the disabling binary logics that have for so long organised
modern thought. Against this, we argue for a focus on *incorporation* – as
metaphor and as process – as a useful way of bringing nature into the body of
social theory and, more literally, into the social body of living organisms,
including ourselves. More specifically, we bring this insight to bear on a subject
of inter-disciplinary concern (and a long-standing concern of ours), which
illuminates "society" and "nature" in their most intimate entanglement: the
political economy of modern agriculture and agro-food networks.

We begin with a critical examination of two important bodies of thought –
environmental history and Marxist political economy – that take issue with the
Enlightenment ontology of nature as external, primordial and mechanistic and
that seek to re-theorize the agency of nature. Additionally, each has focussed
their critical energies on modern agriculture, a primary locus of metabolism

between society and nature. Following a critical appreciation of both bodies of thought, we suggest that the notion of *corporeality* provides an alternative means of theorizing nature's agency, one that evades the pitfalls of the "hostile" dichotomy of nature and society (Williams 1980) and avoids the modernist tendency to nominate either nature or society as ontologically prior (Latour 1993). How are the social practices and institutions of agriculture and food consumption theorized in recent environmental history and Marxist scholarship, and how do these theorizations assert conceptualizations of nature? More specifically, what are the implications for agrarian political economy if, within the metabolic relations which incorporate agricultural production and food consumption, we conceptualize nature as both internal *and* autonomous, as causal *and* contextual, and as always consequential?

Most work in agrarian political economy has been constituted in response to a discourse about capitalism and modernity that exalted manufacturing and addressed agriculture as a secondary locus, presenting the role of nature in agriculture as "exceptional." It has either failed to give theoretical weight to the metabolic relations which distinguish agro-food systems or has dismissed such approaches on the grounds of ecological or technological determinism. Here agriculture is routinely presented as just another industrial sector or branch of economic activity. This abstraction of nature has been applied *a fortiori* to the orthodox, modern disciplines of agricultural economics and rural sociology, but it also permeates more critical analyses based in political economy.

By positing the shared corporeality of agro-food practices – of nature into the human body and of humans into the complex life ways of the natural world – this chapter challenges the abstraction of agriculture as an economic activity. Our explanation of nature's agency places this corporeality at the forefront of the analysis. At any given conjuncture, the practice of agro-food is described by mutually constitutive practices and processes of metabolism and incorporation, which involve the dual and combined agency of biophysical processes and of social labour engaged in production and reproduction. This notion of shared corporeality speaks explicitly to a more general insistence that ecology and social relations, the production and reproduction of nature and society, be located within a unified analytical frame.

After reviewing environmental history (in the second section) and ecological Marxism (in the third section), we turn to our central theme of agro-food networks as the locus of metabolic incorporation between humans and nature (fourth section). The final section extends the notion of incorporation to the ingestion of "food" and considers metabolic, corporeal, social, and symbolic dimensions, both pleasurable and pathological, by focussing on the cases of BSE and anorexia. Although these are preliminary considerations, we hope that by reconceptualizing nature's agency we can begin to find progressive ways of remaking nature, ways which "commonsense" dualistic thinking currently disables.

NATURE'S AGENCY IN ENVIRONMENTAL HISTORY

Nature as historical agent

Environmental historians are attempting to break social theory's long silence on the active material presence of nature in social life and social reproduction by reasserting nature's agency. This new field is a major current in the broad movement to redress the Enlightenment legacy of the ontological and historical abstraction of nature. Environmental historians have responded to this abstraction by reactivating the neglected side of the nature–society dualism to seek a new synthesis, what Worster (1984: 2) described as "an ecological perspective on history."[1] This ecological perspective is founded ontologically on the premise that societies arise within nature, tracing out multiple possibilities upon the grounds and landscapes laid out by nature's processes, nature's practices in time and space. Environmental histories thus begin with the material engagement of people and land. They practice "ecological thinking," constructing:

> Nature as an active partner . . . As parts of the whole, humans have the power to alter the networks in which they are embedded. Nature as active partner acquiesces to human interventions through resilience and adaptation or "resists" human actions through mutation and evolution. Nonhuman nature is an actor; human and nonhuman interactions constitute the drama.
>
> (Merchant 1989: 25)

This project to bring nature into history has been pursued in different ways and with varying narrative strategies. Here, we briefly review the recent evolution of this project, addressing changing views on how it should be prosecuted, to clarify the assumptions and meanings behind environmental historians' claim that nature is an independent actor on the stage. We engage selected canonical texts and commentaries which delineate "moments" of critical self-reflection on the field, its purpose, and prospects.

The first of these moments brings agriculture into the narrative tropes of the new environmental history. We identify this moment with the publication of Donald Worster's *Dust Bowl* in 1979. At this point, environmental history began to loosen its early ties with intellectual and political history, and its associated focus on the conservation movement and wilderness preservation as the antecedents of postwar American environmentalism.[2] Other landmarks of this transition include William Cronon's *Changes in the Land* (1983), and Carolyn Merchant's *Ecological Revolutions* (1989). Each signals a shift of attention, from wilderness to occupied landscape, from nature as primordial to nature as present and active in human affairs. How is this relationship delineated at this juncture, where the necessary human action of agriculture is brought into the narrative of

196

environmental history? Richard White (1985: 335) alludes to two aspects of this relationship that bear on the question of agency: physical limits and reciprocity. "Nature does not dictate, but physical nature does, at any given time, set limits on what is humanly possible. Humans may *think* what they want; they cannot always *do* what they want, and not all they do turns out as planned. It is in the midst of this compromised and complex situation of the reciprocal influences of a changing nature and a changing society that environmental history must find its home."[3]

The question of reciprocity is taken up explicitly in William Cronon's (1983) study of colonial New England. Cronon fully recognizes the reciprocal relations of nature and society in shaping this regional environment, but he also notes that "the period of human occupation in postglacial New England has seen environmental changes on an enormous scale, many of them wholly apart from human influence" (1983: 11). In short, "ecosystems have histories of their own" (p. 12). For Cronon, ecological history assumes that nature and society are entangled in "a dynamic and changing relationship," where "the interactions of the two are dialectical" (p. 13). In this "interacting system," culture and environment are reshaped dialectically in a continuing "cycle of mutual determination." Accordingly, "changes in the way people create and recreate their livelihood must be analyzed in terms of changes not only in their *social* relations but in their *ecological* ones as well" (p. 13, original emphasis).

As a second "moment" of reflexive and programmatic analysis, we have selected a debate between leading environmental historians in *The Journal of American History* (1990). These exchanges revealed profound epistemological divisions and threw into question the moral certitude and political goals which, to paraphrase Cronon (1994), had nurtured "this child of environmentalism." In the lead paper, "Transformations of the earth: toward an agroecological perspective in history," Donald Worster (1990a: 1091) takes the "new" history to task for its "conceptually underdeveloped" analysis of interactions between the structure of production, technology, and the environment, which constitute the "techno-environmental base" of society. To rectify this omission, Worster proposes that environmental history should focus on "modes of production as ecological phenomena, and particularly as they are articulated in agriculture" (ibid.).[4]

Although he concentrates on the rise of the capitalist mode of production, Worster hastens to distance himself from Marxist conceptualizations which, he suggests, have neglected the possible causative role of ecological factors in the capitalist transition. These do not "acknowledge that the capitalist era in production introduced a new distinctive relation of people to the natural world. The *reorganization of nature*, not merely of society, is what we must uncover" (p. 1100, original emphasis). If, as White (1990) suggests, Worster's inspiration is Braudelian rather than Marxian, his program to comprehend the capitalist transformation of nature nevertheless has close parallels with ecological reformulations of Marxist theory, as we see below in the third section.

Rather than giving greater unity and purpose to the field, Worster's attempt to set its agenda evoked strong dissent and exposed wide disagreement on the fundamental issues of causality and agency. Worster's sense of ecology is moral and holistic, following the views of nature of Clements and Odum. But his assertions of the authority of the ecological vision came at a time when this view had lost hegemony in academic ecology, where mathematical modeling of population dynamics and experimental manipulation of two- or three-species model systems now characterizes most research practice and where "scientific" ecology seeks to free itself from "natural history." At the "moment" of the 1990 debate, environmental historians were sharply divided on the epistemological implications of these conceptual disputes in ecology. As perhaps becomes the author of *Nature's Economy* (1977), which examines the cultural context and intellectual history of changing scientific paradigms of nature, Worster (1990a) is unimpressed by the latest "revisionist swing," and even less so by relativist interpretations of their significance (Worster 1990b).

However, Worster's colleagues are less sanguine about taking these ecological disputes in their stride. Thus, White (1990: 1114) believes that environmental historians are left without "a firm basis for their morality and causality." Cronon (1990: 1127) argues that new methodological tools are required to replace such outmoded concepts as equilibrium, community, and successional climax, which "underlie the old idea of the 'balance of nature' that so often supplies the analytical (and moral) scale against which we measure the environmental effects of human societies." White (1990: 1115) goes to the crux of the relativist critique when he states that "although ecology has been reified into nature, ecology is, in fact, only an academic discipline." But it is left to Stephen Pyne (1990: 1139) to spell out the implications of this reification when he argues that Worster "appeals to the laws of ecology to construct a nature that is external to humans and that provides a moral template against which to measure human behavior. This grants him, as author, a privileged omniscient position with which to view the spectacle." In rejecting this "implied claim to privilege," Pyne (1990: 1140) asserts that: "The laws of ecology become functional equivalents to other presumed 'laws' of human behavior and history and have, I believe, no more validity."[5]

Our third "moment" in the development of environmental history is the publication of William Cronon's acclaimed study, *Nature's Metropolis: Chicago and the Great West* (1991). Deservedly recognized as a milestone in the "new" history, this monumental work is considered here for what it has to say on the question of nature's agency, and particularly the distinction Cronon draws between "first" and "second" nature. Much of this theoretical terrain has previously been mapped by contributors to the "production of nature" debates, which followed Alfred Schmidt's *The Concept of Nature in Marx* (1971), and gained further momentum with the publication of Neil Smith's book *Uneven Development: Nature, Capital and the Production of Space* (1984).[6] However, in a more recent reconnaissance, David Demeritt (1994a) engages environmental historians in a conversation with cultural geographers, which reveals the very different and irreconcilable worlds portrayed by their respective "metaphors of nature as agent and landscape as text" (p. 164). These metaphoric strategies, he argues, place environmental history and cultural geography at opposite poles of the nature/culture dualism "that fixes nature and landscape as either autonomous natural actors or absolute social productions" (p. 163).

Environmental history occupies its polar position because to make the analytical connection between the autonomous agency of nature and the consequences of that agency "requires an ability to step outside ourselves . . . to discover and acknowledge another, objective reality that we have not created nor ever fully controlled" (Worster 1990b: 1146); that is, external nature, primordial and originary, pre-human.[7] In short, what environmental historians assert as the "strong" criterion of nature's agency draws its explanatory power from the opposition between an external, pre-human nature and an historically subsequent state when nature and society have materially commingled. As we have just seen, recourse to the ontological categories of ecology, notably Clementsian ecosystems, with their putative "balance" and homeostatic properties, which are reified as nature "before" human disturbance, serves the same methodological purpose.

Similar considerations apply to Cronon's analytical scheme which, in much the same way, counterposes "first" and "second nature." "First nature" is identified with the vegetational geography – grasslands, hardwood and softwood forests, lakes and rivers – of the postglacial landscape of the Great West, in contrast to "A kind of 'second nature,' designed by people and 'improved' toward human ends, (which) gradually emerged atop the original landscape that nature – 'first nature' – had created . . . " (Cronon 1991: 56). "First nature," which is "the result of autonomous ecological processes," is the source of natural wealth or "natural capital" (p. 205), such as soil fertility, abundant forests: "this was the wealth of nature, and no human labour could create the value it contained" (pp. 149–50). Using the recurring metaphor of transmutation, "first" nature is transmuted into "second" nature, constructed nature, at the hands of humans and their technical devices, driven by the "logic" and "geography" of capitalism.

Cronon (1991) is fully aware of Schmidt's work and Neil Smith's critique and further elaboration of the concept of "second nature." However, in *Nature's Metropolis*, he chooses to ignore the warning that to treat "'first Nature,' primordial Nature as analytically and historically prior to 'second Nature,' social Nature" (Fitzsimmons: 1989a: 118, n. 2) risks falling back into the trap of Enlightenment ontology.[8] As Demeritt (1994a) notes, Cronon's use of the binary "first nature/second nature" framework displaces "the culture–nature dualism into a stratigraphy he can excavate" in order to reveal "the autonomous agency of nature" (p. 196). Echoing the misgivings expressed earlier by White (1990) and Pyne (1990) about environmental history's reliance on scientific ecology, Demeritt (1994a) argues that Cronon reifies the concept of ecosystem as nature, "first" nature, and "thus determines the essence of what nature 'really' is. In so doing, he reintroduces the problematic dualism between nature and culture that his book so effectively subverts" (p. 178).

In seeking to find a place for the autonomous agency of nature in its stories, environmental history continues to wrestle with the nagging question of causation and how this is to be resolved without depending on the received ontology of ecological science and its binary categories. The methodological dangers of this position are appreciated by White (1990: 1114) when he observes that: "The field has a tendency to produce cautionary tales. But without a clear demonstration of causality, a teller's cautionary tale becomes a listener's just so story." Later, in the fourth section, we return to discuss the new "metaphoric tools" developed by Michel Callon, Bruno Latour and Donna Haraway that Demeritt (1994a) believes will release environmental history from Enlightenment antinomies by making "it possible to imagine nature as both a real material actor and a socially constructed object" (p. 163). First, however, we move from these general formulations of nature's agency to consider a more specific theorization – agency as ecological rules which set natural limits – as it is deployed in environmental histories of agriculture.

Agricultural narratives

The theorization of nature's agency as ecological processes that surround and constrain human action is a recurrent theme in the new environmental history and a key element in its methodological arsenal of causality. The most comprehensive statement of this view is to be found in the work of Donald Worster. We have already noted his advocacy of the concept of mode of production and his dissatisfaction with Marxist conceptualizations of the capitalist mode (Worster 1990a). These must be reformulated, he argues, to incorporate not only the "restructuring of *human* relations" (1990a: 1098), but also "The *reorganization* of nature" (p. 1100). Worster further recommends agriculture as a particularly appropriate arena for this research agenda, since it is through food production that "every group of people in history" has been "connected in the most vital, constant and concrete way to the natural world" (pp. 1091–2).

To investigate the capitalist reorganization of nature and human ecological relations, Worster (1990a) urges environmental historians to center their analyses on something like Polanyi's "great transformation." Correspondingly, in agriculture, the common bond uniting Worster's analysis of ecological restructuring and nature's agency, and the leitmotif of his narrative strategy, is the agrarian transition. This narrative, framed in vivid declensionist terms, describes the transformation of traditional agricultural landscapes – "carefully integrated, functional mosaics that retained much of the wisdom of nature" (p. 1096) – into highly specialized, ecologically risky and vulnerable capitalist monocultures. In one story, nature appears as a benevolently active partner, in the second, as a capricious, potentially revengeful presence, with the autonomy to expose the hubristic ambition of capitalist agriculture, and its dualistic Baconian–Cartesian foundations.

This analytical framework and narrative trajectory structure Worster's account of the US Dust Bowl in the 1930s, "one of the worst environmental catastrophes in recorded human experience" (p. 1106). This "disaster was due to drought, the most severe drought in some two hundred years. But it was also the result of the radically simplified agroecosystem the Plains farmers had tried to create" (ibid.). The moral to be drawn from this story, whether on the Great Plains or other places, involves humankind "knowing the earth well, knowing its history and knowing its limits" (ibid.).

Worster's "howling world of nature" that "has always been a force in human life" (p. 1096) is revealed when humankind transgresses its limits. In this framework of explanation, the agency of nature is made manifest through the narrative contrast between two 'second' natures or agroecological landscapes: pre-capitalist and capitalist. One is "based on close observation and imitation of the natural order" (ibid.), and the other on "a movement toward the radical simplification of the natural ecological order" (p. 1101). This natural order is represented by the laws of ecology to which all agroecosystems are subject. However, within this compass, since what distinguishes traditional agrarian landscapes from capitalist agroecosystems is precisely their "close observation" of these "laws," they are characterized as being "at once the product of nonhuman factors and of human intelligence, working toward a mutual accommodation" (p. 1097).

Merchant and Cronon also focus strongly on agriculture and the increasing exploitation of farming and natural resource economies under developing capitalism. Merchant (1989: 25) presents this transition as an ecological revolution, defined as "processes through which different societies change their relationship to nature. They arise from tensions between production and ecology and between production and reproduction. The results are new constructions of nature, both materially and in human consciousness." Cronon moves from the European occupation and later abandonment of the New England landscape as a core region of commercial agriculture (in *Changes in the Land*) to the integration and subordination of the West by urban and commercial institutions arising with the growth of capitalist agro-food systems (in *Nature's Metropolis*), setting up the modern dialectic between city and countryside which is mediated by agro-food institutions.

Thus, although ecological science speaks universally, to use Demeritt's phrase, "for what nature 'really' is," two dualistic or binary notions of agency and nature–society dialectics, each heavily charged with both moral and material ecological imperatives, are embedded on either "side" of Worster's analysis of agrarian transition, of Merchant's interrogation of ecological revolution, and of Cronon's city of the merchants of food. These are the premodern dialectic of "mutual accommodation," and the capitalist dialectic of accumulation, degradation and ecological catastrophe as humankind challenges nature's limits. We now turn to theorizations of the agency of nature encountered in "green" or ecological Marxism. Here again we will find that agriculture occupies a special

201

position in the conceptualization of nature–society dialectics and the constraints to human activities of production and reproduction.

NATURE'S AGENCY WITHIN CAPITALISM

Nature as limits, nature as process

If nature as a constraint on human action is a recurrent theme in the new environmental history, it is also, but in quite different ways, a concept crucial to new work in "ecological Marxism." We use this term to review new engagements with the question of nature and society founded in Marxism in several disciplines. These formulations locate the point of analytic entry in society – in the laws of accumulation and the social relations and structures of modes of production, particularly of capitalism – rather than in the laws and processes of nature. This notion of the epochal specificity of nature–society dialectics, which has to be teased out of the analysis of the environmental historians, receives a more explicit, formal treatment in the discussions of ecological Marxism we review here. Using more emphatically the language of dialectics, rather than simply "interaction," nature is brought into the social world through the agency of labour in the extraction and transformation of use values, the industrial internalization of nature's processes, the instrumental knowledge and action structured by science and technology in the service of accumulation, and through the globalization of production and exchange mediated by the market for commodities. And it is here that we can begin to talk of nature as literally "incorporated."

But what "nature" is incorporated? Here we find that contributors to the discussions of ecological Marxism constitute nature quite differently in their theories. Some address nature as eco-physical, reading nature from within society as a physical nature of stocks and flows of potential use values (Martinez-Alier 1987; Altvater 1993) or as "conditions of production" (O'Connor 1988). Here nature remains external, inorganic. Others use a more supple dialectic to envision nature as biological, as life in process (Benton 1989, 1991, 1992, 1993; Harvey, 1982, 1996) within which human life is metabolically embodied through praxis. The implications of these two views for the incorporation of nature in ecological Marxist theory are quite distinct.

The concept of nature as physical nature leads inevitably to issues of limits: limits in nature's services, nature's stocks and flows, nature's reserves. Abstracting from limits in place to a grand global calculation of natural "accounts," this view presents nature through the lens of ecosystem ecology, in which life is a series of pumps and pipes for energy and materials. Thus Martinez-Alier and Altvater invoke the laws of thermodynamics to anticipate an ecological crisis of resource exhaustion and environmental pollution, while O'Connor, who also focusses on crisis, proposes a second contradiction of capitalism – the problem of reproduction of the conditions of production – which connects to and parallels Marx's first contradiction – that of reproduction of the forces of production (human labour). This approach – nature as physical nature – presents a categorical conception which reduces nature to the inorganic. "Nature is man's *inorganic body* – nature, that is, insofar as it is not itself human body. Man *lives* on nature – means that nature is his *body*, with which he must remain in continuous interchange if he is not to die. That man's physical and spiritual life is linked to nature means simply that nature is linked to itself, for man is a part of nature" (Marx, in Benton 1993: 24). Though "man is a part of nature," the metaphor of metabolism that Marx presents here emphasizes action upon, not action within. Embodiment occurs through necessity, necessity mediated by the presence of class relationships of exploitation through which both labour and nature are subordinated to the service of capital. "The exploitation of the worker is simultaneously the exploitation of the soil." Nature's things, like people, are used and abused. As Harvey (1996) points out, this sense of nature as the subject of domination, as the location of exhaustible resources, connects some Marxist positions both to the "pessimism" of the Frankfurt School and to the dangerous straits of Malthusianism, in which nature (and nature in agriculture) becomes the transhistorical and universalized explanation of poverty and suffering.

From the publication of David Harvey's essay "Population, resources and the ideology of science" (1974), he has been a strong voice against this Malthusian position of universal limits. Those familiar with Harvey's writings on urbanization may not know that his early work was on agriculture. This grounds his sense of metabolism in the "form-giving fire" of the labour process in which "Labour is, in the first place, a process in which both man and Nature participate. . . . He opposes himself to Nature as one of her own forces, setting in motion arms and legs, head and hands, the natural forces of his body, in order to appropriate Nature's productions in a form adapted to his own wants" (Marx, in Harvey 1982: 101). The quotation continues: "By thus acting on the external world and changing it, he at the same time changes his own nature" (*Capital* I: 177). The sense of metabolism present in Harvey's (1996) work conceives nature as internal to the social species-being of people, as a part of the organic, not the inorganic body. Thus, the way people act is *within* not simply on nature, in the context of "socio-ecological projects" strongly structured by the modes of production within which that action is mobilized and constrained.

> What exists in "nature" is in a constant state of transformation. To declare a state of ecoscarcity is in effect to say that we have not the will, wit, or capacity to change our state of knowledge, our social goals, cultural modes, and technological mixes, or our form of economy, and that we are powerless to modify either our material practices or "nature" according to human requirements. To say that scarcity resides in nature and that natural limits exist is to ignore how scarcity is socially produced and how "limits" are a social relation within nature (including human society) rather than some externally imposed necessity.
>
> (Harvey 1996: 145)

Harvey's sense of ecological nature returns us to a sense of possibilism, rather than determination. Instead of abstracting a universalized view of society and nature, Harvey recommends an "evolutionary" view, based on a "dialectical and relational schema for thinking through how to understand the dialectics of social–environmental change" (p. 190). This evolutionary view avoids the dissolution of socio-ecological projects into the abstractions of nature and society and thus allows for a necessary heterogeneity of outcomes and possibilities for transformation. As Harvey puts it, "The *general* struggle against capitalist forms of domination is always made up of *particular* struggles in which capitalists are engaged and the distinctive social relations they presuppose" (p. 201).

Agricultural narratives: ecoregulation and the labour process

Agroecological processes figure prominently in Ted Benton's formulation of an ecological Marxism (1989, 1992, 1993). Indeed, the distinctive "materiality" of nature in agriculture plays a pivotal role in the revision of Marxian economic concepts that Benton believes "constitute an indispensable starting point for any theory that would adequately grasp the ecological conditions and limits of human social forms" (1989: 63–4).

Benton's reconstructive analysis focusses on Marx's conceptualization of the labour process, both the abstract concept "as a transhistorical condition of human survival" and the forms labour takes under capitalism. In each case, Benton argues, "Marx under-represents the significance of non-manipulable natural conditions of labour-processes and over-represents the role of human intentional transformative powers vis-à-vis nature" (p. 64). Marx's abstract concept of the labour process, Benton suggests, is represented by a transformative intentional structure, possibly with some handicraft activity as a model, in which the "subject" of labour – products of nature or raw materials, variously defined – is transformed into use value. The difficulty with this conceptualization is its claim to be universally applicable to all types of human need-meeting interaction with nature. Benton (1989) contends that the transformative intentional structure is an inappropriate representation of the labour processes in extractive activities and agriculture. To overcome this lack of "independent conceptual specification," Benton uses the criterion of intentional structure to construct a broader taxonomy that distinguishes between productive, transformative labour processes, primary labour processes (hunting, mining, fishing), which directly appropriate nature, and eco-regulatory labour processes.

The last are exemplified by agriculture, where labour is "primarily deployed to sustain or regulate the environmental conditions under which seed or stock animals grow and develop" (Benton 1989: 67). The distinctive aspects of eco-regulatory labour processes include the deployment of labour to optimize conditions for organic growth and development, the corresponding orientation of labour activities to sustain, regulate and reproduce these conditions, and spatio-temporal distributions of labour that are dependent mainly on these contextual conditions and on the rhythms of organic development (pp. 67–8). This dependence of eco-regulatory labour processes on naturally given contextual conditions means that: "For any specific technical organization of agriculture, these elements in the process are relatively impervious to intentional manipulation, and in some respects they are absolutely non-manipulable" (p. 68).

In the case of specifically capitalist forms, Benton suggests that Marx privileged instrumental transformative labour processes, and thus tended to under-theorize the extent to which capitalist industrial production "remained tied to eco-regulatory and primary-appropriative labours as the necessary sources of energy, raw materials and food, and so, also, to a range of non-manipulable contextual conditions" (p. 70). To capture these "footprints," Benton proposes an eco-logical reconstruction of Marx's conceptualization of capitalist labour processes that would give independent conceptual weight to naturally given enabling contextual conditions. In addition, this reconstruction would emphasize their continuing role in the sustainability of production, and recognize that naturally mediated but unintended consequences of production may undermine the persistence or reproduction of these contextual conditions (pp. 73–4).

Benton's further suggestions for an ecological Marxist political economy have some parallels with Worster's less formalized agenda for environmental history, notably Benton's insistence (which Harvey shares) on the historical, geo-graphical, and social relativity of nature–society interactions, and also of "natural limits" and the associated combination of enablement and constraint (positions with which Harvey 1996, disagrees). Benton suggests that each historical epoch or "form of social and economic life," with its own characteristic mode of production and labour processes, will present a "specific structure of nature–society articulation" (1989: 77). The dynamic of each historically articulated structure, in turn, produces correspondingly distinctive ecological problems and interrelations with its own specific contextual conditions and limits. It follows that "since natural limits are themselves theorized, in this approach, as a function of the articulated combination of specific social practices and specific complexes of natural conditions, resources and mechanisms, what constitutes a genuine natural limit for one such form of nature/society articulation may *not* constitute a limit for another" (p. 79).

Benton thus offers "a genuinely dialectical concept of society–nature relations" (Castree 1995: 24), and an historically contingent theorization of natural limits. He also provides the foundations for a rigorous, historicized analysis of modes of production and their characteristic structures of nature–society articulation. By stressing articulation rather than binary opposition, Benton avoids the pitfalls of Worster's moral rhetoric and teleology, with its normative opposition between a (pre-capitalist) dialectic of reciprocity and a capitalist dialectic of over-exploitation and degradation.

Benton (1989) makes two proposals for further reconstructive work that deserve careful consideration, not least by environmental historians. The first would extend the typology of labour processes and their intentional structures in order to emphasize their character "as modes of social appropriation of nature" (p. 80). The second proposal would "reconceptualize the Marxian typology of modes of production" so that each mode would be specified in terms not only of its social–relational elements, but also its form of nature–society interaction. "Each mode must be conceptualized in terms of its own peculiar limits and boundaries, and its own associated liabilities to generate environmental crises and environmentally related patterns of social conflict" (p. 81).

Benton provides a partial yet significant development of this first proposal in his book, *Natural Relations* (1993), which is concerned primarily to convene an "encounter" between traditions of socialist thought, ecological politics and philosophies of animal rights. In this discussion, Benton argues for the specificity of agricultural labour processes, which he attributes, following Goodman and Redclift (1991), to the constraints associated with biological processes that have created distinctive historical patterns of capital accumulation and labour process organization. Adapting his earlier formulation, he emphasizes the importance of analyzing socio-economic relations and non-human nature, in this case livestock, within a single interpretive framework.

Many Marxist approaches to the capitalist transformation of agriculture have relied heavily on the concept of "formal subsumption" to analyze the processes underlying the emergence of an agricultural wage labour force. Benton (1993: 156) would extend this concept to encompass the "whole complex of wage-labour, stock animals, and ecological conditions, whose interrelations are reorganized and subjected to distinctive forms of regulation under commercial pressures." Suggesting that "we should, perhaps, speak of the ecologico-socio-technical organization of the labour process" (p. 156), he refers to the capitalist reorganization of this complex as "material subsumption." Benton invokes his concept of eco-regulatory labour processes to insist that the distinctive character of agricultural labour practices is preserved even into the intensive regime of the factory farm and animal confinement systems, and "continues to operate as a limit to the full subordination of the labour-process to the prevailing mode of economic calculation" (ibid.).

Benton goes on to discuss the tension between pressures to treat animals instrumentally as mere "things" and countervailing limits "to the full realization of these pressures, which derive from the organic, psychological and social requirements of the animals themselves" (p. 157). Benton's analysis here is limited to different regimes of animal husbandry, and he denies that he is attempting an historical account or periodization of agricultural labour processes and their reorganization. Nevertheless, the promise of this theoretical framework for an ecological history of agro-food systems is clear.

AGRI-CULTURE AS A NATURE–SOCIETY HYBRID

> Relationship more than system should be our starting point.
>
> (Cronon 1990: 1130)

> All of culture and all of nature get churned up again every day.
>
> (Latour 1993: 2)

Many of the theorizations of nature–society relations reviewed above rely heavily on interactive or else dialectical concepts, whether framed in terms of natural limits, reciprocity or mutual accommodation. Yet even posing the issue as one of a dialectical relationship between nature and society – with its Hegelian lineage of dissolving dualisms – seems still to reproduce the dualism which we are seeking to resolve. If nature and society are active partners locked in an irreversible, continuing process of mutual determination, how is the materiality of these conjoined natures to be theorized? On many different temporal and spatial scales, nature and society are active contingent categories, mutually determined and constituted in endless cycles of reproduction and production. To portray these co-productions as nature–society dialectics is to risk, to loosely paraphrase Donna Haraway, trying to balance on both poles at once. It is this balancing act that persuades Demeritt (1994a: 183) that new metaphors are needed "for framing nature as both a real material actor and a socially constructed object."

For this purpose, Demeritt recommends the work of Bruno Latour and Donna Haraway. Latour (1993: 51) closes the Kantian "Great Divide" between nature and culture by analyzing these interactions in terms of symmetrical processes that create hybrids, which, following Michel Serres, he calls "quasi-objects, quasi-subjects" that "are simultaneously real, discursive and social" (p. 64). Haraway (1991) develops the metaphor of the cyborg to reveal the partnerships, though not always equal, between human and nonhuman actors joined in the mutual construction of artifactual nature. Since it is currently more prominent in agro-food studies, we concentrate below on actor network theory (ANT) formulated by Latour in collaboration with Michel Callon.[9]

In seeking to retie the Gordian knot between nature and society, Latour (1993) advances the notions of mediation and network. The work of mediation "creates mixtures between two entirely new types of beings, hybrids of nature and culture" (p. 10), which are mobilized and assembled into networks. The practices of mediation, mobilizing things and assembling hybrids correspond, in Latour's metaphor, to "a delicate shuttle" weaving the natural and social worlds into "a seamless fabric." This "socialization of nonhumans" (p. 42) prompts Latour to "use the word 'collective' to describe the association of humans and nonhumans" (p. 4), which forms the "Middle Kingdom" between the nature–society poles of modernity.

Latour's exploration of this middle territory of collective associations poses an incisive, frontal challenge to the notion of autonomous agency that is so central to the environmental history project. In his account, the ever-problematic modern duality has collapsed under its own weight, unable to withstand the increasing scale and scope in the mobilization of networks and multiplication of hybrids. This "third estate" is not "faithfully represented either by the order of objects or by the order of subjects"; the practice of mediation has, in essence, conflated the two poles (Latour 1993: 49–50). Agency in these associative networks is conceptualized as the collective capacity of humans and nonhumans to act; it is an effect of these heterogeneous networks. Nonhuman agents also are endowed with active properties in the conceptualization of collective agency formulated recently by Michel Callon and John Law, based on the notion of hybrid *collectifs* (Callon and Law 1995).[10] Donna Haraway (1992) develops a related perspective when she argues that nature and society have imploded, and calls for formulations in which nature can be "a social partner, a social agent with a history, a conversant in a discourse where all of the actors are not 'us'" (p. 83).

These conceptualizations of collective agency reveal the limitations of formulations which, in the words of Callon and Law (1995: 502), "localize agency as singularity – usually singularity in the form of human bodies." While environmental historians also are seeking to correct "singular" notions of agency based on human intentionality and praxis, their "strong" formulation of nature as an autonomous actor, for all its usefulness, runs the risk of repeating the same error at the opposite pole. To paraphrase Latour (1993: 106–7), if the concept of agency is not centered on the "productions of natures-cultures" he calls collectives, we risk falling back into the binary categories in which nature is construed as external "things-in-themselves" and society as "men-among-themselves." As Demeritt (1994a: 180) observes, Latour's "metaphors make it possible to follow environmental historians in talking about the agency of nature without appealing to a transcendent nature beyond (society)."

The analytical framework articulated by Latour (1993: 107) of heterogeneous associations "of elements of Nature and elements of the social world" also provides a "common matrix" for a "comparative anthropology" of contemporary societies and others. "All collectives are different from one another in the way they divide up beings, in the properties they attribute to them, in the mobilization they consider acceptable" (ibid.). However, although all collectives obey the common principle of symmetry, that is, of being co-productions of nature–society, they may differ in size, in the scope of the mobilization, which offers an understanding of "the practical means that allow some collectives to dominate others" (pp. 107–8).

Increases in size, scope and power of collectives, "like the successive helixes of a single spiral," thus require increasing enlistments of quasi-objects, of "both forms of nature and forms of society" (p. 108). Respecting the common principle of symmetry, differences in the complexity, intricacy and length of networks are explained by the capacity to recruit these hybrids. It is for this reason that modern "Sciences and technologies are remarkable . . . because they multiply the nonhumans enrolled in the manufacturing of collectives and because they make the community that we form with these beings a more intimate one" (ibid.).

A history of agricultural quasi-objects: the corn collective

This analytical perspective of mediation and the creation of heterogeneous collectives, allied to an understanding of their differences or asymmetries – "The extension of the spiral, the scope of the enlistments . . . the ever-increasing lengths to which (modern science) goes to recruit" quasi-objects – offers some intriguing insights into agro-food systems. (Latour 1993: 108) This potential can be glimpsed if we recognize the intricate ways in which the "seamless fabric" is continually refashioned over space–time in production and reproduction. For example, as hunting and gathering practices, already highly selective, gave way to agrarian communities with "domesticated" animals and plants, those "quasi-objects, quasi-subjects" co-produced over the centuries become the "traditional" livestock bloodlines and landraces, the "heirloom" varieties whose history as quasi-objects is conveyed to the present in genetic materials and situated farmer knowledges. This history clearly can be described by Latour's metaphor of collective associations of humans and nonhumans, differentiated by the "scope of the enlistments," and extending through time and space to form international networks.

In the case of corn, for example, such a history might encompass, selectively, the agricultural origin story of the corn mother told by Indians in eastern North America, and associated with corn planting in forest clearings (Merchant 1989), the place of corn in meso-American cultures, the Columbian exchange (Crosby 1972), botanical expeditions, and the expansion of corn as a commercial crop, supported in the USA by an extensive institutional infrastructure. Subsequent mappings of corn following the rediscovery of Mendelian genetics are perhaps even more revealingly portrayed as collective enlistments of widening scope, emphasizing Latour's contribution as co-founder with Michel Callon of the so-called "Paris School" of science studies (Pickering 1992). These mappings, again selectively, would include early research into hybrid vigor, or heterosis, parent corn lines, hand pollination, the double-cross technique of hybridization, and the development of the Burr-Leaming hybrid (Bogue 1983). This highly innovative quasi-object permitted the production of seed corn on a commercial scale, the enabling condition for the introduction of locally adapted hybrid varieties to the Mid-West.

The increasing scope of this collective mobilized, *inter alia*, the offspring of parent corn lines, private philanthropy, public agricultural research stations, state land-grant colleges, cosseted experimental seedbeds, and a private hybrid seed industry. In this collective, we can see the origins of the modern agro-food complex as mechanical and chemical innovations converged toward, and are inscribed on, the new seeds, transforming the economic, social and techno-ecological foundations of agriculture and accelerating the processes of industrial appropriation and substitution (Goodman *et al.* 1987). These inscriptions produced a new plant architecture as hybrid varieties were designed for yield responsiveness to intensive fertilizer use, greater photosynthetic efficiency, higher plant density and whose strong stems, uniform height and simultaneous ripening facilitated mechanized harvesting. The "translation" of this collective to the tropics and sub-tropics,[11] extending its real, social and discursive elements, enlisted the Rockefeller Foundation, Mexican agricultural research and policy institutions, large-scale public irrigation programs, multinational corporations, and strategies of agricultural modernization, discursively and geo-politically propagated as the "Green Revolution."

This skein later enrolled new human and nonhuman recruits as it extended to the Ford Foundation and a system of international agricultural research centers, subsequently coordinated by the Consultative Group on International Agricultural Research representing governments, private agencies, and multilateral funding institutions. The hybrid corn collective and its multiplying quasi-objects are easily discerned in current practices of gene prospecting, gene banking, the mapping of plant genomes, and the gene-mixing networks of agri-biotechnologies. It may be fruitful to consider these collectives, at least in part, as a productive force (Yoxen 1981), constantly shaped and reshaped by translation and mediation.

Returning to the hybrid corn collective, this quasi-object is also metabolized industrially and in human stomachs as, for example, corn flour, tortillas, bread, corn starch, and as an ingredient of many thousands of food products. Corn as animal feed similarly is metabolized in the double gut of feedlot cattle and dairy cows, and by factory-farmed chickens and hogs, before its secondary metabolism as animal protein for human consumption. This indirect consumption of grain, reflected in the monocultural practices supporting the livestock-feed complex, is the hallmark of modern agro-food networks. So, this collective becomes *corporeal*, incorporated into a social body of multiple quasi-objects.

In effect, the real, social and discursive character of the hybrid corn collective is literally ingrained in the spatio-social structures and practices, the political ecology, for example, of Mid-Western corn-hog production and the broiler industry of the American South, and their international translations. Finally, the increasing scope of collectives enlisting the new quasi-objects of industrial metabolism is exemplified by the biotechnological production of high fructose corn syrups, single cell microbial proteins using corn starch as the feedstock, and a variety of new feed and food byproducts. Again, the notion of translation or network is a useful metaphoric device which, in this case, captures salient characteristics of the industrial substitution processes that are transforming modern agro-food production networks.

This brief discussion of actor network theory, particularly when infused with an awareness of human metabolism, also suggests potentially fruitful avenues for the analysis of food habits and practices. These habits can be seen as incorporations of, constituents of, and constituted by, collectives of varying scope, which enroll practices of food provision. That is, these collectives "translate" or mediate the personal. This approach may thus offer a way to integrate the ethnography of the personal, the familiar, everyday experience, with different collectives of quasi-objects varying in size and scope. Some reflections along these lines are presented in the following section.

FOOD AND EVERYDAY METABOLIC STORIES

The quasi-objects which appear within "food" allow us to examine the location of metabolism and incorporation at multiple sites along a dimension which connects nature, agriculture, the food system and its produced commodities and human needs, and the social constitution of the body as both actor and medical object. "Food has shifted from being something that people see in its nutritional values and its appetite appeal and pleasure, to something that they have to be careful of because it will do them harm. . . . This opens the opportunity to market many, many new foods" (NPR, 1996).

The two short stories we provide to illustrate the possibilities of this approach have various elements in common: both require the decomposition of organisms (humans and others) into elements of nature/science, culture, and power; and both allow the translation of certain material elements, practices and meanings from eaten to eater in ways mediated by particular intervening industrial metabolisms. Yet each is deaf to parallel anti-narratives that communicate power relations more clearly. The first story, that of BSE (bovine spongiform encephalophy) transforms animals into cannibals; the second, that of Redux or fen/phen, recomposes anorexia, a lethal disease of the social body, into "oral anorectics," now a pharmaceutical cure for the metabolic "disease" of obesity.

BSE: a co-production denied

The saga of bovine spongiform encephalophy (BSE) – mad cow disease or *la maladie de la "vache folle"* – is full of metaphorical possibilities and invites multiple "readings." It can be represented as a grotesque consequence of the hubris of modern industrial farming, with its subtext of "nature" striking back against wanton human interference. The story can be told as an archetypal "food scare," those episodic occurrences that expose the co-productions and collective associations of non-human and human "actants" that make up the human food chain. The BSE controversy also illustrates beautifully Latour's notion of the translation of actants into different collectives, which variously struggle to deny, recognize, contain or extend the scope of the mobilization. That is, translation as the creation of links between agents that did not exist previously and which

211

modify the agents involved. The politics of the BSE saga can also be interpreted, somewhat freely, in terms of another meaning Latour (1994: 36) gives to mediation, that of "blackboxing," or "a process that makes the joint production of actors and artifacts entirely opaque." In this paper, a brief chronology of BSE in Britain only hints at the rich analytic and metaphoric material awaiting further exploration.

Spongiform encephalopathies are an extremely rare group of neuro-degenerative diseases that includes scrapie in sheep and several human forms, such as Creutzfeldt-Jakob Disease (CJD), whose incidence is under one case per million people per year. The existence of scrapie has been known since the nineteenth century, and there is no evidence of its transmission to humans through the food chain. Although some experts assert that a very rare spongiform encephalopathy may have previously existed in cattle, the crisis began when BSE was identified in a dairy herd in Kent, England, in November 1986. For the public, the "scare" factor arises from the belief that BSE can be transmitted to humans as a form or variant of CJD as the result of eating infected beef products.

The political and scientific crisis that accompanied and greatly exacerbated the BSE scare was based on the persistent rejection by the public authorities, despite circumstantial evidence to the contrary, of any possible link between BSE and CJD. This position reflected scientific doubts, later modified in the 1990s, that spongiform encephalopathies could cross species barriers: from sheep to cattle and thence to humans. This skepticism was born of ignorance, however, since there is no convincing explanation of how either BSE or CJD is transmitted, although there are hypotheses involving misshapen proteins called prions and viral agents (Radford 1996; *The Economist* 30 March, 1996: 25–7).

While systematically denying any conceivable link between BSE and CJD, the British government took a number of steps that were directly at odds with its public pronouncements. In 1988, the government banned the use of meat and bone meal from rendered sheep and cattle carcasses as protein feed for cattle. This gave consumers a rare glimpse into the "black box" of the food chain to reveal the industrial conversion of ruminants into both carnivores and cannibals. With ruminant tissue used in cattle feed now apparently identified as the proximate cause of BSE, attention then focused on the British rendering industry. For a variety of reasons, rendering processes were modified in the early 1980s as companies abandoned the use of powerful chemical solvents and carcasses were "cooked" at lower temperatures for shorter periods than previously. In the party political arena, these findings prompted attacks on the neo-liberal deregulation policies implemented by the Conservative governments of Margaret Thatcher and John Major. In an extravagantly dubious gesture of patriotic confidence in British beef, John Gummer, the Minister of Agriculture and Fisheries, appeared on television to feed his daughter with a beefburger!

While still vigorously denying the transmissibility of BSE to humans, the government took a further contradictory step in 1990, when it removed specific kinds of bovine offal from the food chain. As Radford (1996) notes, this measure came one year after pet food manufacturers had taken this same step independently. This pattern of rumor and consumer suspicion, government denials and damage limitation continued until late 1995, when ten cases of a new variant of CJD were reported, strengthening the circumstantial evidence of a transmissible encephalopathy. Moreover, the victims of this new variant were young people rather than the old, as with classical CJD. With this news, the "mad cow" collective began its pan-European "translation."

The European Commission, on 17 March 1996, imposed a ban on the international sale of all meat and other products of British cattle, which subsequently provoked the so-called "beef war" following the British government's decision, on 21 May 1996, to adopt a policy of "non-cooperation" with its European partners, urged on by bellicose Euroskeptics and the tabloid press. Awkward questions have followed about shipments of ruminant-based cattle feed to Europe after the 1988 ban in Britain, as well as exports of calves to France and other countries. Although some experts had rejected the possibility of "vertical" or maternal transmission of BSE to calves born to infected cows, circumstantial evidence again indicated that this was indeed conceivable, as the British government acknowledged in August 1996.

As the BSE crisis has deepened, lower beef consumption has brought sharp falls in beef prices throughout Europe, threatening rural livelihoods and raising the specter of widespread farm bankruptcies. French farm unions have organized a series of protests, including a national road blockade on 28 August 1996, to check the provenance of beef carcasses and live cattle in trucks and at abattoirs (*Le Monde* 30 August 1996). In France, beef producers, processors and distributors have instituted an "identity card" scheme to track individual animals through the beef commodity chain. This scheme, under the label "Viande Bovine Française," creates new conventions of quality assurance and a preference for local beef at the expense of imported meat.

Consumers in Europe have been exposed to complex debates over the rates of selective culling and the length of time needed to eradicate BSE from cattle herds. In Switzerland, for example, the government plans to slaughter 230,000 cattle, representing one in eight of its total herd (*Financial Times* 17 September 1996). Between May and October 1996, over 500,000 cattle were slaughtered in Britain following the decision to remove cattle aged over thirty months from the food chain. Newspaper readers have become aware of the shortage of special incinerators for infected cattle and a host of other macabre details, such as an estimated backlog of cows awaiting slaughter in October 1996 of 400,000. Moreover, since the rendered down carcasses of cattle will no longer enter the food chain as animal feed for their fellow species, schemes are afoot to use these remains for fuel in British power stations. As Radford (1996) wryly observes, the nutritionists' adage of food as fuel will now take on a literal but entirely different meaning.

The BSE crisis and its trajectory fits nicely with Latour's notion of agency as "productions of natures-cultures" or collective associations. In this respect, a food scare of this magnitude opens the "black box" of the modern agro-food system to wider scrutiny, revealing the scope of enlistment of human and non-human actants in prosaic decisions of everyday life. Yet, this is a co-production of the most intimate kind: of food and human metabolism. Ultimately, it is a question of identity, and most pointedly so in the case of BSE, since human forms of spongiform encephalopathy take over the victim's proteins bringing loss of memory, dementia, and death.

Anorexia/anorectic redux: hunger and metabolism

Another co-production of food and human metabolism appears internal to the human body. It is incorporated in the intersection of intention, culture and metabolism as expressed in the apparently individual construction of the body itself. Here, food is dangerous because it is food, and may accumulate. Obesity has been formally characterized as a social (as well as a medical) disability. "The US Equal Opportunity Commission, responding to complaints of widespread discrimination against obese persons, has now declared obesity a protected category under the federal Americans with Disabilities Act" (Medical Sciences Bulletin 1997: 1). The stigmatization of obesity is closely socially coupled with the simultaneous construction of social food, offering pleasure, pain, health, and illness as co-productions.

Anorexia nervosa – self-starvation – is increasingly recognized among the eating disorders with which modern society has become concerned. Anorexia and the related disorders of bulimia and binging occur most frequently in young women, though occasionally in men. Anorexia refers to "want of appetite," even to "lack of desire." Current readings of anorexia discover the agency of this phenomenon in both the social construction of women's bodies and the individual's need to master that little left to her own control. In a recent review of two books on anorexia, Susie Orbach reflects on what "is so fundamental in the anorexic's plight: *the struggle for human agency.* For women who starve themselves, there is always the attempt to create out of the material of the body a self that they can find acceptable – a self that, in its very essence, questions the one that has been given" (Orbach, *TLS* 8/9/96, emphasis added).

Susan Bordo (1985) provides us with a remarkable interrogation of the cultural and social axes on which anorexia is constructed: mind–body dualism, control, and gender/power. She writes: "the anorexic's 'protest,' like that of the classic hysterical symptom, is written on the bodies of anorexic women and *not* embraced as a conscious politics, nor, indeed, does it reflect any social or political understanding at all. . . . The anorexic is terrified and repelled, not only by the traditional female domestic role – which she associates with mental lassitude and weakness – but by a certain archetypal image of the female: as hungering, voracious, all-needing, and all wanting. It is this image that shapes and permeates her experience of her own hunger for food as insatiable and out-of-control, which makes her feel that if she takes just one bite, she won't be able to stop" (p. 44).

This "intimate" collective now enlists the quasi-objects of pharmacology, with its technological triumphalism. Metabolism can be defeated; the desire to incorporate food can be overcome. "Molecular biology will ultimately provide the 'cure' for obesity. . . . Obesity is a chronic disease that requires lifelong management. Drug treatment can play an important role in any management strategy, and combination therapy with two anorectic drugs may be the best approach. . . . Physicians who prescribe appetite-control drugs should plan to continue therapy for at least 5 to 10 years' (Http://pharminfo.com/pubs/msb/obesity.html). The world wide web becomes a frame within which to promote this strategy, and its implements, and to contest it. The mechanism for this transformation is a set of drugs called oral anorectics, drugs which induce anorexia by design.

These drugs are revolutionizing the diet industry, which each year collects several billion dollars from consumers of networks of quasi-objects (of drugs, alternative diets, exercise programs, and psychological support) which provide antidotes to the dangers of food. Fen/phen (fenfluramine and phentermine) and Redux (dexfenfluramine, the dextro-isomer) apparently act on the production of serotonin in the brain. Serotonin is an important neurotransmitter, and concerns about the impact of these drugs under long-term administration are now surfacing, but the commercial possibilities of these proprietary medications are immense. Some research suggests that these medications, if taken over a long period, can lead to the death of serotonin-producing cells or to primary pulmonary hypertension, a lethal disease of the lungs. Supporters of these medications find flaws in this research, and argue that the experience of thousands of Europeans who have taken these drugs proves their safety. Enthusiasts, such as Dr Pietr Hitzig (http://www.fenphen.com/hope.html) promote fen/phen as "a major paradigm shift in modern medicine." Hitzig writes: "Beyond weight management, when properly administered, FEN/PHEN can eliminate craving disorders such as severe obesity, nicotine longing, and drug or alcohol addiction. It can manage psychoneuroses, affect immune disorders like asthma and hives, and rein in obsessive/compulsive conditions such as nailbiting and bulimia. It can control attention deficit disorders, even those with hyperactivity, and it can eliminate sexual addiction, rage, and depression." What more could any body want?

Metabolism and incorporation link the corporeality of the social body to the agro-food system and its discontents. Rather than addressing the transgressions of social justice incorporated in (real) starvation at the Malthusian margin of the global market, the agro-food system finds new markets in social fears, incorporating (pharmacological) starvation into the commodities of food. Food is now a drug for which the antidote can be found.

In both of these short stories, incorporation is tied to transgression. Our use of actor-network theory encourages us to take our analysis of the dialectical composition of nature and culture, and of the hybridities that appear reflected in the facets of that dualism, into the social composition of the commodity itself.

For it is in commodities, including the commodity body of a woman (or a steer), that the natural and social worlds are mutually incorporated, through the intertwined agency of "eco-regulatory process" and the "form-giving fire" of labour, not in circumstances of our own choosing.

CONCLUSION

The "greening" of social theory has coincided with a growing awareness that modern, asymmetric formulations of nature–society relations are profoundly problematic. Often, the nonhuman realm is "blackboxed" and agency is conceptualized as the "singular" property of human intentionality, which acts on, and in opposition, to a passive, mechanistic nature. In this chapter, we have focused on environmental history and Marxist political economy as two privileged theoretical sites where the abstraction of nature is taken seriously, and have critically examined the strategies deployed to redress this asymmetric treatment: the autonomous agency of nature as ecological limits, and nature–society dialectics. Yet these proposals are still infused, to varying degrees, with the polarities of modernism, both in their general articulation and more specifically when extended to the case of agriculture.

This ontological "discontinuity problem," in Val Plumwood's (1991) elegant phrase, can be addressed more productively, we suggest, by applying the alternative metaphorical tools of actor-network theory. With interactive symmetry as its basic premise, agency is conceptualized as the collective capacity of associative, and most importantly, relational networks. The principle of symmetry, and its importance in overcoming modern polarities and silences, is that consideration is given to non-human and human entities equally, with no prior assumption of privilege, rank or order. This concept of collective agency, of active relational materiality, offers fruitful new ways to theorize the processes of production and reproduction that describe, conjoin, and mutually constitute the natural and social worlds into Latour's "seamless fabric."

In addition to its promise as a general framework of analysis of nature–society interactions, the actor-network perspective of relational symmetry between nonhuman and human worlds underlies the notion of corporeality developed in this chapter. This concept is used here to theorize agency in agro-food networks and to reject a more local ontological "discontinuity": the abstraction of nature in mainstream agrarian political economy. The notion of corporeality is proposed to capture the relational materiality of ecologies and bodies that characterizes agro-food networks. In other words, those collective associations constituted by the prosaically complex, recursive metabolic exchanges that define the social reproduction and production of agro-food nature and humans.

Explorations of this kind, moving between general and specific theorizations of nature–society relations, illustrate the conceptual challenges and potential pitfalls that lie behind the seemingly simple task of bringing nature "back in." Yet this reconstructive analytical work has much to contribute to the articulation of political projects to overcome the environmental despoliation and social asymmetries of late modernity and create more ecologically sensitive, egalitarian societies.

NOTES

1 As a corrective to the erasure of the materiality of produced nature that he detects in some Marxist geographical accounts, Castree (1995: 21) approvingly quotes Donald Worster that "nature itself is 'an agent and presence in history.'" This reference is more than convenient coincidence since, in this particular respect, Castree's project for Marxist geography has some affinities with the epistemological foundations of environmental history.

2 Worster (1984) observes that in 1972, when the *Pacific Historical Review* devoted an issue to environmental history, "the main themes . . . were conservation, water development, wilderness, national parks, and the Department of the Interior" (2, note 3).

3 The existence of physical limits to human possibility and the elemental role of environment in shaping society are stressed by John Opie (1983) in an essay on method which appeared at the same "moment." Opie cites the *Annales* school and particularly the work of Fernand Braudel in support of this position.

4 Worster advocated this conceptual innovation in his 1982 presidential address to the fledgling American Society of Environmental History, arguing that the new field should give analytical prominence to the ecological origins and impacts of modes of production. This ecological perspective on the development of social institutions and social change, which Worster (1984) attributes to Wittfogel's extension of Marx, is more fully elaborated in Worster's *Rivers of Empire* (1985). Additional discussion of modes of production also can be found in Worster (1987, 1988).

5 With the notable exception of William Cronon (1992, 1994b, 1995), engagement with the social constructionist challenge to the moral authority and normative stance of environmental history has been dismissive and slight. David Demeritt (1994b) provides a fuller formulation of the constructionist position, building in part on Cronon's discussion of the fundamental role of narrative or "stories" and narrative strategy in environmental history, but Cronon (1994b) so far has been his only interlocutor.

6 These debates are reviewed by Redclift (1987) and Castree (1995).

7 Demeritt (1994a) traces this explanatory framework to the ways in which environmental historians have read the narrative strategy of Carl Sauer and H. C. Darby in cultural geography, which uses sequential landscape cross-sections to evaluate the impacts of human–nature interaction on landscape change. However, Sauer never begins with a pre-human landscape, but with "the reality of the union of physical and cultural elements" (1925: 325). He quotes Vidal de la Blache: "Human geography does not oppose itself to a geography from which the human element is excluded; such a one has not existed except in the minds of a few exclusive specialists."

8 In response to critical commentary, Cronon (1994a: 173) acknowledges "his analytical compromises in the service of . . . writerly rhetoric. 'First' and 'second' nature are not completely successful because they appear to reintroduce the very dualism they seek to undermine, but I was unwilling to introduce 'third' nature to label a dialectic that already seemed inescapably implied by 'second' nature." Cronon identifies his central question in *Nature's Metropolis* as being about "human alienation from nature" (p. 171).

9 For actor network theory in agro-food and rural studies, see Murdoch (1995), Long and van der Ploeg (1994, 1995), Lowe and Ward (1997), and Whatmore and Thorne (1997).
10 We are indebted to Sarah Whatmore and Lorraine Thorne for this reference to Callon and Law (1995). The notion of hybrid *collectifs* is used by Whatmore and Thorne (1997) to analyze alternative international circuits of food production, distribution, and consumption.
11 Latour (1994) clarifies the concept of translation by noting that like Michel Serres and following the sociological usage of Michel Callon (1986), "I use *translation* to mean displacement, drift, invention, mediation, the creation of a link that did not exist before and that to some degree modifies two elements or agents" (p. 32).

REFERENCES

Altvater, E. (1993) *The Future of the Market: An Essay on the Regulation of Money and Nature after the Collapse of "Actually Existing Socialism,"* London: Verso.
Benton, T. (1989) "Marxism and natural limits," *New Left Review* 178: 51–86.
—— (1991) "Biology and social science: why the return of the repressed should be given a (cautious) welcome," *Sociology* 25: 1–29.
—— (1992) "Ecology, socialism and the mastery of nature: a reply to Reiner Grundmann," *New Left Review* 194: 55–74.
—— (1993) *Natural Relations: Ecology, Animal Rights and Social Justice*, London: Verso.
Bogue, A. G. (1983) "Changes in mechanical and plant technology: the corn belt, 1910–1940," *Journal of Economic History* 43(1) 1–25.
Bordo, S. (1985) "Anorexia nervosa: psychopathology as the crystallization of culture," in D. Curtin and L. Heldke (eds) *Cooking, Eating, Thinking: Transformative Philosophies of Food*, Bloomington: Indiana University Press.
Callon, M. (1986) "Some elements of a sociology of translation: domestication of the scallops and the fishermen of St. Brieuc Bay," in J. Law (ed.) *Power, Action, and Belief: A New Sociology of Knowledge?*, London: Routledge & Kegan Paul.
Callon, M. and Law, J. (1995) "Agency and the hybrid *collectif*," *South Atlantic Quarterly* 94: 481–507.
Castree, N. (1995) "The nature of produced nature: materiality and knowledge construction in marxism," *Antipode* 27: 12–48.
Cronon, W. (1983) *Changes in the Land: Indians, Colonists, and the Ecology of New England*, New York: Hill and Wang.
—— (1990) "Modes of prophecy and production: placing nature in history," *Journal of American History* 76(4): 1122–31.
—— (1991) *Nature's Metropolis: Chicago and the Great West*, New York: W. W. Norton.
—— (1992) "A place for stories: nature, history, and narrative," *Journal of American History* 78: 1347–76.
—— (1994a) "On totalization and turgidity," *Antipode* 26: 166–76.
—— (1994b) "Cutting loose or running aground," *Journal of Historical Geography* 20: 38–43.
—— (1995) (ed.) *Uncommon Ground: Toward Reinventing Nature*, New York: W. W. Norton.
Crosby, A. W. (1972) *The Columbian Exchange: Biological and Cultural Consequences of 1492*, Westport, CT: Greenwood Press.

Demeritt, D. (1994a) "The nature of metaphors in geography and environmental history," *Progress in Human Geography* 18: 163–85.

—— (1994b) "Ecology, objectivity and critique in writings on nature and human societies," *Journal of Historical Geography* 20: 22–37.

FitzSimmons, M. (1989a) "The matter of nature," *Antipode* 21: 106–20.

—— (1989b) "Reconstructing nature," *Society and Space* 7: 1–3.

Goodman, D. and Redclift, M. (1991) *Refashioning Nature: Food, Ecology and Culture*, London: Routledge.

Goodman, D., Sorj, B. and Wilkinson, J. (1987) *From Farming to Biotechnology: A Theory of Agro-Industrial Development*, Oxford: Blackwell.

Haraway, D. (1991) *Simians, Cyborgs, and Women: The Reinvention of Nature*, New York: Routledge.

—— (1992) "Otherworldly conversations; terran topics, local terms," *Science as Culture* 3: 64–98.

Harvey, D. (1974) "Population, resources, and the ideology of science," *Economic Geography* 50: 256–77.

—— (1982) *The Limits to Capital*, Chicago: University of Chicago Press.

—— (1996) *Justice, Nature, and the Geography of Difference*, Oxford: Blackwell.

Latour, B. (1993) *We Have Never Been Modern*, Brighton: Harvester Wheatsheaf.

—— (1994) "On technical mediation, philosophy, sociology, genealogy," *Common Ground* 3: 29–64.

Law, J. (1992) "Notes on the theory of actor-network: ordering, strategy, and heterogeneity," *Systems Practice*, 5: 379–93.

Long, N. and van der Ploeg, J. D. (1994) "Heterogeneity, actor, and structure: towards a reconstitution of the concept of structure," in D. Booth (ed.) *Rethinking Social Development, Research and Practice*, Harlow: Longman.

—— (1995) "Reflections on agency, ordering the future and planning," in G. E. Frerks and J. H. B. den Auden (eds) *In Search of the Middle Ground*, Wageningen: Wageningen Agricultural University.

Lowe, P. and Ward, N. (1997) "Field-level bureaucrats and the making of new moral discourses in agri-environmental controversies," in D. Goodman and M. Watts (eds) *Globalising Food: Agrarian Questions and Global Restructuring*, London: Routledge.

Mann, S. (1990) *Agrarian Capitalism in Theory and Practice*, Chapel Hill, NC: University of North Carolina Press.

Mann, S. and Dickenson, J. (1978) "Obstacles to the development of a capitalist agri-culture," *Journal of Peasant Studies* 5: 466–81.

Martinez-Alier, J. (1987) *Ecological Economics: Energy, Environment and Society*, Oxford: Blackwell.

Merchant, C. (1989) *Ecological Revolutions: Nature, Gender, and Science in New England*, Chapel Hill, NC: University of North Carolina Press.

Murdoch, J. (1995) "Actor-networks and the evolution of economic forms: combining description and explanation in theories of regulation, flexible specialization and networks," *Environment and Planning A* 27: 731–57.

NPR (National Public Radio) (1996) "Morning edition, September 27", interview with George Rosenbaum (Bob Edwards, interviewer).

O'Connor, J. (1988) "Capitalism, nature, socialism: a theoretical introduction," *Capitalism Nature Socialism* 1: 11–38.

Opie, J. (1983) "Environmental history: pitfalls and opportunities," *Environmental Review* 7: 8–16.

Pickering, A. (1992) "From science as knowledge to science as practice," in A. Pickering (ed.) *Science as Practice and Culture*, Chicago: University of Chicago Press.

Plumwood, V. (1991) "Nature, Self, and Gender: Feminism, Environmental Philosophy, and the Critique of Rationalism" in K. J. Warren (ed.) Ecological Feminism, *HYPATIA* 6, Special Issue: 3–27.

Pyne, S. (1990) "Firestick history," *Journal of American History* 76: 1132–41.

Radford, T. (1996) "Poor cow," *London Review of Books* 5 September: 17–19.

Redclift, M. (1987) "The production of nature and the reproduction of the species," *Antipode* 19: 222–30.

Sauer, C. O. (1925) "The morphology of landscape," in J. Leighly (ed.) (1963) *Land and Life: A Selection from the Writings of Carl Ortwin Sauer*, Berkeley: University of California.

Schmidt, A. (1971) *The Concept of Nature in Marx*, London: New Left Books.

Smith, N. (1984) *Uneven Development: Nature, Capital, and the Production of Space*, Oxford: Blackwell.

Unger, R. (1987) *False Necessity: Anti-Necessitarian Social Theory in the Service of Radical Democracy*, Cambridge: Cambridge University Press.

Walker, R. (1994) "William Cronon's *Nature's Metropolis*: a symposium," *Antipode* 26: 113–76.

Whatmore, S. and Thorne, L. (1997) "Nourishing networks: alternative geographies of food," in D. Goodman and M. Watts (eds) *Globalising Food: Agrarian Questions and Global Restructuring*, London: Routledge.

White, R. (1985) "American environmental history: the development of a new historical field," *Pacific Historical Review* 54: 297–335.

—— (1990) "Environmental history, ecology, and meaning," *Journal of American History* 76: 1111–16.

Williams, R. (1980) *Problems in Materialism and Culture*, London: Verso.

Worster, D. (1977) *Nature's Economy: A History of Ecological Ideas*, San Francisco: Sierra Club Books.

—— (1979) *Dust Bowl: The Southern Plains in the 1930s*, New York: Oxford University Press.

—— (1984) "History as natural history: an essay in theory and method," *Pacific Historical Review* 53: 1–19.

—— (1985) *Rivers of Empire: Water, Aridity, and the Growth of the American West*, New York: Oxford University Press.

—— (1987) "New West, true West: interpreting the region's history," *Western Historical Quarterly* 18: 141–56.

—— (1988) "Appendix: doing environmental history," in D. Worster (ed.) *The Ends of the Earth: Perspectives on Modern Environmental History*, New York: Cambridge University Press.

—— (1990a) "Transformations of the earth: toward an agroecological perspective in history," *Journal of American History* 76: 1087–1106.

—— (1990b) "Seeing beyond culture," *Journal of American History* 76: 1142–7.

Yoxen, E. (1981) "Life as a productive force," in L. Levidow and R. Young (eds) *Science, Technology and the Labour Process*, London: CSE Books, pp. 66–122.

10

TO MODERNISE OR ECOLOGISE? THAT IS THE QUESTION

Bruno Latour (translated by Charis Cussins)

WILL POLITICAL ECOLOGY PASS AWAY?

This chapter explores the destiny of political ecology.[1] It is very much influenced by the French political situation and the continuing marginality of the country's various green parties. It relies on three different strands: first, a very interesting model to understand political disputes devised by two French sociologists, Luc Boltanski and Laurent Thévenot, in a book that is not yet available in English (Boltanski and Thévenot 1991); second, a case study by the author on the recent creation by law of what could be called 'local parliaments of water' (Latour and Le Bourhis 1995);[2] third, a long-term project in philosophy to develop an alternative to the notion of modernity (Latour 1993) and to explore the political roots of the notion of nature. The point of the chapter can be stated very simply: political ecology cannot be inserted into the various niches of modernity. On the contrary, it requires to be understood as an alternative to modernisation. To do so one has to abandon the false conceit that ecology has anything to do with nature as such. Disabused of this notion, political ecology is understood here as a new way to handle all the objects of human and non-human collective life.[3]

For the last ten years or so, the question has arisen as to whether the eco-movement is in fact a new form of politics or a particular branch of politics. This uncertainty is reflected in the difficulty that the environmental parties have experienced in carving out a niche for themselves. On track for rapid integration into people's everyday concerns, environmentalism could well follow in the footsteps of the nineteenth-century hygiene movement – a movement with which, obvious differences notwithstanding, it greatly resembles[4] – with the defence and protection of the environment becoming a feature of everyday life, rules, regulations and goverment policy, just as preventive vaccination, the scientific analysis of water quality and health records did. One would no more drop litter in the woods than spit on the floor, but that does not make habits of good manners and civility into an entire political project. Just as there is no 'hygienists' party' today, there will soon be no green party left. All political parties, all goverments and all citizens will simply add this new layer of behaviour and regulations to their everyday concerns. A good indicator of this progressive normalisation of ecologism will be the creation of specialised administrative bodies, like those for bridges and highways or water and forests, which would be all the more effective since they would be cast in the mould of the well-established depoliticising tradition of public sector administration (Lascoumes 1994)

The inverse solution consists of making ecology responsible for all of politics and all of the economy, on the basis of the argument that everything is interrelated, that humankind and nature are one and the same thing and that it is now necessary to manage a single system of nature and of society in order to avoid a moral, economic and ecological disaster. But this 'globalisation' of environmentalism, even if it constitutes the common ground of numerous militant activities and of the public imaginary at large, still does not seem to replace the normal domain of political action.

As convinced as its adherents might be, this submersion of all politics and all of society into nature seems unrealistic. It would appear to lack political sense and plausibility, for at least two reasons that are easily understood.[5] In the first place, the nature whole into which politics and human society would supposedly have to merge transcends the horizons of ordinary citizens. For this Whole is not human, as is readily seen in the Gaia hypothesis (Lovelock 1979). Second, the only people who would be capable of defining these connections and revealing the infinitely complex architecture of this totality would be specialists, whose knowledge and breadth of view would remove them from the lot of common humanity (Lafaye and Thévenot 1993). In any case, these scientific demigods would not belong to the ordinary rank and file of county councils, administrative boards and local organisations. Accepting that ecology bears on every type of connection would be thus to lose sight of humanity twice: first to the advantage of a unity superior to humankind, and second to the advantage of a technocracy of brains that would be superior to poor, ordinary humans.

Consequently, on the one hand, ecology integrates itself into everyday life without being able to become the platform for a specific party and, on the other, it becomes inflated to the point of assuming responsibility for the agendas of all the other parties, while handing the pen to men and women who do not belong to the world of politics and who speak of a global unity which no longer has the political domain as its horizon.

However, practical experience does not confirm either of these two extreme hypotheses.[6] Militant action remains both far more radical than one would believe if the hypothesis of ecology becoming a fact of everyday life was correct – nothing to do, in this respect, with hygiene which was always the concern of a few prominent administrators – and far more partial than it should be if one were to accept the hypothesis of globalisation. It is always *this* invertebrate, *this* branch of a river, *this* rubbish dump or *this* land-use plan which finds itself the subject of concern, protection, criticism or demonstration.

In practice, therefore, ecological politics is much less integrable than it fears, but a lot more marginal than it would like. To express this paradox of totality in the future and present marginality, there is no shortage of formulae which enable it to get out of the problem: 'think globally, act locally', integrated management, new alliance, sustainable development, and so on. According to political ecology, it should not be judged by its modest electoral results.[7] It begins with individual cases, but it will soon, slowly but surely, incorporate them all into a general movement that will end up embracing the whole earth. According to political ecology, the courage to address itself to small causes rightly comes from the certain knowledge that it will soon have to assume responsibility for all the major issues.

If this were indeed the case, we should be witnessing the rise, perhaps hesitant but certainly irreversible, of a political ecology taking up, day after day, the whole task of political life. Yet the scenario of ecology becoming a synonym for politics *tout court* seems increasingly improbable. This is certainly the case in France where, although the number of environmental parties is increasing, they still do not account for more than 5 per cent of the votes, and even this total appears to be declining. In spite of the presence of three candidates in the 1995 French presidential elections, green parties could well go out as they came in, like any other passing trend. For a party that must take responsibility for Mother Earth herself, there is more than one problem in this continuing marginalisation. It is a challenge that is making it necessary to rethink the very basis of its aspiration to become global.

In this chapter, I would like to advance the hypothesis that the rise in power of political ecology is hindered by the definition it gives itself, as both politics and ecology! As a result of this self-definition, the practical wisdom acquired after years of militant action is incapable of expression by a principle of classification and ordering – about which I shall say more below – that would be politically effective. As the prophet Jonah said of the Hebrew people, 'it can't tell its left from its right'. Without this principle of ordering, political ecology makes little impact upon the electorate and does not manage, using all the arguments that it nevertheless so effectively reveals, to develop lasting and consistent political viability.

IS POLITICAL ECOLOGY AN ORIGINAL TYPE OF JUSTIFICATION?

In their pioneering work, Boltanski and Thévenot (1991) have offered us the ideal acid test to see whether or not political ecology can survive as an original form of politics, or if, on the contrary, it can easily be dissolved into very ordinary regimes which have been put in place during the last century or so.

By studying the details of how ordinary people engaged in disputes over right and wrong justify their action, these authors have been able to identify six different 'regimes of justification' (which they call 'Cités' in French). The novelty of their approach is to have proven that each of those regimes is complete, although utterly contradictory with the others. In other words, it is possible to demonstrate that in contemporary French society people engaged in disputes may ascend to six different overarching principles ('principe supérieur commun'), each of them engaging a full-fledged and coherent definition of what humanity should be ('principe de commune humanité'). Each regime is the result of a long history of political philosophy, and has now become an everyday competence activated easily by every member of the society. Each of them defines through trials a scale of right and wrong ('grandeur' et 'petitesse'/'greatness' and 'smallness'), that allows one to pass judgement and to settle disputes. Each of them – and this is the great strength of the model – has the capacity to denounce the others because they lack morality or virtue.[8]

We do not need to go into the details of this magisterial theory. For the present chapter, the great interest of this model is that it allows us to test whether or not political ecology offers a new principle of justification, or if it can be reduced to the six others which have been sedimented through the course of time. Is political ecology old wine in new bottles, or, on the contrary, new wine in old bottles?[9]

At first glance, the answer is clear. There can be no 'ecological regime' since it is very easy to show that any of the empirical cases tackled by green politics borrows its principle of justification from one of the six 'Cités' already in place – in fact we will limit ourselves here to the Domestic, Civic, Industrial and Commercial regimes of justification.

The majority of issues considered – in the case of the landscape, water and waste, natural parks, etc. – can be related easily to what Boltanski and Thévenot call the 'domestic regime', the principle of which is to justify the worth of a human by the quality of his lineage and the solidity of his roots. And it is true that many practical disputes in ecology are always a question of defending a particular territory, a particular aspect of national heritage, a particular tradition or a territory against the de-sensitised, de-territorialised, stateless, monstrous character of an economic or technical enterprise. Starting from these principles of justification, one can denounce the 'industrial regime' and, more recently, the 'civic regime' without scruple. This is probably why political ecology appeared so original in the beginning. In short, it gave back value to the 'domestic regime' which two centuries of republican and revolutionary spirit had reduced to a mere 'domesticity', to the domain of the home. Thanks to ecology, the domestic domain became once more what it was before the revolutionary ethos.

The curious alliance between conservatives, conservationists of heritage and nature conservationists would thus be easily explained. Against the 'civic' and 'industrial regimes', another justification has been revived after centuries of piti-less denunciation. By attacking a bullet-train line, by protecting a garden, a rare bird's nest or a valley spared by the suburbs, one could finally be simultaneously reactionary and modern. In short, the originality of ecology would only last long enough partially to rehabilitate the quality of the private domain. Nature, it is easy to see, is becoming as 'domestic' in the Vallée de Chevreuse as among the Achuars.[10] In this revamped 'domestic regime' the state of highness is achieved by ancientness, by durability and by familiarity; the state of smallness, by the anonymity of people without roots or attachments.[11]

If many burning issues of political ecology can be reduced to the 'domestic regime', other issues can be reduced even faster within the 'industrial regime' (Barbier 1996). This is notably the case in all the battles over waste, pollution and the like.[12] Here again, the originality of ecology disappears rapidly in favour of equipment and regulations designed to end waste and reduce pollution. After the initial cries of horror at the accounts to be balanced, the costs to be met and the equipment to be installed, it is 'business as usual' for ecology in the 'industrial regime'. Domestic waste is becoming a raw material that is managed like any other raw material by simply extending the production process. Pollution rights are traded on a market in environmental goods which is fast ceasing to be exotic. The health of rivers is now monitored like the health of the workforce. It is not worth treating ecology as a separate concern; it is more a question of using it to explore new and profitable business opportunities. There was a waste problem. We put an end to it. There was a pollution problem. We put an end to it. It is now only a question of controlling, monitoring and managing. That's all there is to it. Exit the bearded and hairy ecologists: they have become obsolete.

Are the ecological issues that cannot be reduced to the 'domestic' or 'industrial' regime a proof that there is something original in political ecology? No, because they can appear – although it is slightly less straightforward – reducible to a third regime, the one that Boltanski and Thévenot call the 'civic regime' and that is defined by 'general will'. In this regime, worth is defined by the ability of one agent to disentangle himself from particular and local interests so as to envision only the General Good. In its aspirations to globality, ecology encounters in the definition of the general will an opponent which is all the more formidable since it has the support of almost all mainstream political institutions since the mid-eighteenth century.

Here again, it seems, ecologists do not manage to establish their justifications for long and cannot claim to represent more than one lobby among many. Although some green party may speak in the name of the common good, it is always the elected mayor who signs the land-use plan and not the association that is defending, often for its own petty reasons, some end of a garden, some bird, some snail or other (Barbier 1992). It is the local goverment that closes a polluting factory and not the manufacturer who, in the name of efficiency, is exploiting employees. It is the Water Board that protects resource for everyone and not the angling association which has its own fish to fry. Rehabilitating domestic traditions and extending efficiency to include natural cycles is one thing; directly opposing the general will on such terrain is quite another and an extremely delicate issue.[13]

The new compromise that enables the 'civic regime', without modifying itself in any lasting way, to absorb most ecological issues consists in extending the electorate deemed to participate in the expression of the general will to include future generations of citizens.[14] Future generations are indeed mute, but no more so than the minors who have just been born, the ancestors who are already dead, the abstainers who are said to 'vote with their feet', or the incompetents who have rights through various sorts of stewardships. At the cost of a slight enlargement in the number of electors, the 'civic regime' can absorb most of the issues pending. At the cost of a delicate compromise with the 'domestic regime', it could even reconstruct this 'community of the dead and the living', which would permit it to be of both on the right and on the left, thus casting its net wide and thereby diluting the green vote even further.

On the basis of these various reductions, there would therefore be no 'ecological regime' since the issues that it raises can all be resolved in the 'domestic,' 'industrial' and 'civic regimes'. What is left could easily be pigeonholed into the 'commercial regime,' as can be witnessed in the unashamed processing of the numerous 'green products', 'green labels' and other 'natural' products.[15] With this hypothesis one could account for the necessarily ephemeral vogue for ecology.

If we follow this not very charitable reduction, we could say that there is no durable originality in the political philosophy of ecology. To be sure, on seeing the irruption in debates of waterways, landscapes, noise, dustbins, the ozone layer and unborn children, it was some time before civil society recognised its ancient preoccupations.[16] This is why for several years, many have believed in the originality of this new social movement before realising that it did not, underneath it all, pose any real threat. We remain humans, after all, despite taking nature into account. Consequently, as the old regimes regain their importance, the originality of ecology is being gradually eroded and its electoral favour dwindles with each election.

Another reason would make the failure of the environmental parties inevitable. Outside the 'civic regime', a party has no chance of situating itself within the classic framework of the left–right scenography. Trying to define a super-will is at once accepting the classic framework of political life, but hurtling toward defeat if one can only oppose the habitual spokespersons and electors with mute entities – birds, plants, ecosystems, catchment areas and biotopes – or specialists – scientists, fanatics, experts, activists – speaking in their name but on their own authority. Without a new type of spokesperson, natural entities have no voice or are only represented by a specialist knowledge that is incommensurable with public life.[17] By becoming a party, political ecology was forging ahead. But by rejecting party life, it would run the risk of becoming either a branch of the associated movements for domestic community or else a specific sector of industrial or market production.

226

SHOULD WE ABANDON THE PRINCIPLE OF COMMON HUMANITY?

To escape this horrible fate it would seem that there is but one solution, and that is to depart from the model of Boltanski and Thévenot by abandoning its principal axiom, that of common humanity. All the regimes developed by the six types of political philosophy have humanity as their measure. They disagree on how to rank humanity and about the yardstick that allows one to order smallness and greatness in each of the six 'Cités', but they all agree that 'humanity is the measure of all things'. This is what makes these six principles of justification, no matter how contradictory with one another, all completely incompatible with the racist or eugenic or social Darwinist reactionary politics developed during the last century. How is it possible to abandon the notion of common humanity, without immediately falling into the danger of 'biopolitics'? The standard answer is that ecology is no longer about humans – even extended to include future generations – but about nature, a higher unity which would include humans among other components associated with other ecosystems.

We saw above the political incoherence of this solution. How can political life be mixed up with a total unity – nature – which is only known by the science of complex systems? At best, one would arrive at a sort of super-Saint-Simonism, a government of experts, of engineers and of scientists who would abolish the difference between the 'civic regime' and 'industrial regime' by the controlled management of natural cycles. At worst, it would lead to an organicism which would abolish the difference between the 'domestic regime' and all the other regimes, and which would be prepared to sacrifice 'mere humans' to maintain the only truly worthy object: Mother Earth. Perish humanity so long as elephants, lions, snails, ferns and tropical rainforests recover their 'equilibrium' of yester-year: the permanently disequilibriating state of intense natural selection.[18]

It is difficult, one would imagine, to present oneself in front of one's electorate with a programme that envisages the possibility of making them disappear in favour of a 'congress of animals' who don't even vote or pay taxes! As for abandoning the framework of elections altogether, one could certainly do that, but it would be in the name of a fundamentalism that would abandon democracy once and for all. And to whose advantage? Leaders directly inspired by nature? Or mad scientists versed in the sciences of complexity? Faced with such an alternative, the reaction of the ordinary citizen is understandable: 'I would rather live a shorter life in a democracy than sacrifice my life today – and that of my descendants – to protect a mute nature represented by such people.' One can see the difficulty of discovering the 'seventh regime', which now resembles those cities, lost in the jungle, that the 'raiders of the lost ark' hoped to find.

Either one accepts the principle of common humanity, and then there is no longer the slightest originality in political ecology which reduces, with more or less difficulty, to the three (or six) other regimes. Alternatively, by retaining the originality of political ecology, i.e. its equal concern for non-humans and humans, one departs from the framework of the most elementary morality and the healthiest of democracies. Faced with such intellectual dilemmas, one can understand why the environmental parties have considerable difficulty explaining to themselves, to their members and to their electors the meaning of their fight.

WHAT IF ECOLOGY DID NOT CONCERN ITSELF WITH NATURE?

Perhaps we've taken the wrong route. Perhaps we have misunderstood the model that has guided us thus far. Perhaps we have too slavishly followed what political ecology says about itself without paying enough attention to its practice which, happily, differs greatly from its explanations of itself. It seems, in fact, that the originality of political ecology is a lot more subtle than we have so far imagined it to be.

Let us reconsider things by measuring the distance that separates practice from self-representation by setting up two contrasting lists: the first states what political ecology believes it ought to do without really managing to do; and the second sets out the advantages of *not* following the ideals that it flaunts with so much obstinacy.

What ecology believes it ought to do without managing to

Political ecology claims to talk about *nature*, but it actually talks about endless *imbroglios* which always involve some level of human participation:

- It claims to protect nature and shelter it from humans but, in all the empirical cases that we have read or studied, this actually amounts to greater human involvement and more frequent, increasingly subtle and more intimate interventions using increasingly invasive scientific equipment (Chase 1987; Western and Pearl 1989; Western *et al.* 1994).
- It claims to protect nature for its own sake – not as a substitute for human egotism – but at every turn the mission it has set itself is undertaken by men and women who see it through, and it is for the welfare, pleasure or conscience of a small number of carefully selected human beings that one manages to justify it.
- It claims to think with systems known by the laws of science, but every time it proposes to include everything in a higher cause it finds itself drawn into a scientific controversy in which the experts are incapable of coming to agreement.[19]
- It claims to take its scientific models from hierarchies regulated by cybernetic control systems, but it is always displaying surprising heterarchic assemblages whose reaction times and scales always catch off balance those who think

228

they are talking of fragility or of solidity, of the vast size or of the smallness of nature.

- It claims to talk about everything, but only succeeds in shaking up opinion and modifying power relations by attaching itself to particular places, biotopes, situations and events: two whales trapped in the ice, one hundred elephants in the Amboseli National Park (Cussins 1998) or thirty platane trees on the Place du Tertre in Paris.
- It claims to be becoming more powerful and to embody the political life of the future, but it is everywhere reduced to the smallest share of the electoral ejector and jump seats. Even in countries where it is a little more powerful, like Germany, it only brings to bear a secondary force.

One could despair at this severe appraisal. But one can also seize all the advantages that there would be if political ecology were to disabuse itself of its own illusions. Its practice is worth infinitely more than its utopian ideals of a natural super-regime, managed by scientists for the exclusive benefit of a Mother Earth who could at any moment become a cruel or unnatural mother.

Let us return to the list of its miscontruals, now considering the 'defects' of its practice as just so many positive advantages. The encrypted message which permits the discovery of the lost regime is immediately illuminated by a new meaning.

What ecology (happily) does extremely well

- Political ecology does not and has never attempted to talk about nature. It bears on complicated forms of associations between beings: regulations, equipment, consumers, institutions, habits, calves, cows, pigs and broods that it is completely superfluous to include in an inhuman and ahistorical nature. Nature is not in question in ecology; on the contrary, ecology dissolves boundaries and redistributes agents and thus resembles premodern anthropology much more than it thinks.[20]
- Political ecology does not seek and has never sought to protect nature. On the contrary, it wants to take control in a manner yet more complete, even more extensive, of an even greater diversity of entities and destinies. To the modernism of world domination, it adds modernism squared.[21]
- Political ecology has never claimed to serve nature for its own good, since it is totally incapable of defining the common good of a dehumanised nature. It does better than protect nature (either for its own sake or for the good of future generations). It suspends our certainties with regard to the sovereign good of human and non-human beings, of ends and means.
- Political ecology does not know what an eco-political system is and does not rest on the insights of a complex science whose model and methods would, anyway, if it existed, totally escape the reach of poor thinking and (re)searching humanity. This is its great virtue. It does not know what makes

and does not make up a system. It does not know what is and is not connected. The scientific controversies in which it becomes embroiled are precisely what distinguish it from all the other politico-scientific movements of the past. It is the only one that can benefit from another politics of science. Neither cybernetics nor hierarchy make it possible to understand the agents that are out of equilibrium, chaotic, Darwinian, as often as they are global, sometimes rapid, sometimes slow, that it brings into play via a multitude of original experimental devices whose mixed unity precisely does not – and this is the point – form an exact and definitive science.

- Political ecology is unable and has never sought to integrate all its very meticulous and particular actions into a complete and hierarchised unity. This ignorance with regard to totality is precisely its saving grace since it can never rank small human beings and vast ozone layers, or small elephants and middle-sized ostriches, into a single hierarchy. The smallest can become the largest. 'The stone that was cast aside has become the corner stone.'

Political ecology has, unfortunately, remained marginal until now because it has not yet grasped either its politics or its ecology. It believes it is speaking about nature, the system, a hierarchised totality, a world without human beings, a certain science, and it is precisely these too well-ordered statements that marginalise it, while the hesitant statements of its practice would perhaps permit it finally to attain political maturity if only it could grasp their meaning.

By comparing those two lists, one can see the new solution towards which we can now turn. If we leave aside the over-lucid explanations that ecology gives of itself, and focus solely upon its embroiled practical application, it becomes a completely different movement, a wholly other destiny. Political ecology makes no mention of nature, it does not know the system, it buries itself in controversies, it plunges into socio-technical imbroglios, it takes control of more and more entities with more and more diverse destinies, and it knows less with any certainty what they all have in common.

WHAT IS COMMON IN THE EXPRESSION 'COMMON HUMANITY'?

Before crying 'paradox!', an attempt should be made to explore this new avenue. Messages, even decoded, can have a double meaning. Now, if we return to the regimes model, we can see that, at the price of a fundamental but miniscule reinterpretation of the central axiom, the 'seventh regime', which had escaped our looking for so long, suddenly emerges like Merlin's castle.

What in fact is 'common' humanity? Boltanski and Thévenot were content with the usual reading offered by the canonical commentators of political philosophy they chose to consider. They took for granted the detached human offered to them by the humanist tradition, the human whose ultimate risk would be to be confused with a-human nature.[22] But non-human is not inhuman. If ecology has nature as its goal and not humans, it follows that there can be no regime of ecology. But if the aim of ecology is to open up the question of humanity, it conversely follows that there is a 'seventh regime'.[23] The meaning of the adjective 'common' in the expression 'common humanity' changes totally if the non-humans are not 'nature'.[24]

The question opened up by the 'seventh regime' is to know what would a human be without elephants, plants, lions, cereals, oceans, ozone or plankton? A human alone, much more alone even than Robinson Crusoe on his island. Less than a human. Certainly not a human. The regime of ecology does not at all say that we should shift our allegiance from the human realm to nature. That is why it has taken so long to find it, for that requirement appeared too absurd. The regime of ecology simply says that we do not know what makes the common humanity of human beings and that, yes, maybe, without the elephants of the Amboseli, without the meandering waters of the Drôme, without the bears of the Pyrenees, without the doves of the Lot or without the water table of the Beauce they would not be human.

Why don't we know? Because of the uncertainty concerning the relationship between means and ends. To define ecology, it might be sufficient, strangely enough, to return to the definition that Kant gives of human morality, a definition that is so well known that people forgot to see that it is in fact wonderfully apposite for non-humans. Let us get back to this most canonical of all definitions:

> Everything in creation which he wishes and over which he has power can be used merely as a means; only man, and, with him, every rational creature, is an end in himself. He is the subject of the moral law which is holy, because of the autonomy of his freedom. Because of the latter, every will, even the private will of each person directed to himself, is restricted to the condition of agreement with the autonomy of the rational being, namely, that it be subjected to no purpose which is not possible by a law which could have its origin in the will of the subject undergoing the action. This condition requires that the subject never be used simply as a means but at the same time as an end in itself.
>
> (Kant 1956: 90)[25]

The style is abominable, but the thought is clear. In this definition of morality only the first sentence which presupposes a creation composed of mere means presented to human ingenuity needs to be modified. Let us generalise to all the beings of the creation the aspiration to the kingdom of ends. What do we find? An exact definition of the practical connections established by ecologists with those they are defending: rivers, animals, biotopes, forests, parks and insects. They do not at all say that we should not use, control, serve, dominate, order, distribute or study them, but that we should, as for humans, never consider them as simply means but always also as ends. What does not hold together in Kant's definition is the truly incredible idea that simple means could exist and that the principle of autonomy and freedom would be reserved for man in isolation. On the other hand, what does not hold together in ecology's theories is the improbable belief in the existence of a nature external to humans and threatened by the latter's domination and lack of respect.[26]

Everything becomes clear if one applies this admirable Kantian sentence to elephants, biotopes and rivers: 'that [they] be subjected to no purpose which is not possible by a law which could have its origin in the will of the subject undergoing the action [say, the actor itself]. This condition requires that the subject [the actor] never be used simply as a means but at the same time as an end in itself.' It is this conjunction of actors who can never take each other as simple means which explains the uncertainty into which we are plunged by the 'seventh regime'. No entity is merely a means. There are always also ends. In other words, there are only mediators.

Let us come down from the heights of moral philosophy to listen to what the actors engaged in the defence of, for example, a river have to say. 'Before, water went its own way', says an elected representative, 'it was part of the furniture, it was part of the environment.' This paradoxical statement gives a clear indication of the status of water which, contrary to ecological myth, passes from the outside to the inside of the social world. Whereas it was a simple means, part of the furniture, it has now become the subject of political concern. To enter the realms of ecology, it must leave the environment. But the paradox is resolved by ecologists themselves: 'We are defending the fulfilment of the river, the river outside any human context, the river-river', says one activist, seeming to justify the outrage of the moralists and seeming to follow to the letter the mythologies of this social movement. But then he immediately adds: 'When I say the river outside of its human context, I mean the aggressive human context that treats the river solely as a tool.' And here he is applying Kant's slogan to the letter. He is not defending the river for its own sake, but he does not want it to be treated simply as a means.[27]

By adopting this perspective, one understands that the ambiguous phrases that seemed to be easily reducible above to the 'industrial regime' – because that regime does not take account of nature solely for itself but also for the good of humans – explores in fact a 'seventh' type of regime, by applying the (slightly rewritten) Kantian law. As one water authority engineer explained:

> You have to be extremely humble when dealing with a river. You pay for work which takes you the next thirty years to complete. In work carried out to increase productivity it's necessary to *get rid* of the water, to straighten, clean and calibrate – that was the watchword. We didn't know that rivers took their revenge by regressive erosion that we corrected with pseudo-natural sills. It's a slow process, there are still local agricultural authorities where a river after land consolidation

232

appears as a drainage ditch on the map! Fortunately, there is a great deal of pressure from anglers and nature conservationists. There is a clear generation gap; they all talk about the natural environment but, in the same corridor, you can have a bloke who makes everything straight and consolidates land with a vengeance, while another puts back in meanders and 'chevelus'.[28]

Such an analysis does not confirm either the notion of nature saved for its own sake by sacrificing human interests or that of free human beings dominating nature to promote their own freedom alone. A canalised river is seen as something bad and undesirable within the 'seventh regime', not because this futile development will be seen as expensive – taking thirty years to complete and being quickly eroded – but because the river has been treated as merely a means, instead of also being taken as an end. By conspiring with a 'law which could have its origin in the will of the subject undergoing ther action', according to the Kantian expression, rivers are allowed to meander again, to keep their dishevelled network of rivulets, to have their flood zone.[29] In short, we leave the mediators partially to deploy the finality which is in them.[30]

AN ALTERNATIVE TO MODERNISATION

This suspension of certainty concerning ends and means speaks to the question of how smallness and greatness are scaled in the 'green city'. In the 'green city' what is small is knowing for sure that something has or, conversely, has not a connection with another, and knowing it absolutely, irreversibly, as only an expert knows something. Something has value in the 'green city', something is great when it leaves open the question of solidarity between ends and means. Is everything interrelated? Not necessarily. We do not know what is interconnected and woven together. We are feeling our way, experimenting, trying things out. Nobody knows of what an environment is capable.[31]

One of the advantages of this definition of the scaling inside the green regime is that it removes an obstacle that had slowed everyone down in the march towards the lost city. In spite of its claims, fundamentalist ecology, or 'deep ecology,' occupies the state of Worthlessness in the 'seventh regime'. The more certain an ecology is that everything is interrelated, seeing humans simply as a means of achieving Gaia, the ultimate end, the more worthless that ecology. The more strident, militant and assured it is, the more wretched it is. Conversely, the state of greatness peculiar to this 'seventh regime' presupposes a deep-rooted uncertainty as to the nature of attachments, their solidity and their distribution, since it only takes account of mediators, each of which must be treated according to its own law.

One can understand how such an outcome has, for a long time, concealed the lost regime under a thick camouflage of foliage. Political ecology can only come to fruition on condition that those who have terrorised it thus far are reduced to their rightful place. Fundamentalist ecology has, for a long time, fulfilled the same role *vis-à-vis* political ecology as the Communist Party *vis-à-vis* socialism: a raising of the bidding so well justified that it paralysed its adversary/ally into believing it was too soft, too compromised, too much of a 'social traitor'. And yet there is no outbidding, no gradation of virulence in the political courage or radicality of the different movements, since deep ecology simply does not have a place in the regime of ecology – just as, conversely, there is no place for the tranquil certainty of the modernists who have, until now, released into external nature objects with no other purpose, no other risk than those they thought they knew all about it.[32]

One might be surprised that, to define the 'seventh regime', it is necessary to invoke the practice of the ecological movements and set it in opposition to the theoretical justifications of their followers. Nevertheless, the reason for this shortcoming seems clear to me. To justify the regime of ecology, it is necessary to be able to speak about science and politics in such a way as to suspend their certainties twice: with regard to subjects, on the one hand, and objects, on the other. All the other regimes clearly belong to the world of political philosophy. They are all anthropocentric. Only the 'seventh regime' forces us to speak about science and to plunge human beings into what makes them humans. But since enthusiasts of the sciences are loathe to undertake the task of justification, which would force them to throw out their epistemology, and since the partisans of the political sciences find that they need to know far too much science and need to be too interested in non-humans in order to give an account of these debates which completely escape the usual framework of public life, one cannot find authors who are interested in both.[33] In order to disentangle the 'green city', one has to deal at once with science and with politics and to disbelieve epistemology as much as political philosophy. This is why the regime of ecology is still waiting for its Rousseau, its Bossuet, its Augustin or its Hobbes.

In the new regime, everything is complicated and every decision demands caution and prudence. One can never go straight or fast. It is impossible to go on without circumspection and without modesty. We now know, for example, that if it is necessary to take account of everything along the length of a river we will not succeed with a hierarchised system that might give the impression, on paper, of being a wonderful science with wonderful feedback loops but which will not generate new political life. To obtain a stirring up of politics, you have to add uncertainty so that the actors, who until now knew what a river could and could not tolerate, begin to entertain sufficient doubts. The word 'doubt' is in fact inadequate, since it gives the impression of scepticism, whereas it is more a case of enquiry, research and experimentation. In short, it is a collective experimentation on the possible associations between things and people without any of these entities being used, from now on, as a simple means by the others.[34]

Political ecology, as we have now understood it, is not defined by taking account of nature, but by the different career now taken by all objects. A planner for the local agricultural authority, an irrigator, a fisherman or a concessionaire for drinking water used to know the needs of water. They could guarantee its form by assuming its limits and being ignorant of all the ins and outs. The big difference between the present and the previous situation does not lie in the fact that, before, we did not know about rivers and now we are concerned about them, but in the fact that we can no longer delimit the ins and outs of this river as an object. Its career as an object no longer has the same form if each stream, each meander, each source and each copse must serve both as an end and a means for those claiming to manage them.

At the risk of doing a little philosophising, we could say that the ontological forms of the river have changed. There are, literally speaking, no more things. This expression has nothing to do with a sentimentalism of Mother Earth, with the merging of the fisherman, kingfisher and fish. It only designates the uncertain, dishevelled character of the entities taken into account by the smallest river contract or the smallest management plan. Nor does the expression refer to the inevitable complexity of natural milieux and human–environment interactions, for the new relationships are no more complex than the old ones; if they were, no science, management or politics could be done on their behalf, as Florian Charvolin (1993) demonstrated so well. It solely refers to the obligation to be prepared to take account of other participants who may appear unforeseen, or disappear as if by magic, and who all aspire to take part in the 'kingdom of ends' by suddenly combining the relationships of the local and global. In order to monitor these quasi-objects, it is therefore necessary to invent new procedures capable of managing these arrivals and departures, these ends and these means – procedures that are completely different from those used in the past to manage things.

In fact, to summarise this argument, it would have to be said that ecology has nothing to do with taking account of nature, its own interests or goals, but that it is rather another way of considering everything. 'Ecologising' a question, an object or datum, does not mean putting it back into context and giving it an ecosystem. It means setting it in opposition, term for term, to another activity, pursued for three centuries and which is known, for want of a better term, as 'modernisation.' Everywhere we have 'modernised' we must now 'ecologise.' This slogan obviously remains ambiguous and even false, if we think of ecology as a complete system of relationships, as if it were only a matter of taking everything into account. But it becomes profoundly apposite if we use the term ecology by applying to it the principle of selection defined above and by referring it to the Kantian principle for the justification of the green regime.

'Ecologising' means creating the procedures that make it possible to follow a network of quasi-objects whose relations of subordination remain uncertain and which thus require a new form of political activity adapted to following them. One understands that this opposition of modernisation and ecologisation goes much further than putting in place a principle of precaution or prudence like that of Hans Jonas. Or rather, in defining the regime of ecology, we manage to select – from among the arguments of the principle of precaution – those which belong to the new political life and those which are part of the old repertoire of prudence. In ecology, it is not simply a matter of being 'cautious' to avoid making mistakes. It is necessary to put in place other procedures for politico-scientific research and experimentation.[35]

In contrasting modernisation and 'ecologisation' (it will obviously be necessary to find another term, which is less unwieldy and more inspirational and mobilising), one could perhaps escape the two contrary destinies with which we began. Political ecology can escape banalisation or over-inflation. It does not have to take account of everything and especially not nature, and in any case not nature for nature's sake. Nor does it have to limit its designs to the existence of a body of administrators responsible for the environment, just as other bodies are responsible for school health or for monitoring dangerous factories. It is very much a question of considering everything differently, but this 'everything' cannot be subsumed under the expression nature, and this difference does not reduce to the importation of naturalistic knowledge into human quarrels. To be precise, starting from the green regime and according to the Boltanski–Thèvenot method, the interplay of denunciations of the other regimes and the inevitable compromises to be agreed with them, one could perhaps drag political ecology from its present state of stagnation and make it occupy the position that the left, in a state of implosion, has left open for too long.

ACKNOWLEDGEMENTS

This chapter is an English version of an article originally published in French: Latour, B. (1995) 'Moderniser ou écologiser. A la recherche de la septième Cité', *Ecologie politique* 13: 5–27. It is part of a larger project by the Centre de sociologie de l'innovation on the novelty of political ecology. It is thus very dependent on the many case studies pursued there on water politics, waste management, the history of ecology and of political science. I owe a special debt to Charis Cussins and David Western who have shaped most of the arguments here presented (for which of course they are in no way responsible).

NOTES

1 This term does not have the same specific meaning in this essay as it does in Anglophone academic debates surrounding the political-economy of environmental use, as in, for example, Bryant (1992) and Peet and Watts (1996). Rather, it serves two purposes here. First, it is used as a general term used to signify the environmental movement as such or the green parties and groups who in various ways have sought to politicise environmental issues. Second, though, its meaning is reconfigured as the chapter proceeds: see note 7 below [the editors].

2 All the quotations by officials and activists on water used in this chapter are taken from a study by the Centre de sociologie de l'innovation on the novelty of political ecology. The new law of 1992 on water requires catchment of sensible rivers to be represented in 'Commissions locales de l'eau' (CLE), which are a very original experiment in the French context since they aim in part to make politically visible the river's health and sustainable well-being.

3 'Non-human' is my technical term to designate objects freed from the obligation to do politics through nature. Nature is here considered as what assembles all entities into one whole. It is thus a political definition that is sometimes opposed to human politics or, as is the case here, merged with politics. On the genealogy of this bizarre way of doing politics through the notion of a nature cast away from all human politics see Latour (1997).

4 For a comparison of health and ecology see D. S. Barnes (1994), W. Coleman (1982) and R. J. Evans (1987). The anthropocentrism of the nineteenth-century health movement clearly distinguishes it from ecology. Nobody championed the cause of miasmas and microbes.

5 Apart from the many reasons specific to France developed in A. Roger and F. Guéry (1991), France is interesting because the idea of a nature untouched by human hands does not have the evocative strength that it has in the USA or Germany.

6 Bryan Wynne in England, Charis Cussins in the USA, Camille Limoges and Alberto Cambrosio in Quebec, Rémi Barbier and Laurent Thévenot in France, and several others, have begun to collect detailed analyses on the practical work of militant ecologists. It would be interesting to make a systematic comparison which, to my knowledge, has not be attempted. But see Western et al. (1994) for the case of 'community based conservation'.

7 I have used the term 'political ecology' patterned out of the very well-known term 'political economy' to designate not the science of ecosystems – ecology – nor the day-to-day political struggle – green parties – but the whole intersection of political philosophy of human and non-humans. In the course of this chapter the meaning is going to shift from a concern for nature to a concern for a certain way of handling associations of human and non-humans that would be an alternative to modernisation. Hence the rather idiosyncratic sense of the expression. For two militant but directly opposed classifications, see M. W. Lewis (1992) and C. Merchant (1992).

8 The book thus offers a general grammar of indignations that accounts for one of the most puzzling features of contemporary societies: the intensity of moral disputes, the absence of one overarching principle that would include all the others, the ease with which, none the less, every member passes judgement as if there existed one such unique principle. The work of Boltanski and Thévenot is the first in sociology to take seriously the work of justification that is a central part of social action. But they do not simply add moral and political considerations to the study of social forces. They have found a very original and productive way to compare moral and political actions.

9 I was inspired by similar attempts to use the same model by Barbier (1992), Lafaye and Thévenot (op. cit.) and O. Godard (1990)

10 See P. Descola (1986, English translation 1993) and all the work carried out by that author since 1986 on the appropriation of the social world, especially his articles on the non-domestication of the peccary in Latour and Lemonnier (1994) and in Descola and Pallson (1996).

11 It should be remembered that the regimes model makes it possible to classify human beings, from the most lowly to the most elevated, according to a principle that is constant inside each of the 'Cités' but which varies from one regime to the next. 'Smallness' and 'greatness' ('petitesse' and 'grandeur') are thus at once both ordered and multiple. Someone 'small' in one regime maybe 'great' in another. This is the

source of most denunciations and what allows the grammar of indignations to be mapped out.

12 In the industrial regime greatness is achieved by efficiency, and smallness by waste. Here is a typical comment by a Department of Agriculture representative concerning the treatment of the River Gardon: 'The river has been completely destroyed by flood channels, which were cleared with the approval of government departments. This complete destruction serves no purpose in the event of flooding, and destabilises the river – to the point that ground sills have had to be constructed – by causing part of the water table to disappear: this is an absurd system.' This high offcial does not pit the river *per se* and its interest against the human needs for order and efficiency. On the contrary, he takes the new respect for the river's own impetus as one way to gain a faster, less expensive and less wasteful leverage on the other agents. The appeal to the river is here clearly reducible to the ancient industrial order as in this excerpt with another high official – a polytechnician in charge of one of the water basins: 'Engineers only think about the anthropic aspect of things; they can't realize that on the long range the respect for Nature will be beneficial; it does not cost more to be soft or to be hard, except that the soft approach requires much more work and attention at the beginning before the companies are fully trained.' This engineer adds an automat to all the automats that make up the world as in this sentence where he explains why he has been converted to the softer sustainable development approach: 'I have been converted by the aesthetic aspect of things, by the protection of the landscape, then by ecology; in terms of long-term management, it is better with a river that self-regulates itself than with a river that is degradating itself all the time.'

13 Two opposite points of view are clearly expressed, the first by a staunchly militant ecologist, and the second by an elected – communist – representative and teacher: 'Elected representatives protect their electors, we are protecting a population in its environment, in its totality, everyone else is protecting their own interests, their own particular clique, even fishermen protect their fish, only ecologists are disinterested.' The other replies: 'When you create facilities, you automatically make enemies, it's part of being a statesman, it's what politics is all about. I am not an enemy of the ecologists, but there is a collective interest that must come *before* individual interests.'

14 This is the solution explored by Godard (*op. cit.*). See also the classic work of E. Weiss-Brown (1989). Witness the increase in generality on the part of the mayor of a tiny village in the Côte d'Or region of France who is addressing a local meeting on water. He turns to a Cistercian monk – who is present in the local parliament of water because his monastery has been diverting water from the river since the twelfth century! – to call him to witness: '"Be fruitful, and multiply and control the Earth." That's in the Bible! Father Frédéric will not say otherwise, it is essential for our grandchildren to have clean water.' (We can note in passing that the theological theme of the creation is interpreted here in a somewhat contradictory manner since, in giving freedom to his creature, God gave man a level of control that he denies himself to his fellow creatures. We only have to treat nature as our Creator treated us, to completely overturn the supposed link between Christianity and control over nature.)

15 Witness this remark by one of the few French elected representatives who is an ecologist, and boldly combines a concern for nature with civic concern for the region and concern for the market economy: 'Upstream the region Limousin wants the most natural river water and environment possible, not for itself but for economic development. The preserved part of the environment is our trump card, we cannot make up for thirty years of heavy industry, we must not oppose ecology and economy, we are not yet polluted, we have 700,000 inhabitants, we can play the quality-of-life card.'

16 How long will it be before the self-interest anthropocentrism behind this phrase will

be recognised: 'The river Gardon is an umbilical cord, we are all very much attached to it, in the final analysis we have neither the right to pollute it, nor to harness it, so as not to deprive others of an element that they need, we will invitably have to work out a way of sharing'? Or behind this other phrase that gives the river free rein while at the same time draining European Community funds: 'On the lower river Doubs farmers wanted to keep the river in check with stone pitching, but the policy was blocked in favor of creating a free meandering section of the river, where farmers change their crops in order to receive subsidies under the European Community article 19 on agro-environmental measures'?

17 Scientific knowledge continues to remain, with extremely rare exceptions, a black box in the eco-movements, where the social sciences rarely serve as a point of reference for opening controversies between experts. See Latour *et al.* (1991).

18 For a detailed criticism of the theory of natural balance see D. B. Botkin (1990). For its history, see J-M. Drouin (1991).

19 For a caricature of an appeal to scientism that is none the less unable to eliminate scientific controversies, see Ehrlich and Ehrlich (1997).

20 See P. Descola (*op. cit.*) and, for a recent analysis, M. Strathern (1992). See also Western *et al.* (1994) on 'community based conservation' and the recent work of Charis Cussins (*op. cit.*).

21 A position which is particularly clear in Lewis (1992). See also Latour (1994b) on this constant involvement.

22 This is what Luc Ferry did with great efficacy, successfully killing much of the French intellectuals' interest in ecology (Ferry 1995).

23 As we will see below, deep ecology is no more part of ecology than the Cartesian forms of humanism because it does close off the question that was just reopened, by stating unequivocally that 'humanity is obviously part of nature'.

24 In fact 'nature' is merely the uncoded category that modernists oppose to 'culture' in the same way that, prior to feminism, 'man' was the uncoded category opposed to 'woman'. By coding the category of 'natural object', anthropological science loses the former nature/culture dichotomy. Here, there is obviously a close link with feminism. See D. Haraway (1991). Nothing more can be done with nature than with the older notion of man.

25 L. Ferry (1995) rightly wanted to refer to Kant, but chose the wrong critique, opting for the aesthetics of the third rather than the morality of the second.

26 Since the classic work of C. D. Stone (1985), lawyers have gone much further than political philosophers in the invention of partial rights that turn simple means into partial ends. See, for example, M-A. Hermitte (1996) on the tainted blood scandal which is much more typical of 'ecological' issues in France than anything related to 'nature'.

27 Rivers are a wonderful source of conflict between the 'civic' and 'green' regimes. Since large towns and cities are usually situated on their lower reaches, the general will rapidly reach an agreement to sub-represent the depopulated, rural upper reaches.

28 'Chevelus' is the technical term used in French to describe the network of rivulets that have the shape of dishevelled hair and are visible either in flood zones, in deltas or near the sources.

29 There is no anthropomorphism in the reference to the river taking its revenge, merely the sometimes painful revelation of a being in its own right with its own freedom and its own ends. A surprising remark from a water specialist, trained from his youth in the culture of the waterpipe and who admits: 'Nobody imagined that their isolated actions would have repercussions, nobody thought we could dry up the river, nobody thought that removing the gravel in one place would lay bare the foundations of the

bridge in the village of Crest twenty kilometres away. You have to experience extreme situations before you realize.'

30 We must obviously return to the difference between necessity and freedom and invest the sciences with a role that is both more important and more anthropological. See B. Latour (1996).

31 An important advantage of this regime is that it can absorb Darwinism which, of course, has nothing to do with social Darwinism, that is only too well acquainted with the distinction between ends and means, as well as understanding all too easily how to create a hierarchy of the strong and the weak, a ranking that is impossible when all forms of teleology are abandoned. See S. J. Gould (1989).

32 Witness this remark by a technician: 'My predecessor was very much a "harnesser" ... we were technicians, we harnessed water, full stop.' He adds, to emphasise the complexity of a regime that now only has mediators and can no longer simplify life by going 'straight ahead': 'Now things have gone too far in the other direction and you can't do anything any more.'

33 Ethics and law, on the other hand, are extremely well developed but leave the question of scientific objects intact. Even those who, like Stone (*op. cit.*), are interested in things, do not include the production of facts and the emergence of objects in their analyses. Only Serres has tried, in his own idiosyncratic way, to make the connection between the scientific status of objects and the legal status of people: M. Serres (1995). Ulrich Beck (1995) is one of the very few thinkers of the ecological crisis to take into account the sociology of science.

34 This is the great interest of the work developed by Beck (see, for example, 1995), because he extends risk very far, away from 'nature', and makes it a whole theory of what he calls 'reflexive modernity' and that I would prefer to call 'non-modernity'.

35 This argument is developed in B. Latour (1994a).

REFERENCES

Barbier, R. (1992) 'Une cité de l'écologie', DEA thesis, EHESS.

Barnes, D. S. (1994) *The Making of a Social Disease: Tuberculosis in 19th Century France*, Berkeley: California University Press.

Beck, U. (1995) *Ecological Politics in the Age of Risk*, Cambridge: Polity Press.

Boltanski, L. and Thévenot, L. (1991) *De la justification: Les économies de la grandeur*, Paris: Gallimard.

Botkin, D. B. (1990) *Discordant Harmonies: A New Ecology for the 20th Century*, Oxford: Oxford University Press.

Bryant, R. (1992) 'Third World political-ecology', *Political Geography* 11: 12–36.

Charvolin, F. (1993) 'L'invention de l'environnement en France (1960–1971)', Les pratiques documentaires d'agrégation à l'origine du Ministère de la protection de la nature et de l'environnement, Ecole nationale supérieure des mines de Paris.

Chase, A. (1987) *Playing God in Yellowstone: The Destruction of America's First National Park*, New York: Harcourt Brace.

Coleman, W. (1982) *Death is a Social Disease: Public Health and Political Economy in Early Industrial France*, Madison: University of Madison Press.

Cussins, C. (1998) 'Elephants, biodiversity and competing models of science: Amboseli National Park, Kenya', in J. Law and A-M. Mol (eds) *Complexity in Science, Technology and Medicine*, Durham, NC: Duke University Press.

Descola, P. ([1986]1993) *In the Society of Nature: Native Cosmology in Amazonia*, Cambridge: Cambridge University Press.

Descola, P. and Palsson, G. (eds) (1996) *Nature and Society: Anthropological Perspectives*, Routledge: London.

Drouin, J.-M. (1991) *Réinventer la nature. L'écologie et son histoire*, Paris: Desclée de Brouwer.

Ehrlich, P. R. and Ehrlich, A. H. (1997) *Betrayal of Science and Reason: How Anti-Environmental Rhetoric Threatens Our Future*, Washington, DC: Island Press.

Evans, R. J. (1987) *Death in Hamburg: Society and Politics in the Cholera Years 1830–1910*, Harmondsworth: Penguin Books.

Ferry, L. (1995) *The New Ecological Order*, Chicago: University of Chicago Press.

Godard, O. (1990) 'Environnement, modes de coordination et systèmes de légitimité: analyse de la catégorie de patrimoine naturel', *Revue économique* 2: 215–42.

Gould, S. J. (1989) *Wonderful Life: The Burgess Shale and the Nature of History*, New York: W. W. Norton.

Haraway, D. J. (1991) *Simians, Cyborgs, and Women: The Reinvention of Nature*, New York: Routledge.

Hermitte, M.-A. (1996) *Le sang et le droit. Essai sur la transfusion sanguine*, Paris: Le Seuil.

Kant, I. (1956) *Critique of Practical Reason*, trans. L. W. Beck, New York: Liberal Arts Press.

Lafaye, C. and Thévenot, L. (1993) 'Une justification écologique? Conflits dans l'aménagement de la nature', *Revue Française de Sociologie* 34: 495–524.

Lascoumes, P. (1994) *Eco-pouvoir. Environnements et politiques*, Paris: La Découverte.

Latour, B. (1993) *We Have Never Been Modern*, Cambridge, MA: Harvard University Press.

—— (1994a) 'Esquisse du parlement des choses', *Ecologie politique* 10: 97–107.

—— (1994b) 'On technical mediation', *Common Knowledge* 3: 29–64.

—— (1996). 'On interobjectivity – with discussion by Marc Berg, Michael Lynch and Yrjo Engelström', *Mind, Culture and Activity* 3: 228–45.

—— (1997) 'Socrates' and Callicles' settlement or the invention of the impossible body politic', *Configurations* 5: 189–240.

Latour, B. and Bourhis, J.-P. L. (1995) *Comment faire de la bonne politique avec de la bonne eau? Rapport sur la mise en place de la nouvelle loi sur l'eau pour le compte de la Direction de l'eau*, mimeo, Centre de sociologie de l'innovation, Paris.

Latour, B. and Lemonnier, P. (eds) (1994) *De la préhistoire aux missiles balistiques – l'intelligence sociale des techniques*, Paris: La Découverte.

Latour, B. *et al.* (1991) 'Crises des environmements: défis aux sciences humaines', *Futur antérieur* 6: 28–56.

Lewis, M. W. (1992) *Green Delusions: An Environmentalist Critique of Radical Environmentalism*, Durham, NC: Duke University Press.

Lovelock, J. E. (1979) *Gaia: A New Look at Life on Earth*, Oxford: Oxford University Press.

Merchant, C. (1992). *Radical Ecology: The Search for a Livable World*, London: Routledge.

Peet, R. and Watts, M. (eds) (1996) *Liberation Ecologies*, New York: Routledge.

Roger, A. and Guéry, F. (eds) (1991) *Maîtres et Protecteurs de la Nature*. Le Creusot, Champ Vallon (diffusion La Découverte).

Serres, M. (1995) *The Natural Contract*, Ann Arbor: University of Michigan Press.

Stone, C. D. (1985) '"Should trees have standing?" revisited: how far will law and morals reach? A pluralist perspective', *Southern California Law Review* 59: 1–154.

Strathern, M. (1992) *After Nature: English Kinship in the Late 20th Century*, Cambridge: Cambridge University Press.

Weiss-Brown, E. (1989) *In Fairness to Future Generations*, New York: Transnational Publishers.

Western, D. and Pearl, R. (eds) (1989) *Conservation for the 21st Century*, Oxford: Oxford University Press.

Western, D. *et al.*, (eds) (1994) *Natural Connections. Perspectives in Community-Based Conservation*, Washington, DC: Island Press.

11

NATURE AS ARTIFICE AND ARTIFACT

Michael Watts

The Ogoni people were being killed all right, but in an unconventional way.... The Ogoni country has been completely destroyed by the search for oil.... Oil blowouts, spillages, oil-slicks, and general pollution accompany the search for oil.... Oil companies have flared gas in Nigeria for the past thirty years causing acid rain.... What used to be the bread basket of the delta has now become totally infertile. All one sees and feels around is death. [Petrolic] degradation has been a lethal weapon in the war against the indigenous Ogoni people.

(Ken Saro-Wiwa, cited in Nixon 1996: 44–5)

INTRODUCTION

On 10 November 1995, Ken Saro-Wiwa and eight other prisoners – all residents of oil-rich, and oil-devastated Ogoniland in Rivers State in the southeast of Nigeria – were awakened at dawn, shackled at their ankles, and transported from Bori military camp where they had been held during their murder trial, to Port Harcourt Central Prison. Saro-Wiwa, an internationally recognized novelist, environmental activist and leader of the Movement for the Survival of Ogoni People (MOSOP) was granted his last rites by a sobbing priest, and surrendered his remaining property including his trademark pipe. In a moment of darkest farce, the executioners had presented themselves at the prison only to be turned away because their papers were not in order. Dressed in a loose gown and a black headcloth, Saro-Wiwa was, after this interregnum, led to the gallows. The pit into which Saro-Wiwa fell was shallow and the fall failed to break his neck. It took him twenty minutes to die. A videotape of the hanging was sent by courier to General Abacha, head of the Nigerian military junta, as proof of Saro-Wiwa's death. Seven others, who were also found guilty of the murder of four prominent Ogoni leaders by a kangaroo court hastily convened by the military government, suffered a similar fate. The executioners were said to have poured acid on the corpses to speed decomposition and to discourage Ogoni activists from taking possession of the bodies. Within hours of the hanging, 4,000 troops were deployed throughout Ogoniland – a Lilliputian area of 400 sq. miles containing half a million people and almost one hundred oil wells. Nigeria's Kuwait was in effect under military occupation. Special military forces beat any person caught mourning MOSOP's deceased leadership in public and embarked upon a systematic attempt to erase any trace of Saro-Wiwa's influence.

Saro-Wiwa's murder and the oppositional movement which he spoke for were framed by post-colonial Nigeria's tense and difficult relationship to petroleum, nature's black gold. At independence in 1960, Nigeria depended upon oil of an altogether different sort: peanut oil from the lowly groundnut. Peanut production, like so much else in Nigeria, occurred within a profound set of regional sensibilities and loyalties. The national political economy consisted of three semi-autonomous regions – each associated with a primary export commodity – and a marketing board which taxed agriculture to fund political elites who mobilized support largely through forms of ethnic identification. The advent of a turbo-charged petrolic capitalism turned this world upside down. By 1980 petroleum dominated Nigeria's economy more than peanut oil ever had. Moreover, oil unleashed an unprecedented boom. Federal revenues grew at 26 per cent per annum during the 1970s and state investment expanded vastly, creating a torrent of imports and urban construction. The proliferation of everything from stallions to stereos produced a commodity boom that one commentator aptly called a Nigerian "cargo cult."

Fifteen years later, the luster of the oil boom had tarnished. The collapse of the boom in 1981 and a foreign exchange crisis were compounded by mounting external debt obligations. By 1982 the president and his advisors talked of the need for sacrifice and denial; the 1970s, they said, had been a time of "illusion." They were promptly ousted in a military coup in 1983; but the boom was over and the watchword was austerity. Two years later Nigeria signed a structural adjustment program (SAP) with the IMF and the World Bank. The medicine was bitter: the economy contracted, the naira collapsed from $1.12 to 10 cents, and the real wage of industrial workers was savaged. Money was scarce and popular discontent widespread, especially in the anti-SAP riots in 1988 and 1989. In the 1990s debt service ran on average at 25 per cent of export revenues, the currency (the naira) collapsed in value, and inflation galloped ahead at 57 per cent annually over the period 1992 to 1995. Manufacturing output fell from an index of 100 in 1990 to 89 in 1995, while average industrial capacity stood near 29 per cent.

The "oil fortress" had been rocked (*Le Monde* 2 October 1994: 15). As the *Economist* (8 June 1996: 48) noted: "People complain fiercely, even by Nigerian standards, that their wealth and their future have been stolen."[1] Today, at least half the population lives below the poverty line and almost one-third of children die before the age of 5. What began with petro-euphoria and bountiful money in 1973 ended two decades later with scarcity, a huge debt, the deepening of economic inequalities, urban looting and bodies in the streets (Lewis 1996). This roller-coaster economy had fundamentally shaped the everyday life of all sectors of Nigerian society – seemingly for the worst.[2]

Saro-Wiwa's murder and the history of MOSOP is inexplicable outside the local rhythms of petroleum boom and bust in Nigeria. Furthermore, this history is one of what I shall call natural and political economy. Natural in the sense that the black gold pumped unadulterated from the swamps of the Niger Delta was quite literally part of the Ogoni natural world; and also in the sense that petroleum exploitation devastated the environmental basis of longstanding forms of Ogoni livelihood. And political because the key actors in the local petroleum economy – Shell and the federal state apparatus – became the oppositional reference points for the Ogoni movement and its struggles for compensation, local political autonomy, self-determination and joint revenue sharing. Running through these natural and political economies was the irreducible fact that oil itself is – from the vantage point of the Ogoni – a complex and ambiguous entity. As a "natural" resource of Ogoniland, it is identified with the core cultural claims and representations of what Ogoniland represents; yet as a source of unimaginable power, wealth and development, it also brings with it unprecedented destructive urges. It is the simultaneity of these qualities – oil as both natural and artifactual – which renders the role of petroleum such a difficult and contested presence in Ogoni lives.

My concern also is to locate these ambivalences, tensions and contradictions on the larger landscape of environmental movements in the South, and their central relation to so-called anti- or alternative development[3] (Escobar 1995; Pieterse 1996). In invoking opposition and alterity, I wish to engage with a substantial and diverse body of work which takes conventional development as a failed modernist project: "You must be either very dumb or very rich if you fail to notice," says Gusavo Esteva (1992: 7), "that 'development' stinks." The alternatives to development reside in the resistance to modernity generated by social movements – typically environmental in some profound sense – which also figure centrally in the post or alternative development imaginary. Nature is one source, then, of thinking about other social orders.

Standing at the heart of this chapter is an argument about green movements and, specifically, their polyvalency. In the case of the Ogoni, the movement is environmental, ethnic, political, and cultural, and this permits us to explore the complexity of the politics of produced natures as well as the politics of writing theories of produced nature. Standing at the center of the Ogoni struggle is oil (Watts 1994) which is necessarily and unavoidably artifactual – a product of science, technologies and social relations – and natural (crude black gold). If oil as nature is in a strong sense constructed – oil as a set of discourses, as a form of wealth and value, as an embodiment of social relations in the form of the state and transnational capital – it is also the case that some fundamental social identities – the Ogoni people, the Nigeria nation state, Shell oil company – cannot be understood apart from nature, that is, apart from oil as a natural resource. It is not simply that these central forms of identification are contested on the social and ecological landscapes of late twentieth-century Nigeria, it is that all these identifications are, as it were, channeled through nature, through the oil nexus. One might say that social identities and nature constitute what Latour (1993) calls an "imbroglio," networks of hybrid identities and quasi objects. But *pace* Latour, I have chosen to emphasize the ways in which objects, networks and identities are built and how such construction matters, not the least of which for the Ogoni. This seems to me important because at a certain point the complex identities and natural relations passed through the body of Saro-Wiwa – hence why I began with his hanging – which becomes not only a site within these networks where cultural politics is materialized, but the object of state violence. It is here that the politics, and the costs, of some sorts of hybrid identity become brutally clear.

In taking apart the Ogoni movement, and in exploring how nature, nation, statehood, citizenship and ethnicity are constituted within the contested social and ecological landscape of oil in late twentieth-century Nigeria, I shall also touch upon four central issues in postcolonial politics and theory (both of which have been surprisingly silent on the question of the relations between nature and identity): first, how nature is constructed in relation to the movement itself; second, the purportedly hybrid nature of the movements, and how hybridity might be understood; third, the new ways of doing politics and the centrality of the community in such political doings; finally, the relations between green movements and modernity.

NATURE, SOCIAL MOVEMENTS AND THE POST-DEVELOPMENT IMAGINARY

A striking consequence of radical economic restructuring in the South in the wake of the 1980s debt crisis, and in the post-socialist bloc after 1989, has been "the resurrection, reemergence and rebirth of . . . civil society" (Cohen and Arato 1992: 29). Partly in response to state contraction, and partly as an outcome of an uneven democratization process, various forms of local and community movements have emerged in the interstices of the state–society nexus. Fred Buttel (1992) sees these new social movements (NSM) as "new" insofar as they represent a sort of postmodern politics outside of, and in many respects antithetical to, class or social-democratic party politics. Many are also new in that they are an integral part of a widespread "environmentalization" of institutional practices. Enormously heterogeneous in character and scale (anti-dam movements, squatter initiatives, minority cultural movements), these grassroots movements often focus on efforts to take resources out of the marketplace, to construct, as Martinez-Alier (1990) would have it, a sort of moral economy of the environment. In much of this literature the label "environmental" is misleadingly narrow, since the proliferation of grassroots and NGO movements often focus more broadly on livelihoods and justice. Indeed, it is striking how indigenous rights movements, conservation politics, food security, the emphasis on local knowledges, and calls for access to and control over local resources (democratization broadly put) course through environmental struggles. This multi-dimensionality is, according to Arturo Escobar (1995), indicative of a new mode of doing politics, involving so-called "autopoietic" (self-producing and self-organizing) movements which exercise power outside the state arena and which seek to create "decentred autonomous spaces."

The local community and grassroots environmental initiatives loom very large in post-structural approaches to development (Parajuli 1991; Shiva 1993; Booth 1994; Escobar 1995). What they represent is certainly a form of collective action, but more specifically and profoundly a "resistance to development" (Escobar 1995: 216; also Routledge 1994) which attempts to build new identities. The implication is, of course, that these identities fall outside of the panoptic gaze of the hegemonic development discourse as new forms of subjectivity which stand opposed to – or ambiguously with respect to – modernity itself. As Escobar puts it (1995: 216), these movements are not cases of "essentialized identity con-struction" but are "flexible, modest, mobile, relying on tactical articulations arising out of the conditions and practices of daily life." To the extent that these movements are "environmental" or "green," very substantial claims have been made on their behalf: they are a "revolt against development" (Alvarez 1992: 110); "a new economics for a new civilization" (Shiva 1989: 24); and "learning to be human in a posthuman landscape" (Escobar 1995: 226). What then is the new content of such movements and what are their relations to anti-development?

At least five fundamental and overlapping properties have been attributed to environmental movements as vehicles of another development. First, they purportedly contain new sorts of politics and new sorts of political subjectivity. They are typically local, outside of the organized state sphere and "without one particular ideology or political party" (Escobar 1992b: 422). They are "self-organizing and self producing" (ibid.), exercising non-state forms of power. Second, "cultural difference is at the root of postdevelopment" (Escobar 1995: 225) and hence the movements are, above all, examples of popular cultural discourse. Minority cultural communities figure centrally in both green and anti-development movements. Indeed, the "indigenous" becomes the lodestar for the "unmaking of the Third World." Indian confederations in Latin America or "ethnic" green movements in Africa often turn on the ways in which cultural identity is mobilized as "a transformative engagement with modernity" (Escobar 1995: 219) in which "the greatest political promise for minority cultures is their potential for resisting and subverting the axiomatics of capitalism and modernity in their hegemonic forms" (p. 224). Third, the movements employ, in creative ways, local or subaltern reservoirs of knowledge. The proliferation of the field of "indigenous technical knowledge" (ITK) and the so-called actor-oriented interface analysis is another (see Arce *et al.* 1994). In singing the praises of this subaltern science position, women's knowledge and nature are often central. In

Shiva's (1989: xvii) words: "women as victims of violence of patriarchal forms of development have risen against it to protect nature" and by virtue of their organic relationships to things natural have a "special relationship with nature" (p. 43). Indeed for Shiva, feminine/ecological ways of knowing are "necessarily participatory" (p. 41). Fourth, local community and "tradition" are neither erased nor preserved as the basis for alternative development but are refashioned as a hybrid: hybridity entails "a cultural (re)creation that may or may not be (re)inscribed into hegemonic constellations" (Escobar 1995: 220). This is the heart of the new political subjectivity which speaks to a "transcultural in-between world reality" (ibid.). And finally, these movements produce a defense of the local: such a defense is a "prerequisite to engaging with the global . . . [and represents] the principal elements for the collective construction of alternatives" (p. 226).

To the extent that these movements from the South are claimed to be both environmental and new, they should be subject to careful scrutiny. Clearly, there is a long history of grassroots initiatives, and these have been emboldened by the new spaces opened up in the 1980s by the contraction of the state and the democratic opening worldwide (Seabrook 1995). Equally, some of these movements in the South are "new" in the sense that they do not conform to a simple Eurocentric model of single-issue struggles. As an Indian worker quoted in Seabrook (1995: 103) says: "You cannot separate work from all other aspects of life: all must be integrated into a single struggle." Escobar properly points out that, the new "green" movements are more than "environmental" insofar as they seem to link in complex ways a number of social justice and cultural issues. Many of these non-governmental organization (NGO) activities are contributing to a new sort of internationalism through global electronic networking and solidarity activity (Keck 1995). But a central weakness of the social movements as an alternative approach is precisely that greater claims are made for the movements than the movements themselves seem to offer. Indeed, the sorts of claims made by Shiva – that ecological knowing (and local knowledge in general) is "necessarily participatory" or that women have a special relationship with nature which organically produces conservation and protection of it – express exactly the sorts of essentialisms that she (and others) attribute to retrograde Eurocentric discourses. In much of this work, culture and popular discourse from below is privileged uncritically; identity politics is championed by Escobar because it represents part of an alternative reservoir of knowledge and because such ideas stand against the "axiomatics of capitalism." But there is surely nothing necessarily anti-capitalist, anti-modern or particularly progressive about cultural identity: calls to localism can produce Hindu fascism as easily as Andean Indian co-operatives. Running through much of the social movements as alternative to development literature is an uncritical appeal to the "people" – that is to say populist rhetoric – without a sensitivity to the potentially deeply conservative aspects of such local particularisms.

A striking feature of much of the work on environmental movements in the South is therefore the uncritical appeal to the local, to the popular, and to the cultural (where cultural is synonymous with a local sense of community or simply ethnicity). Yet as Pierre Bourdieu has noted, in discussion of "the people" and "popular" discourse, what is at stake is the struggle between intellectuals (1990: 150). These debates among intellectuals celebrate, in a quasi-mystical way, the efficacy of all action/knowledge from below. They contain a rejection of a *fin-de-siècle* modernity rooted in the losses and reaffirmations of local particularisms which are and always have been the accompaniment of capitalism. Moreover, they often forget that the "local" is never purely local, but is created in part by extra-local influences and practices over time. And to this extent there is the danger that the alternatives to development school has, to quote Marshall Berman (1982: 17), "lost touch with the roots of its own modernity."

By interrogating the case of MOSOP and the Ogoni struggle – a movement which has been lauded like Chipko in India as a compelling case of a hybrid, multi-dimensional "environmental" movement rooted in the powers and knowledges of the subaltern – I seek to engage with the particular ways in which nature is interpolated in the practical politics of the movement.

OIL, OIL POLITICS AND DEVELOPMENT IN POST-COLONIAL NIGERIA

While the promise of the piece of the "oil" pie keeps Nigeria together as a nation, the very nature of its distribution has destroyed the social, economic and political fabric from within.

(Sara Ahmad Khan 1994: 8)

Prospecting for oil began in Nigeria in 1908 when a short-lived German corporation commenced drilling along the southeastern coast, but the first commercial oilfield was not discovered until 1956 in tertiary sediments in the Niger Delta basin at Oloibiri, 90km west of Port Harcourt. Eight years later the first offshore oilfield was located off the Bendel state coast. By 1996 Nigeria was producing just less than two million barrels per day, about 2.9 per cent of the world's total from 176 oilfields and some 2,000 producing wells. While Nigerian petroleum is typically seen to be relatively high cost (roughly two to three times higher than the Middle East), the quality of its marker crude, so-called Bonny Light, is high due largely to its low sulphur content. The Niger Delta basin contains 78 fields (including the largest field at Forcados) spanning 75,000 sq km across Rivers, Edo, Imo, Abia, Akwa-Ibom, Cross River and Ondo states, and remains the most prolific and important oil-producing area in the country (Figure 11.1).

As of January 1996 there were fourteen oil-producing companies in Nigeria. All production has Nigerian participation (usually through the state oil company founded in 1971 as the Nigerian National Oil Company) in which eleven foreign oil companies operate through a complex concession system. During the

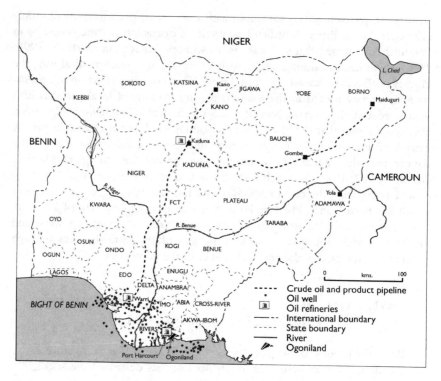

Figure 11.1 The geography of Nigerian oil production

Shell–BP concession era (1921–55), oil companies determined price and production in the industry (Ikein 1990). Until the mid-1960s, the state confined its activities to the collection of taxes and rents and royalties. In effect, throughout the 1970s there was a gradual process of nationalization undertaken via the state petroleum corporation (renamed the Nigerian National Petroleum Corporation [NNPC] in 1977) which increased its equity stake to 35 per cent in 1971 and finally to 60 per cent in 1979. The current leasing situation, in which oil-prospecting licences and oil mining releases are dispensed to foreign companies, primarily takes the form of joint-venture contracts in which NNPC and the company shares the costs of investment, exploration and production in direct proportion to their participation stakes.[4] From a peak production of 2.3 million barrels/day in 1979, the output and price levels have fallen – dramatically so in the mid-1980s when oil exports receipts were about one-fifth of 1980 levels. In 1995 the oil sector accounted for $9.15 billion, roughly 96.6 per cent of all Nigerian exports by value and close to 80 per cent of all state revenues.

Oil has produced an unequivocal centralizing dynamic in the Nigerian political economy – indeed it converted Nigeria into a strikingly single-commodity economy on which the state directly depended. But oil's wider consequences can only be grasped in terms of the complex political geography into which the first petroluem export revenues were inserted. Nigeria is a huge and heterogeneous country. Among its 96 million inhabitants, there are an estimated 370 ethnic groups. Three ethno-linguistic groups – the Hausa-Fulani in the north, the Ibo in the East and the Yoruba in the West – each with distinctive religious loyalties and identifications, are disproportionately dominant, and their relative size, autonomy and resource base contribute to the instabilities of Nigeria's federal structure. As Tom Forrest (1995) notes, colonial Nigeria was ruled by the British less as a single political unit than as distinct regional entities each with local authority structures and highly regionalized judiciaries, fiscal institutions (the marketing boards which taxed export commodities) and legislative assemblies with responsibility for education, health and local government. Colonial indirect rule – the policy of divide and rule in which local systems of authority and community sentiments were upheld and encouraged – entrenched a regional tripartite structure (the North, the West and the East) in which ethnic loyalties were created, refashioned and deepened. Within each region a single ethnic group predominated while federal authority (established formally in 1954) and nationalist sentiment were weak in the face of strong and fissiparous regional subnationalisms.

If the regions were the source of identification and political loyalty, they were also marked by patterns of unequal development. The northern region, while larger in population and area than the other regions combined, was the poorest and least exposed to Western education. The West conversely was by virtue of cocoa, coastal access, industrial development and early education, the wealthiest region and captured 38.3 per cent of the statutory (i.e. federal) revenue allocation by 1954–5. Educational inequities contributed to regional tensions as southern-ers (Yoruba and Ibos) dominated federal posts and attempted to penetrate northern government. As a consequence, northerners attempted to slow down the transition to independence in the 1950s and promoted a northernization policy to limit Yoruba and Ibo incursions.

In such a highly charged, competitively regionalized polity, mobilization built upon ethnic and religious loyalties from below. As public institutions within the regions grew in the run-up to independence, patronage was distributed through contracts, public office, loans and so on. Patronage and clientelism were, in the context of intense regionalism, the legitimate perquisites of political and public office. Not surprisingly, in the wake of the federation in 1954, the establishment of political parties assumed an irreducible regional cast, and state patronage developed apace. Furthermore, in the transition from the British to the first generation of Nigerian leaders, a number of key issues – public service employment, the national census – were cast and contested, in strongly regional/ ethnic terms. But in the political competition for power in post-colonial Nigeria, it was the economically weak and "backward" northerners, under the careful orchestration of the British, who assumed gradual federal supremacy in an atmosphere of deepening animosity and parochialism.

The emergence of petroleum as the centerpiece of the Nigerian export economy and the mainstay of state revenues had enormous consequences for the political development of post-colonial Nigeria. First, the geography of oil mattered (Figure 11.2). Close to 80 per cent of the petroleum was located in the eastern region – more precisely in the delta which represents roughly 8 per cent of the country – and not infrequently in the territories of ethnic minorities (i.e. non-Ibo). While the civil war (1967–70) – the attempt by the Ibo to secede from the federation and establish the independent state of Biafra – was not in any simple sense caused by the discovery of oil, the control of oil revenues was the central issue which precipitated the crisis of February 1967 in which the governor of the eastern region, Colonel Ojukwu, passed the Revenue Collection Edict #11 by which all revenues collected by the federal government would be paid to the treasury of the eastern government.

Second, petroleum rewrote the political script of the relations between the regions and the federal center. Growing nationalization of the petroleum sector and the establishment of a national oil company in 1970 channeled petroleum rents directly to federal coffers. Centrally controlled oil revenues superseded the regionally based revenues derived from the commodity marketing boards. As a consequence of the stunning growth of state revenues in the 1970s, the political center possessed a newfound fiscal capacity by which petrodollars could be used to manufacture a sort of political compliance. Conversely the regions discovered a new interest in gaining access to the seemingly infinite wealth provided by centrally controlled black gold. Petroleum enhanced the capacities of the historically weak center.

Nature's black gold is therefore the key to understanding the two fundamental properties of Nigerian politics in the period since 1970: state creation and revenue allocation. One of the first acts of the postwar military government under

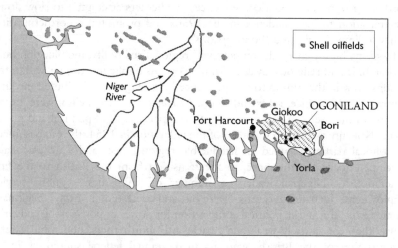

Figure 11.2 Location of Ogoniland and Shell oilfields

Gowon was to create twelve new states in 1967 from the existing four regions. Designed to balance north and south with six states, and thereby break the power and pathological competitiveness of the large regional blocs, the new state system had the effect of increasing minority access to federal funds while simultaneously making the entire state structure dependent on central (oil) revenues. Of course the demand for new states to meet the local needs for access to government resources, especially in deprived areas, was in practice difficult to halt. More states were created in 1976 and in 1991, while the number of local government areas (LGAs) within each state also proliferated for similar reasons. The result was the genesis of small states with little or no fiscal basis, totally dependent on what each state saw as "their share" of the national cake (i.e. the oil monies), and a profusion (there are 589) of corrupt, ineffective and hugely expensive LGAs driven by the logic of patronage politics. The multiplication of states from twelve in 1967 to thirty in 1996 did have the effect of irrevocably breaking some aspects of the old pattern of regional power, and accordingly increased the power of minorities who came to hold some form of political representation and economic autonomy. To this extent petrodollars permitted a certain degree of political cohesion within the federation to be quite literally purchased. The cost, however, has been an undisciplined federal structure driven by massive inflationary costs – the proliferation of state bureaucracies driven by prebendal politics – largely without a robust material base. As Khan (1994: 32) notes, the states have abandoned any pretense of a productive identity and rely unashamedly on federal handouts. The result is "power untempered by responsibility. . . . [The states are] . . . miniature versions of their free-spending federal paymasters" (*Economist* 1993: 12).

Embedded in this complex multi-state military federalism, however, the obdurate tensions of colonial Nigeria remain. First, the old regional blocs have certainly not disappeared[5] because major ethnic groups broke the old regions into smaller entities in which they could claim oil revenues from minority states (the Hausa, for example, created eight new states in which they became ethnically dominant). Second, the multiplication of states generated new demands by minorities and other political entities on a federal center dependent on a single and volatile source of revenue.

State creation could not be separated from the revenue allocation question. New states created more and complex demands on the distribution of government revenues.[6] In practice, the allocation rules have always been hotly contested. Fifteen commissions and decrees have refashioned the criteria for allocation over the period 1945–6, in effect defining the relative rights of federal and local authorities to petroleum resources, and the criteria by which petroleum-based government revenues are to be allocated (Furro 1992; Ikporupko 1996). To simplify an enormously complex picture, prior to 1959 statutory revenue was allocated on the basis of a "derivation principle" by which states received allocations from the federal pool in strict proportion to their contribution to these revenues (Ashwe 1996). This generally benefited northern and western regions but in the face of growing oil revenues in the 1960s they sought to change the principles of allocation. Monies to be allocated to the states came to be deposited in a Federation Account (formerly the Distributable Pool Account), the vast proportion of which was (and is) derived from oil. As the size of this account grew, new criteria were developed, largely to amend and supplant the derivation principle. By the 1960s population, need and equity principles were invoked; by the 1980s social development and internal revenue were added. This new horizontal allocation system obviously privileges more populous and larger states.[7]

Amid the shifting sands of revenue politics and allocative criteria, several patterns are evident. First, the proportion of revenues flowing to the north increased substantially from 35 per cent of the total in 1966–7 to 52 per cent in 1985. Second, the proportion of statutory revenues as a part of the local states' budget grew disproportionately. Third, the change in the derivation principle meant that oil-producing states in particular saw their share of statutory revenues fall; Bendel and Rivers states' share fell from 23.1 per cent and 17.1 per cent in 1974–5 to 6.4 per cent and 6.2 per cent respectively in 1989–90. To paint the revenue allocation picture, in short, is to depict northern hegemony in a weak and fissiparous federal system in which the oil-producing states in particular have experienced a sort of fiscal (and political) deprivation.

Running through the recent political history of petroleum in Nigeria is a strong dialectical current. Petrodollars on the one hand became the medium by which a historically weak federal center weakened regional power and manufactured a degree of political consent between center and region while simultaneously preserving northern hegemony (and reaffirming ethnic and regional loyalties). This is the heart of the Nigerian centralized patrimonial (redistributive) state (Naanen 1995: 53). Yet on the other, petroleum revenues vastly expanded patronage politics in a way which incapacitated federal and state governments and "pulverized" the social and civic life of the polity. As Khan (1994: 6) says, there is a "vicious circle" linking the lack of effective governance to regional, ethnic and parochial interests which in turn links these special interests to the vacuum of central political authority. Unregulated patrimonialism stands as the antithesis of effective, legal–rational state action. Petroleum rewarded inefficiency, corruption and despotism[8] in a way that has torn asunder the very social, political and economic fabric of contemporary Nigeria.

BLACK GOLD: THE NATURAL AND POLITICAL ECONOMY OF OGONILAND

The paradox of Ogoniland is that an accident of geological history (and histories of "geology") conferred upon the Ogonis nothing more than ecological destruction and economic backwardness. As a representative from British Petroleum put it: "I have explored for oil in Venezuela . . . in Kuwait [but] I have never seen an oil-rich town as completely impoverished as Oloibiri [where Shell first found oil in 1958]" (*Village Voice*, 21 November 1995: 21). In this sense, the Ogoni hated modernity because they could not get enough of it. They were angry because they could afford neither the cars nor use the roads criss-crossing their homeland which were the icons of petrolic success.

The Ogoni are a distinct ethnic group, consisting of three subgroups and six clans.[9] Their population of roughly 500,000 people is distributed among 111 villages dotted over 404 sq. miles of creeks, waterways and tropical forest in the northeast fringes of the Niger Delta (Figure 11.3). Located administratively in Rivers state, a Louisiana-like territory of some 50,000 sq km, Ogoniland is one of the most heavily populated zones in all of Africa. Indeed, the most densely settled areas of Ogoniland – over 1,500 persons per sq km – are the sites of the largest wells. Its customary productive base was provided by fishing and agricultural pursuits until the discovery of petroleum, including the huge Bomu field, immediately prior to independence. Part of an enormously complex regional ethnic mosaic, the Ogoni were drawn into internecine conflicts within the delta region – largely as a consequence of the slave trade and its aftermath – in the period prior to the arrival of colonial forces at Kono in 1901. The Ogoni resisted the British until 1908 (Naanen 1995) but thereafter were left to stagnate as part of the Opopo division within Calabar province. As Ogoniland was gradually incorporated during the 1930s, the clamor for a separate political division grew at the hands of the first pan-Ogoni organization, the Ogoni Central Union, which bore fruit with the establishment of the Ogoni Native Authority in 1947. In 1951, however, the authority was forcibly integrated into the Eastern region. Experiencing tremendous neglect and discrimination, integration raised long-standing fears among the Ogoni of Ibo domination. As constitutional preparations were made for the transition to home rule, non-Ibo minorities throughout the Eastern region appealed to the colonial government for a separate Rivers state. Ogoni representatives lobbied the Willink Commission in 1958 to avert the threat of exclusion within an Ibo-dominated regional government which had assumed self-governing status in 1957. Minority claims were ignored, however, and instead the newly independent Nigerian government passed the Niger Development Board Act in 1961 to look into the special problems of the delta and to improve social services and infrastructure. But the federal government, in keeping with its history of bad faith toward the delta, "failed to develop one square inch of the territory" (Okilo 1980: 14).

Politically marginalized and economically neglected, the delta minorities feared the growing secessionist rhetoric of the Ibo and consequently led an ill-fated secession of their own in February 1966. Isaac Boro, Sam Owonaro and Nottingham Dick declared a Delta Peoples Republic but were crushed and subsequently, in a trial reminiscent of the Ogoni tribunal in 1995, condemned to death for treason. Nonetheless, Ogoni antipathy to what they saw as internal colonialism at the hands of the Ibo continued in their support of the federal forces during the civil war. While Gowon did indeed finally establish a Rivers state

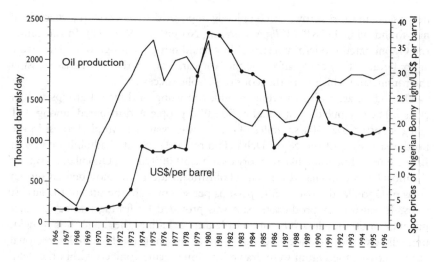

Figure 11.3 Nigerian annual oil production and prices (Bonny Light), 1966–96

in 1967 – which compensated in some measure for enormous Ogoni losses during the war – the new state recapitulated in microcosm the larger "national question." The new Rivers state was multi-ethnic but presided over by the locally dominant Ijaw, for whom the minorities felt little but contempt.[10] In Saro-Wiwa's view (1992), the loss of 10 per cent of the Ogoni people in the civil war was for nought as federal authorities provided no postwar relief, seized new on and offshore oil fields, and subsequently sold out the minorities to dominant Ijaw interests.

In short, Ogoni "nationalism" long predated the oil boom, but was deepened as a result of it. Ogoni fears of what Saro-Wiwa (1992) called "monstrous domestic colonialism," were exacerbated further by federal resistance to dealing with minority issues[11] in the wake of the civil war and by the new politics of post-oil boom revenue allocation. Rivers state saw its federal allocation fall dramatically in absolute and relative terms. At the height of the oil boom, 60 per cent of oil production came from Rivers state but it received only 5 per cent of the statutory allocation (roughly half of that received by Northern States and the Ibo heartland, East Central state). Between 1970 and 1980 it received in revenues one-fiftieth of the value of the oil it produced. In what was seen by the Rivers minorities as a particularly egregious case of ethnic treachery, the civilian Shagari regime reduced the derivation component to only 2 per cent of revenues in 1982, after Rivers state had voted overwhelmingly for Shagari's northern dominated National Party of Nigeria. The subsequent military government of General Buhari cut the derivation component even further at a time when the state accounted for 44.3 per cent of Nigeria's oil production. In 1992 the Oil Mineral Producing Areas Commission (OMPADEC) was established to develop projects in oil-producing communities (funded by 1.5 per cent of oil revenues) but it has been marred by corruption, inefficiency and fierce conflicts among oil-producing communities (Osaghae 1995: 333; Welch 1995).

Standing at the margin of the margin, Ogoniland appears (like Chiapas in Mexico) as a socio-economic paradox. Home to six oilfields, half of Nigeria's oil refineries, the country's only fertilizer plant, a large petrochemical plant, Ogoniland is wracked by unthinkable misery and deprivation. During the first oil boom Ogoniland's 56 wells accounted for almost 15 per cent of Nigerian oil production[12] and in the past three decades an estimated $30 billion in petroleum revenues have flowed from this Lilliputian territory; it was, as local opinion had it, Nigeria's Kuwait. Yet according to a government commission, Oloibiri, where the first oil was pumped in 1958, has no single kilometer of all-season road and remains "one of the most backward areas in the country" (cited in Furro 1992: 282). Few Ogoni households have electricity, there is one doctor per 100,000 people, child mortality rates are the highest in the nation, unemployment is 85 per cent, 80 per cent of the population is illiterate and close to half of Ogoni youth have left the region in search of work. Life expectancy is barely fifty years, substantially below the national average. In Furro's survey of two minority oil-producing communities, over 80 per cent of respondents felt that economic conditions had deteriorated since the onset of oil production, and over two-thirds believed that there had been no progress in local development since 1960. No wonder that the systematic reduction of federal allocations and the lack of concern by the Rivers government was, for Ogoniland, part of a long history of "the politics of minority suffocation" (cited in Ikporukpo 1996: 171). Petroleum was, in the local vernacular, being pumped from the veins of the Ogoni people. Systematic neglect not only deepened the sense of possession over local oil resources – "Leave us our oil for us . . . Rivers State oil for Rivers State only" read an editorial in *The African Guardian* (1991: 12) – but affirmed the belief that exclusion was primarily a result of ethnic and minority discrimination.

If Ogoniland failed to see the material benefits from oil, what it did experience was its devastating ecological costs – what the European Parliament has called "an environmental nightmare." The heart of the ecological harms stem from oil spills – either from the pipelines which criss-cross Ogoniland (often passing directly through villages) or from blow-outs at the wellheads – and gas flaring. A staggering 76 per cent of natural gas in the oil-producing areas is flared (compared to 0.6 per cent in the USA). As a visiting environmentalist noted in 1993, in the delta, "some children have never known a dark night even though they have no electricity" (*Village Voice* 21 November 1995: 21). Burning 24 hours per day at temperatures of 13–14,000 degrees Celsius, Nigerian natural gas produces 35 million tons of CO_2 and 12 million tons of methane, more than the rest of the world (and rendering Nigeria probably the biggest single cause of global warming). The oil spillage record is even worse. According to Claude Ake, there are roughly 300 spills per year in the delta and in the 1970s alone spillage was four times more than the much publicized *Exxon Valdez* spill in Alaska. In one year alone almost 700,000 barrels were soiled according to a government commission. Ogoniland itself suffered 111 spills between 1985 and 1994 (Hammer 1996: 61). Figures provided by the NNPC document 2,676 spills between 1976 and 1990, 59 per cent of which occurred in Rivers state (Ikein 1990: 171), 38 per cent of which were due to equipment malfunction. Between 1982 and 1992 Shell alone accounted for 1.6 million gallons of spilled oil, 37 per cent of the company's spills worldwide.

The consequences of flaring, spillage and waste for Ogoni fisheries and farming have been devastating. A recent spill in 1993 flowed for 40 days without repair, contaminating large areas of Ogoni farmland. Petroleum residues appear in the rivers at levels of 60 ppm and in the sediments around the Bonny terminal reach lethal levels of 12,000 ppm. In the ecologically delicate mangrove and estuarine regions of the delta, oil pollution has produced large-scale eutrophication, depletion of aquatic resources and loss of traditional fishing grounds (see NEST 1991: 44; Benka-Cocker and Ekundayo 1995) which now threaten customary livelihoods. The loss of farmland, dating back to massive blow-outs at Bomu in 1970 and in Ibobu in 1973, has further eroded the subsistence capabilities of Ogoni communities (*Newswatch* 18 December 1995: 10). Indeed, it is the direct threat to the means of subsistence by petrolic destruction that led Saro-Wiwa to talk of Ogoni genocide.

In almost four decades of oil drilling, then, petroleum ushered in a modernist nightmare. It had brought home the worst fears of ethnic marginalization and minority neglect: of northern hegemony, of Ibo neglect, and of Ijaw local dominance. The euphoria of oil wealth after the civil war has brought ecological catastrophe, social deprivation, political marginalization, and a rapacious company capitalism in which unaccountable foreign transnationals are granted immunity by the state.

DRILLING FIELDS: CONSTRUCTING NATURE, COMMUNITY AND ALTERITY IN OGONILAND

My Lord, we all stand before history . . . I and my colleagues are not the only ones on trial. Shell is on trial here . . . [and] the ecological war the company has waged in the [Niger] delta will be called into question sooner or later . . . On trial also is the Nigerian nation . . . I call upon the Ogoni people, the peoples of the Niger Delta, and the oppressed minorities of Nigeria to stand up and fight fearlessly and peacefully for their rights . . . For the Holy Quran says in Sura 42, verse 41: "All those who fight when oppressed incur no guilt."

(Ken Saro-Wiwa 1995)

The hanging of the Ogoni nine in November 1995 – accused of murdering four prominent Ogoni leaders who professed opposition to MOSOP tactics – and the subsequent arrest of nineteen others on treason charges, represented the summit of a process of mass mobilization and radical militancy which had commenced in 1989. The civil war had, as I have previously suggested, hardened the sense of external dominance among Ogonis. A "supreme cultural organization" called Kagote which consisted largely of traditional rulers and high-ranking functionaries, was established at the war's end and in turn gave birth in 1990 to MOSOP. A new strategic phase began in 1989 with a program of mass action and passive resistance on the one hand (the language is from MOSOP's first President, Garrick Leton) and a renewed effort to focus on the environmental consequences of oil (and Shell's role in particular) and on group rights within the federal structure. Animating the entire struggle was, in Leton's words, the "genocide being committed in the dying years of the twentieth century by multinational companies under the supervision of the Government" (cited in Naanen 1995: 66).

A watershed moment in MOSOP's history was the drafting in 1990 of an Ogoni Bill of Rights (Saro-Wiwa 1992; Greenpeace 1994; UNPO 1995; UN 1996). Documenting a history of neglect and local misery, the Ogoni Bill took head-on the question of Nigerian federalism and minority rights. Calling for participation in the affairs of the republic as "a distinct and separate entity," the Bill outlined a plan for autonomy and self-determination in which there would be "political control of Ogoni affairs by Ogoni people . . . the right to control and use a fair proportion of Ogoni economic resources . . . [and] adequate representation as of right in all Nigerian national institutions" (Saro-Wiwa 1989: 11). In short, the Bill of Rights addressed the question of the unit to which revenues should be allocated – and derivatively the rights of minorities. Largely under Saro-Wiwa's direction, the Bill was employed as part of an international mobilization campaign. Presented at the UN Sub-Committee on Human Rights, at the Working Group on Indigenous Populations in Geneva in 1992 and at UNPO in The Hague in 1993, Ogoni became – with the help of Rainforest Action Network and Greenpeace – a cause célèbre.

Ken Saro-Wiwa played a central role in the tactical and organizational transformations of MOSOP during the 1990s. Born in Bori as part of a traditional ruling family, Saro-Wiwa was already, prior to 1990, an internationally recognized author, a successful writer of Nigerian soap operas, a well-connected former Rivers state commissioner, and a wealthy businessman (Nixon 1996). Saro-Wiwa was also President of the Ethnic Minorities Rights Organization of Africa (EMIROAF) which had called for a restructuring of the Nigerian federation into a confederation of autonomous ethnic states in which a federal center was radically decentralized and states were granted property rights over onshore mineral resources (Osaghae 1995: 327). Under Saro-Wiwa, MOSOP focused in 1991 on links to pro-democracy groups in Nigeria (the transition to civilian rule had begun under heavy-handed military direction) and on direct action around Shell and Chevron installations. It was precisely because of the absence of state commitment and the deterioration of the environment that local

Ogoni communities, perhaps understandably, had great expectations of Shell (the largest producer in the region) and directed their activity against the oil companies after three decades of betrayal. There was a sense in which Shell was the local government (*Guardian* 14 July 1996: 11) but the company's record had, in practice, been appalling. In 1970, Ogoni representatives had already asked Rivers state government to approach Shell – what they then called "a Shylock of a company" – for compensation and direct assistance (a plea which elicited a shockingly irresponsible response documented in Saro-Wiwa 1992). Compensation by the companies for land appropriation and for spillage have been minimal and is a constant source of tension between company and community.[13]

In an atmosphere of growing violence and insecurity, MOSOP wrote to the three oil companies operating in Ogoniland in December 1992 demanding $6.2 billion in back rents and royalties, $4 billion for damages, the immediate stoppage of degradation, flaring and exposed pipelines, and negotiations with Ogonis to establish conditions for further exploration (Osaghae 1995: 336). The companies responded with tightened security while the military government sent in troops to the oil installations, banned all public gatherings, and declared as treasonable any claims for self-determination. Strengthening Ogoni resolve, these responses prompted MOSOP to organize a massive rally – an estimated 300,000 participated – in January 1993. As harassment of MOSOP leadership and Ogoni communities by state forces escalated, the highpoint of the struggle came with the decision to boycott the Nigerian presidential election on 12 June 1993.

In the wake of the annulment of the presidential elections, the arrest of democratically elected Mashood Abiola and the subsequent military coup by General Abacha, state security forces vastly expanded their activities in Ogoniland. Military units were moved into the area in June 1993 and Saro-Wiwa was charged with, among other things, sedition. More critically, inter-ethnic conflicts exploded between Ogoni and other groups in late 1993, amid accusations of military involvement and ethnic warmongering by Rivers state leadership.[14] A new and aggressively anti-Ogoni military governor took over Rivers state in 1994 and a ferocious assault by the Rivers state Internal Security Task Force commenced. Saro-Wiwa was placed under house arrest, and subsequently fifteen Ogoni leaders were detained in April 1994. A series of brutal attacks left 750 Ogoni dead and 30,000 homeless; in total, almost 2,000 Ogonis have perished since 1990 at the hands of police and security forces. Ogoniland was in effect sealed off by the military. Amid growing chaos, Saro-Wiwa was arrested on 22 May 1994 and several months later was charged with the deaths of four Ogoni leaders.

In evidence that has come to light during and after the military tribunal, the trial of Saro-Wiwa and eight others was marked by massive irregularities, including witnesses paid by government falsely to implicate MOSOP leaders. The defense team, faced with a kangaroo court, withdrew in June 1995, and Saro-Wiwa and the other defendants were sentenced to death on 31 October 1995.

Despite international outcry and enormous political pressure on the Abacha regime by the Commonwealth, Nelson Mandela and the human rights community, Abacha refused to rescind the death sentence. Throughout 1996, military operations in Ogoniland resulted in more deaths and the arrest of eighteen MOSOP leaders in March to prevent them from meeting with representatives of a UN mission which investigated the murder trials in the first two weeks of April.

In spite of the remarkable history of MOSOP between 1990 and 1996, its ability to represent itself as a unified pan-Ogoni organization was a central issue, particularly for Saro-Wiwa. There is no pan-Ogoni myth of origin (characteristic of many delta minorities), and a number of the Ogoni subgroups engender stronger local loyalties than any affiliation to Ogoni nationalism. The Eleme subgroup has even argued, on occasion, that they are not Ogoni. Furthermore, the MOSOP leaders were actively opposed by elements of the traditional clan leadership, by prominent leaders and civil servants in state government,[15] and by some critics who felt Saro-Wiwa was out to gain "cheap popularity" (Osgahae 1995: 334). Finally, the youth wing of MOSOP, which Saro-Wiwa had used, had a radical vigilante constituency which the leadership were incapable of controlling. Throughout the twentieth century, the sense of pan-Ogoni identity and self-interest has been replete with fractures and tensions along axes of generation, tradition/class, and political strategy in regard to the Nigerian federation.

POST-COLONIAL GREEN

> Ecological movements are not creating a new economics for a new civilization, they are not presenting a solution for the crisis of the modern world, and they do not have the capacity . . . for ending development. But they can show the difficulties, shortcomings and limited scopes of the dominant as well as the alternative models for development at the level of action.
>
> (Antje Linkenbach 1994: 81–2)

Ken Saro-Wiwa presided over a social movement which was, and remains, irreducibly environmental and yet its horizons and claims far transcend even the most catholic sense of environment. MOSOP was, in this regard, multi-dimensional and polyvalent, embracing in equal measure political ambitions which included self-determination, ethnic recognition, cultural rights, social justice, citizenship and political decentralization. But it is precisely this multi-dimensionality which presents the greatest stumbling block for Escobar, Shiva and others who celebrate the ostensibly local and popular–cultural character of post-colonial environmental movements. In the same way, if the coherence of the local or the community or the ethnic group cannot be taken for granted – to

261

employ Mallon's (1995) language a cultural hegemony has to be secured under quite heterogeneous and dispersive circumstances – the purported newness of southern social movements must also be treated with some skepticism. Efforts at creating a pan-Ogoni identity run throughout the twentieth century (long preceding Saro-Wiwa's mass organization), at various times associated with oil and the Biafran secession, at others with multi-ethnic secession for the federation, at others with claims for political autonomy within the interstices of colonial indirect rule.

The importance of rethinking both the historicity and polyvalence of Ogoni politics, is that it presents a more complex and differentiated sense of environmental movements qua movements. In this sense David Harvey's (1996: 177) recent taxonomy of environmental movements – the "incredible political diversity to which environmental opinion is prone" – is helpful but fails to convey a sense of the ways in which these opinions are contained within purportedly unified movements. "Ogoniness" had to be created (socially constructed) within considerable social, political and cultural diversity. This construction is certainly a discursive problem in part – interpellating a particular hegemonic sense of Ogoniness and of its "naturalized" qualities – but it suggests that contained within such movements are tensions and contradictions which threaten any simple sense of "the local" and any self-evident sense of "alternatives."

Interestingly, the new revisionist work on another iconic southern green movement – Chipko, the tree hugging movement of the Garawhal in north India – is also reframing its character in ways not unlike my account of the Ogoni. Priya Rangan (1995) has examined the earlier ur-history of Chipko which points to the role of Gandhian and left organizing in the region on the one hand, and the tensions within Chipko (and among its nominal leadership) on the other. Gadgil and Guha (1995) suggest in fact that there is not (and has never been) one Chipko, but rather three tendencies are represented within it. A Gandhian trend associated with Sunderlal Bahuguna (a claim incidentally which Rangan contests), a Marxist trend represented by a more militant and regional separatist wing, the Uttarakhand Sangarsh Vahini, and an appropriate technology section associated with Dasholi Gram Swarajya Mandal aimed at ecological restoration. Implicit in this account is that the tendencies within Chipko – which do not correspond to neatly bounded subgroups within the movement – are shot through with quite different "imaginaries": each tendency proffers markedly different accounts of science, development, nature/environment, social hierarchy and so on. Linkenbach (1994) suggests that Chipko contains at least three "models" within it: a survival economy, a selective modernization model, and an eco-development model. To each corresponds an invocation, a "calling," of nature which presents distinctive visions of social order.

The parallels with Ogoniland are striking. The history of Ogoni identity and its relation to the environment was a fractured but not incoherent set of narratives. The defense of the local, to employ Escobar's language, and the defense of local resources and local environments, was rooted in tensions over Ogoni development itself, a trajectory which is related to the natural and political economy of oil. The axes of this contention embraced youth militancy and the desire (a desire with a much longer history) for separatism, the efforts of local traditional rulers (in some measure a creation of colonial rule and local despotism) to capture oil revenues, and the bill of rights articulated by Saro-Wiwa which posited a new sense of federation in Nigeria and another sense of citizenship.

The parallels between Ogoni and Chipko suggest several conclusions. The first concerns hybridity and its meanings. There is a sense in which Saro-Wiwa was a sort on "transcultural," in-between activist, moving easily between global environmental movements, Ogoni lineages, and Nigerian political elites. But perhaps the hybridity that matters more is the ambivalent and contrary ways in which various environmentalisms are read into the political project of constructing a coherent and unified local identity. In Chipko it is the hybrid ways in which Gandhian, Marxist and sustainability narratives are constructed and manipulated around particular sorts of interests (logging, feminist oppression, state neglect). In Ogoniland, nature was something to be compensated for in the face of corporate pollution and local territorial identifications, but this compensation was read in hybrid ways into competing claims by a secessionist and militant youth wing, a local "traditional" ruling class anxious to preserve their entitlements through expanded petroleum revenue allocation, and by ethnic separatists talking the language of self-determination and decentralization. Age, class and ethnicity cross-cut any simple sense of local or community hegemony. This raises a related point, namely that there is a need to be more attentive to the fractures and fissures within such movements and a need to grasp the modalities by which local hegemony is (or is not) achieved. It is the active ways in which natures, identities and politics are constructed and contested – there are representations of and representations for – which needs to be grasped.

Second, the relations between environmental movements and modernity appear to be much more refracted and ambivalent than Escobar and others seem prepared to admit. It is one thing to suggest that tradition and modernity are redeployed in creative ways; it is quite another to acknowledge that there are explicit calls for more modernity – whether in the forms of schools, roads, pharmacies and so on which communities have been deprived of, or as a liberal charter of political rights – in which a sense of the local, the indigenous, or the subaltern is occluded. In this sense, to suggest that minority cultures carry the promise of "resisting and subverting the axiomatics of capitalism and modernity" (Escobar 1995: 224) is to prejudge the political tendencies of cultural movements which are surely as likely to produce Hindu nationalism or ultra-modernism as much as "flexible, modest [resistance to capitalism] relying on tactical articulations arising . . . out of daily life" (p. 216). Indeed it is precisely because they do arise out of the conditions of daily life that such movements may be forms of emergent modernism as much as models of alternative development.

Finally, it needs to be asserted that in movements in which claims over nature and culture are linked, the struggle turns on questions of social order. The Ogoni Bill of Rights is in many respects a unequivocal (and modernist) project which seeks to imagine a social order in which the unit of revenue allocation in Nigeria, local rights over resources, the definition of citizenship, and rights of self-determination stand in sharp contrast to the entire post-colonial history of Nigeria. Insofar as this post-colonial history is predicated upon the fiction of the Nigerian nation – what Nigerian leader Awolowo called "a mere geographical expression" – then such movements necessarily represent enormous challenges to the status quo, and to those whose interests are predicated upon often fragile and delicate senses of nationhood and nationalism. Only in this way can one begin to understand why such cultural–environmental movements elicit the ferocious forms of state violence which left Ken Saro-Wiwa and nine others hanging from the gallows of a Port Harcourt prison.

ACKNOWLEDGEMENT

I am especially grateful for the very constructive advice and commentary offered by Bruce Braun, especially in regard to clarifying some of the theoretical connections to the work of Latour and to post-colonial studies. He is of course in no way responsible for the content of the chapter.

NOTES

1 "Watching a minister sweep through Lagos in a twelve-vehicle convoy, complete with an armoured car, a journalist asked bystanders what would happen if they got out and walked: 'We'd kill them they replied.'" (*Economist* 8 June 1996: 8).
2 According to Pat Uyomi of Lagos Business School, personal income has grown at 0.02 per cent since the oil boom and most Nigerians lived better before the 1973–4 oil price rise. (*Economist* 8 June 1996: 48).
3 This field is not of a piece analytically or politically. Its key figures – Vandana Shiva, Claudio Alvares, Shiv Visvanathan, Wolgang Sachs, Gustava Esteva, Rajni Kothari, Arturo Escobar, Ashish Nandy and others – occupy a large territory from reactionary anti-modernism to rampant cultural relativism to feminist essentialism. I wish to make two points however. First, there is an important set of core propositions, especially with regards to environmental and social movements and their post-development imaginaries – and to this extent I shall focus especially on Escobar's important book *Encountering Development* (1995) which has by far the most sophisticated and detailed elaboration of these issues. Second, the intellectual genealogy of these new environmental movements has four points of derivation. One is post-colonial (and discourse) theory in its concern with the vestigial persistence of colonialisms of various sorts. A second is the debate over the "impasse" of Marxism (the meltdown of the Althusserian reactor) and its "post-Marxist" emphasis on non-class reductionism, anti-economism and heterogeneity in resistance politics. The third, is political ecology and its concern with deepening our understanding of what, environmentally speaking, is contested and how. Finally, the rise of global environmentalism as a set of

transnational discourses and practices which encompass the corporate managerialism of the Global Environmental Facility on the one hand and the transnational NGO networks on the other. These four points of reference have as their confluence, the anti-development/new environmental movement axis – what Escobar calls imagining post-development and post-structural political ecology.

4 The largest joint venture between NNPC and Shell accounts for 47.4 per cent of total crude production. Elf, Agip, Mobil and Chevron account collectively for another 49.2 per cent. For the most part there have been few changes in the foreign concession holders in the Nigerian oil industry and, with the exception of the civil war, production has largely been insulated from the volatile political climate (seven military governments, two elected civilian regimes, and one interim civilian government in the last thirty years).

5 In the 1990s roughly twelve of the nineteen positions in the Armed Forces Ruling Council (the former Supreme Military Council) have been held by northerners (Osaghae 1991: 257).

6 In the period between the end of World War II and Biafran secession, the revenues received by the old Northern and Eastern regions fell (respectively from 40.7 per cent to 35.3 and from 34.6 per cent to 29.3); conversely the Western state grew sharply from 24.7 per cent to 34.9 (Forrest 1995: 22). After 1967, however, the allocation process changed six times over a thirteen-year period, an instability which reflects precisely the struggles over a new political order in the context of the oil boom. Indeed, revenue allocation was central to the growing conflicts between federal leaders and the Eastern region prior to secession (over the so-called Binns allocation scheme), and in particular the manipulation of the "rules of the game" by the northerners and westerners (Rupley 1981).

7 Hence, the five oil-producing states which account for 90 per cent of the oil receive only 19.3 per cent of the allocated revenues (Ikporukpu 1996: 168). Five northern non-oil-producing states conversely absorb 26 per cent of allocated revenue.

8 These antinomies of oil are no better captured than in the state petroleum company (NNPC). The Irikefe Report in 1980 which discovered "vast irregularities" in the awarding of contracts, contributed to the NNPC's history of personnel turnover which resembles Italian politics. An estimated $14 billion disappeared from NNPC between 1979 and 1983; according to the 1994 Okigbo Report, $12.5 billion was unaccounted for between 1988 and 1994 (*Africa Confidential* 4 November 1994: 6). In 1990–91 when oil prices leapt as a result of the Gulf crisis, there was no recorded increase in NNPC or government revenue. In 1992, the World Bank estimated that this gap between official and unofficial oil earnings was $2.7 billion, roughly 10 per cent of GDP (*Economist* 1993: 8).

9 Ogoniland consists of three local government areas and six clans which speak different dialects of the Ogoni language. MOSOP is in this sense a pan-Ogoni organization.

10 The Ogoni and other minorities petitioned in 1974 for the creation of a new Port Harcourt state within the Rivers state boundary (Naanen 1995: 63).

11 What Rivers state felt in regard to federal neglect, the Ogoni experienced in regard to Ijaw domination. While several Ogoni were influential federal and state politicians, they were incapable politically of exacting resources for the Ogoni community. In the 1980s only six out of forty-two representatives in the state assembly were Ogoni (Naanen 1995: 77). It needs to be said however – and it is relevant for an understanding of state violence against the Ogoni – that the Ogoni have fared better than many other minorities. In 1993, 30 per cent of the commissioners in the Rivers state cabinet were Ogoni (the Ogoni represent 12 per cent of the state population) and every clan has produced at least one federal or state minister (Osaghae 1995: 331) since the civil war.

12 According to the Nigerian government, Ogoniland currently (1995) produces about 2 per cent of Nigerian oil output and is the fifth largest oil-producing community in Rivers state. Shell maintains that total Ogoni oil output is valued at $5.2 billion before costs.

13 Shell, which was deemed the world's most profitable corporation in 1996 by *Business Week* (8 July 1996: 46) and which nets roughly $200 million profit from Nigeria each year, by its own admission has only provided $2 million to Ogoniland in forty years of pumping. Ogoni historian Loolo (1981) points out that Shell has built one road and awarded ninety-six school scholarships in thirty years. According to the *Wall Street Journal*, Shell employs 88 Ogonis (less than 2 per cent) in a workforce of over 5,000 Nigerian employees. Furthermore, the oft-cited community development schemes of the oil companies only began in earnest in the 1980s and have met with minimal success (Ikporukpo 1993). In some communities, Shell only began community efforts in 1992 after twenty-five years of pumping, and then only providing a water project of 5,000 gallons capacity for a constituency of 100,000 (*Newswatch* 18 December 1995: 13).

14 It seems clear that the conflicts between Ogonis, Andonis and Okrikas were effected by state authorities, disgruntled community elders and the oil companies (Osaghae 1995: 337; Human Rights Watch 1995).

15 This was raised of course in the murder trial, where Saro-Wiwa was accused of referring to some of the elders and chiefs as "vultures" (Human Rights 1995: 29).

REFERENCES

Alvarez, C. (1992) *Science, Development and Violence*, Delhi: Oxford University Press.

Arce, A., Villarreal M. and de Vries, P. (1994) "The social construction of rural development," in D. Booth (ed.) *Rethinking Social Development*, London: Longman.

Ashwe, C. (1988) *Fiscal Federalism in Nigeria*, Monograph 46, Canberra: Australian National University.

Benka-Cocker, M. and Ekundayo, J. (1995) "Effects of an oil spill on soil physico-chemical properties of a spill site in the Niger Delta," *Environmental Monitoring and Assessment* 30: 93–104.

Berman, M. (1982) *All that is Solid Melts into Air: The Experience of Modernity*, New York: Penguin.

Booth, D. (ed.) (1994) *Rethinking Social Development*, London: Methuen.

Bourdieu, P. (1990) *In Other Words*, London: Polity.

Buttel, F. (1992) "Environmentalization," *Rural Sociology* 57: 1–27.

Cohen, J. and Arato, A. (1992) *Civil Society and Political Theory*, Cambridge: MIT Press.

Escobar, A. (1992a) "Imagining a post-development era? Critical thought, development and social movements," *Social Text* 31/32: 20–56.

—— (1992b) "Culture, economics, and politics in Latin American social movements theory and research," in A. Escobar and S. E. Alvarez (eds) *The Making of Social Movements in Latin America*, Boulder: Westview Press.

—— (1995) *Encountering Development*, Princeton: Princeton University Press.

Esteva, G. (1992) "Development," in W. Sachs (ed.) *The Development Dictionary: A Guide to Knowledge as Power*, London: Zed Books.

Forrest, T. (1995) *Politics and Economic Development in Nigeria*, Boulder: Westview.

Furro, T. (1992) "Federalism and the politics of revenue allocation in Nigeria," unpublished PhD dissertation, Clark Atlanta University.

Gadgill, M. and Guha, R. (1995) *Ecology and Equity*, London: Routledge.

Greenpeace (1994) *Shell Shocked*, Amsterdam: Greenpeace International.

Hammer, J. (1996) "Nigerian crude," *Harpers Magazine* June: 58–68.

Harvey, D. (1996) *Justice, Nature, and the Geography of Difference*, Oxford: Blackwell.

Human Rights Watch (1995) *The Ogoni Crisis*, Report 7/5, New York: Human Rights Watch.

Ikein, A. (1990) *The Impact of Oil on a Developing Country*, New York: Praeger.

Ikporukpo, C. (1993) "Oil companies and village development in Nigeria," *OPEC Review* 83–97.

—— (1996) "Federalism, political power and the economic power game: control over access to petroleum resources in Nigeria," *Environment and Planning C* 14: 159–77.

Keck, M. (1995) "Social equity and environmental politics in Brazil," *Comparative Politics* 27: 409–24.

Khan, S. A. (1994) *Nigeria: The Political Economy of Oil*, London: Oxford University Press.

Latour, B. (1993) *We Have Never Been Modern*, Cambridge, MA: Harvard University Press.

Lewis, P. (1996) "From prebendalism to predation: the political economy of decline in Nigeria," *Journal of Modern African Studies* 24: 79–104.

Linkenbach, A. (1994) "Ecological movements and the critique of development," *Thesis Eleven* 39: 63–85.

Mallon, F. (1995) *Peasant and Nation*, Berkeley: University of California Press.

Martinez-Alier, J. (1990) "Poverty as a cause of environmental degradation," report prepared for the World Bank, Washington, DC.

Naanen, B. (1995) "Oil producing minorities and the restructuring of Nigerian federalism," *Journal of Commonwealth and Comparative Politics* 33: 46–58.

NEST (1991) *Nigeria's Threatened Environment*, Ibadan: Nigerian Environmental Study Action Team.

Nixon, R. (1996) "Pipe dreams," *Black Renaissance* Fall: 39–55.

O'Donnell, G. (1993) "On the state, democratization and some conceptual problems," *World Development* 21: 1355–69.

Ogbonna, D. (1979) "The geographic consequences of petroleum in Nigeria with special reference to Rivers State," unpblished PhD dissertation, University of California, Berkeley.

Okilo, M. (1980) *Derivation: A Criterion of Revenue Allocation*, Port Harcourt: Rivers State Newspaper Corporation.

Osaghae, E. (1991) "Ethnic minorities and federalism in Nigeria," *African Affairs* 90: 237–58.

—— (1995) "The Ogoni uprising," *African Affairs* 94: 325–44.

Parajuli, P. (1991) "Power and knowledge in development discourse," *International Social Science Journal* 127: 173–90.

Pieterse, J. (1996) "My paradigm or yours?," Working Paper 229, The Hague: Institute of Social Studies.

Rangan, P. (1995) "Contesting boundaries," *Antipode* 27: 343–62.

Routledge, P. (1994) *Resisting and Shaping the Modern*, London: Routledge.

Rupley, L. (1981) "Revenue sharing in the Nigerian federation," *Journal of Modern African Studies* 19: 252–77.

Saro-Wiwa, K. (1989) *On A Darkling Plain*, Port Harcourt: Saros International Publishers.

—— (1992) *Genocide in Nigeria*, Port Harcourt: Saros International Publishers.

—— (1995) Closing statement for presentation on 1 September to the military tribunal, Port Harcourt, Nigeria. Not published. Available from Ogoni Website: www. oneworld.org/oca

Seabrook, J. (1995) *Victims of Development*, London: Verso.

Shiva, V. (1989) *Staying Alive*, London: Zed Books.

—— (1993) "The greening of the global reach," in W. Sachs (ed.) *Global Ecology: A New Arena of Political Conflict*, London: Zed Books.

UN (1996) *Report of the Fact-Finding Mission of the Secretary-General to Nigeria: Summary of Information and Views Received*, New York: United Nations.

UNPO (1995) *Ogoni: Report of the UNPO Mission to Investigate the Situation of the Ogoni*, The Hague: Unrepresented Nations and Peoples Organization.

Watts, M. (1994) "The devil's excrement," in S. Corbridge, R. Martin and N. Thrift (eds) *Money, Power and Space*, Oxford: Blackwell.

Welch, C. (1995) "The Ogoni and self determination," *Journal of Modern African Studies* 33: 635–50.

Part 4

AFTERWORD

12

NATURE AT THE MILLENIUM

Production and re-enchantment

Neil Smith

Come on nature,
I don't want to read about or talk about the world;
Come on nature,
Let me show the way that I've been feeling all along.
(The Proclaimers, © Proclaimers/Chrysalis Records, Inc.)

INTRODUCTION

It was Angela Carter who, amid the celebratory hype of 1980s postmodernism, wryly remarked that "the *fin* is coming a little early this *siècle*." With the end of the present century coinciding with the end of the millenium this is perhaps not unexpected. Yet the response by Christians, fellow travellers, and indeed the rest of us, for whom this arbitrary moment in time has been fixed with epochal significance, has been somewhat muted. It could well be argued that the revelatory confidence of biblical millenialism is largely overshadowed by symptoms of *angst* concerning identity and being, yet even this more cautious response is so far milder than might have been expected. Mega year-end celebrations for 1999 were planned well in advance with many fully booked years ahead, to be sure, but the excitement remains much more abstract than that which captured the last *fin de siècle*. The *angst* and optimism will surely quicken in the last moments of the second Christian millenium, and they may seem more intense in hindsight, but I suspect too that something else is at work. Assumptions of social power and the power of social construction are to a greater or lesser extent hegemonic in a way they never were a hundred years ago. The abstract time marked by the *fin de millenium* has yielded some of its power of determination.

The geographical and biological determinism of social events and relationships was at its ideological peak a century ago. Today the former has fizzled in the face of dramatically expanded economic, social and technological power over nature, and the latter, while enjoying a distinct comeback as part of the *angst* of century's end – from a resurgent Malthusianism to the wildest fantasies concerning what the Human Genome Project will explain – is nonetheless circumscribed by social ambition. The lesson projected by Scottish Dolly, the cloned sheep, was that however deeply engrained the biological patterning of nature, it is accessible to social engineering. Nature more generally is far more malleable than ever it was in the nineteenth century.

There is cause for celebration, then. Rigidly fixed conceptions of nature have been a major reservoir feeding numerous ideologies of the social and, while this is still true today, increasing acknowledgment of the malleability of nature sharpens the contradictoriness and undercuts the ideological power of "nature" as social explanatory. Environmental activists and Marxists, feminists and work-place safety and health activists, science critics and social theorists of race and sexuality have all contributed to this achievement. But so too has the fury of capital accumulation. We have won a major victory by putting nature squarely and ineluctably on the popular political agenda, but we have also suffered a major defeat insofar as the agenda of politics as normal has largely digested, institutionalized and marketized the politics of nature. Compared with the late 1960s and 1970s when the politics of nature erupted, *fin de millenium angst* about nature is widespread but of low intensity; we're all environmentalists now. The radical genie of the environmental challenge to late capitalist nature destruction has been stuffed back into the bottle of institutional normality just in time to calm millennial jitters about nature. The challenge for the twenty-first century is to start again, to make environmental politics subversive again.

All the chapters in this book, with their commitment to critique contemporary forms of environmental destruction together with institutionalized environmental discourses, point in different ways toward this project of starting again. Part of the project involves the continual ploughing and reploughing of familiar political terrain. But it also involves new intellectual and political departures, with which this volume is especially fertile. This involves, in part, scandalizing contemporary appropriations of environmentalism, but it also involves the more difficult task of eking out an alternative political vision. In an attempt to get the genie back out of the bottle, I would like to conclude with my own rehearsal of the known and the novel as they might contribute to a reformulated environmental politics for the new millennium: the production and simultaneous re-enchantment of nature.

BEYOND SOCIAL CONSTRUCTIONISM: SOKAL'S TAUNT AND THE PRODUCTION OF NATURE

When in the early 1980s theories of "the production of nature" first crystallized as a simultaneous critique of capitalist exploitation and environmental romanticism, a broader social constructionism was beginning to seep into the theoretical air of the English-speaking academy. Since then, of course, constructionism has become de rigueur, even passé, and the claim that even nature is socially constructed is anything but shocking. In retrospect, the notion of the production of nature was already pregnant in the environmental movement, even in subatomic physics albeit in radically different form, before it came to geography and cultural studies. Leaving aside for the moment the world of difference between production and construction, it should now be clear that at one level there is little startling at all about this insistence on an historical reading of nature, or indeed about the seeming oxymoron of the production of nature; it is part of the general ethos of our newly environmental society. The global warming debate only provides the sharpest instance of the extent to which historical explanations invoking social agency in natural change have become normalized, but at the same time it highlights the crucial importance of a critical constructionism. While the causes of global warming are generally traced back to specific kinds of chemical emission from the direct production process and from the consumption of produced goods, policy resolutions focus on the technical (and geographically selective) curtailment of these emissions without questioning either the specific social relations that organized prevailing production and consumption choices or even the global social restructuring implied by technical emission abatement policies. "Environmental management" is little more than a technocratic, neo-liberal rendition of the construction of nature, albeit evoking an especially narrow productionism which keeps nature and culture in separate spheres.

To take a second less obvious example, even among the purest physical scientists the search for order in the universe, the revelation of chaos, the ambition for a unified physical theory have all recently been cast in resolutely historical terms. To understand the micro-structure of physical or biological nature or the macro structure of the universe is today to seek the cosmological origins of matter and events. The uncertainty principle and its outgrowths notwithstanding, the history of nature in physical science is often theologized rather than socialized,[1] to be sure, but this contemporary resort to divine creation itself underscores rather than questions the social embeddedness of some kind of "production of nature" perspective. Who is doing the production and under what circumstances becomes the crucial question even as the history of nature radically excludes social agency.

In cultural and social theory, insistence on the discursive construction of nature, arguably the paramount vehicle in the 1990s for rendering nature complicit with social history, accomplishes something parallel. The central and undeniable insight here is that the authoritative appeal to reality as the ground of truth claims is always filtered through the social muslin of representations gathered into discourses (no matter how liberal and permissive the muslin may be) and that no kind of purely extra-social authority is available for arbitrating the shape and dynamism of nature. Discourse filters all we are able to claim. If this is all that is implied in the claim, to take a particularly eloquent case, that "nature is discursively constructed all the way down" (Escobar 1996) then most scientists and post-structuralists would have little to disagree about. "No reality without representation" was more than a vital corrective slogan in the 1980s; it revitalized the micro political investigation of socio-cultural practices across the board – never a bad thing. But if the further step is taken, as it too often is, that for all practical purposes there is therefore only representation – that nature is only constructed discursively – then it seems to me that all the careful caveats that of course "nature is real" have more rhetorical than practical meaning; the autologic of the post-structural critique has dissolved its reason for being in the first place. Conceptual deconstruction by the philosopher-god substitutes for the active investigation of nature, as discourse flatters itself as its own universal. Turning nature into history does not necessarily require its transformation *tout court* into discourse, and the exclusion of natural "agency," and to this extent at least a viable politics of nature has to get beyond easy resorts to social constructionism.

Alan D. Sokal, who hoaxed *Social Text* (and himself to boot, which has so far escaped comment) with a faux postmodern rendering of contemporary physics (Sokal 1996a), is therefore wrong for all the right reasons.[2] Whether itself an outburst of millenial *angst* or whether a more terrestrial response to declining scientific budgets and career opportunities, Sokal's defensive scientism actually displays an uncomfortable homology with the exclusionary discursivism he takes as his target. This is nowhere clearer than in his taunt: "Anyone who believes that the laws of physics are mere social conventions is invited to try transgressing those conventions from the windows of my [twenty-first storey] apartment" (Sokal 1996b: 62). I am not so interested in prodding the rhetorical silliness of this sneer – not even the most committed discursive constructionist denies the existence of gravity, and Sokal might have learned something by asking why this is so and what it might mean – but rather what does such a claim say about science? In the first place although it is the discourse of science that Sokal defends – "the laws of physics" – his invocation of "real" gravity divulges precisely the flaw that critical scholars of science highlight, namely that the creative translation by the scientist between the reality and the description of nature is fully erased; the discourse and the reality are rendered interchangeable. What irony that the scientist defending the objectivity of science/nature repeats precisely the fault he finds in others.

But apart from erasing the translation between discursive laws and real processes, law bound or otherwise, Sokal's taunt makes a much more hubristic claim for science. In the end it is a claim about ownership, not just of discourse but of nature, and as such gives away a lot more about certain claims of contemporary science than was intended. The central message is that physics is responsible for keeping us from the disaster of abrogating gravity; without physics we would somehow be susceptible to lunging from Sokal's apartment in blissful ignorance of natural forces. But the arrogance of such a claim is matched only by its falsity. Long before modern science, human beings generally managed to avoid throwing themselves from great heights, even if they did not have Sokal's twenty-first storey apartment to tempt them. Even if, as many scientists do, one assumes the historical universality of a single scientific project back to

Euclid or Pythagoras, Eratosthenes or Aristotle, or if one appeals to less exclusively Greco-centric versions of scientific origins, it is difficult to avoid the conclusion that while science has achieved unprecedented insight in describing, understanding, and mobilizing natural forces and processes, there is a learned practical knowledge of the world – including the hazards of gravity – that pre-exists science and remains to some extent extra-scientific. (For that matter it also pre-exists discourse, regardless of misguided claims denying any existence to the extra-discursive world.) This practical knowledge is not even exclusively human: even goats respect gravity, although it is a fair assumption that they know nothing of Sokal's laws of nature. So good are animals at avoiding the accidents of gravity, that when the exception occurs – as allegedly with lemmings – it is an event of extraordinary scientific and social curiosity for us.

Just as discourse does not own nature – and it is especially sad that we needed Sokal to remind us of this – nor does science. Sokal may be the scientist-god of his own apartment, but his hubristic taunt reveals a much more extensive ambition of property claims over nature. This fact was brought home to me several years earlier when I found myself involved in a very odd discussion, the full significance of which only came to me later. In 1991 I had organized a conference entitled "Metaphor and Materiality: The Politics of Space and Nature" that was designed to further an emerging discussion of the politics of metaphor in social and cultural theory. It featured various scholars in the social sciences and humanities. As publicity for the conference went out, I received a telephone call from a professor in the university's mathematics department. He was audibly upset that neither he nor other mathematicians had been invited to present papers at the conference and was not about to be consoled by my assurances that our discussion of the politics of space and nature lay fairly far afield from his concerns. Mathematics was all about space, he insisted, and nature, but especially space. A mathematics student when I first went to university, my own hubris bubbled up as I relished a discussion of mathematical space that did not entirely leave me behind, but I was about to be completely outdone. Readily admitting that mathematical space is wholly divorced from the politics of space, "whatever that might be," our maths professor readily conceded that while his expertise was space he could talk about politics too. I demurred, repeating simply that the discussion was unlikely to make the bridge to his mathematical concerns but that he was more than welcome to participate from the audience. He was not about to be put off in this way and finally raised his voice in frustration to insist that I could not "exclude" him: "Mathematicians own space!"

The voice, I later realized, issued from Norman Levitt who two years later published a co-authored attack on what he presented as a dangerous, anti-intellectual conspiracy against science and who, with the millenium fading, became a vocal cheerleader for Sokal (see Gross and Levitt 1994). In fact *Higher Superstition* is a desperate defense of an eroded authority and a transparent ownership claim over not just space and nature but culture too, insofar as any discussion of whatever science presumes to "own" is to be mercilessly policed by scientists themselves. More than anything this vignette from the science wars convinced me of the old Frankfurt School truth that technological intent is already written into what C. H. Waddington (1941) depicted years ago as "the scientific attitude." The critiques of science and modernism are to this extent correct and should not have to be rediscovered on a generational cycle.

Nonetheless, these are extreme claims from a few scientists and for all that they have succeeded in framing much of the "science wars," it would be a mistake to allow such responses to displace a more subtle critical approach to science. If it is important to critique the scientific idealism that confuses science with reality and the dangerous mobilization of social power that such a misguided science could effect, it is equally important to avoid its virtual image in discursive universalism. While not so intimately linked to social power, such an extreme discursive constructionism similarly exacts its own domestication of nature: if nature is discursive "all the way down," we can bypass nature and make a new world by changing the discourse. But modern science is itself a procedure for producing nature, materially cum discursively, and even the most withering political critique of science has to open up the contestable dialectical translation between discourse and that which is represented – reality – rather than close it off under the sign of discursive construction.

Are less extreme forms of constructionism more sustainable? Sismondo's (1993) identification of artifactual constructionism, which attempts to hold an ontological realism consistent with an epistemological agnosticism, and to locate the question more squarely in the middle ground of social practice is, philosophically, a much more robust argument. It depicts our scientific understanding of nature as the result of social practices in laboratories; the results are as artifactual as the apparatuses that produce them; nature is known through social practice "all the way down," we might say. In many ways this is the approach that critics such as Latour (1987), Haraway (1991) and Barad (1996) have taken. There are two limitations to this perspective, however. In the first place, the focus is very much scientific knowledge and the laboratory, and much less attention is paid to other kinds of social practice. Second, as Demeritt (this volume) has pointed out, there is an inherent danger that such a delicate artifactual constructionism will devolve into some kind of neo-Kantian idealism, and although it is rarely identified as such, it is this approach that is the presumed target of anti-constructionist attacks. This is no minor danger. It has so far attracted little comment, but social theory at the end of the twentieth century seems to be going through something of a neo-Kantian revival in much the same way as at the end of last century, especially in Germany.

The original emphasis on the production of nature rather than its construction was therefore deliberate and in the wake of the science wars seems to be more urgent. It suggests a rather different valence of meaning and a differently inflected politics that might help inspire a twenty-first century "revolutionary environmentalism." For the impetus behind the deliberately provocative language of "the production of nature" has been multifold: it seeks first, not simply to highlight the fact that even nature was somehow socially constructed, quixotic as that might sound, and thereby to try and deflate the vocabulary of wilderness and pristine nature; but second, it is intended to preclude precisely the Kantian cul de sac of an exclusively discursive constructionism. Third, and more positively, the intent was to insist that social labour lay at the heart of our comprehension of the social relation with nature, and in so doing the emphasis on production is consistent with artifactual constructionism but it retains a broader perspective than the laboratory, involving the whole range of social, economic and cultural production. Fourth, the achievements of the highly influential Frankfurt School theory and critique of nature notwithstanding, the "domination of nature" and the cleave of second from first nature represent an insufficiently dialectical foundation for a left comprehension of the mutual implication of society and nature. Finally, via these critiques the "production of nature" thesis is intended to shift the intellectual fulcrum from a Kantian to a Hegelian "constructionism."[3] Thus there was a vital corollary to the project: to learn from environmental politics and thereby to rehabilitate the notion of social production itself, to expose and explore the dialectical tension of nature and production. This in turn would help open up the profound and optimistic possibility that radically different social environments and environmental societies are possible and can be deliberately as much as indeliberately produced.

In Bourdieu's definition, the habitus is where social history is "turned into nature" (1987: 4), but the corollary is equally true. The transformation of received nature into history is general to all societies but the scale, rigour and destructiveness with which nature is historicized are an especial hallmark of capitalism. While it pervades the habitus, the economic and ideological fulcrum of this transformation is the social production process. In the broadest terms, the political intent of analyses of the production of nature is to open up the history of nature both to retrospective examination and to future political agency.

The complaint that "the production of nature" squeezes our theoretical and political comprehension of nature into much too narrow a "productionist" framework is precisely the target of, as much as a common response to, this move. Such an aversion to production springs in part from a sincere conviction that there is much in the world beyond production but it equally emanates from a misguided empiricism that allows "production" none of the liberal expansiveness we currently associate with, for example, "culture." It restricts production to those acts of manual and not imaginative work, economic and not cultural creation, individual labour rather than social accomplishment, and the making of objects rather than productive consumption by subjects. Whatever the uses of such a narrow conception of production, it makes better sense to follow Marx's (1973) more complicated and nuanced analysis. Formulating a conception of production in *Grundrisse*, he argues in the first place against any individualist model, embedding production instead in social relations: "production by an isolated individual outside society" – the Robinson Crusoe myth that founds so much doctrine of economic production – "is as much of an absurdity as is the development of language without individuals living together and talking to each other" (p. 84). In the second place, therefore, production is not arbitrarily distinguished from the other elements of social and economic circulation: "production is also immediately consumption," and "consumption is also immediately production" (p. 90). Thus, production is not "encased in eternal natural laws independent of history" – a widespread if largely implicit assumption which allows "bourgeois relations" to be "quietly smuggled in as the inviolable natural laws on which society in the abstract is founded" (p. 87).

Henri Lefebvre (1991) is only one of many social theorists who has embraced and developed this complexity inherent in Marx's notion of production, and in his pioneering effort to reframe politics as a struggle over the production of space, he explicitly seeks to rehabilitate production from the narrow empiricist rebuff. For Lefebvre the richness of "production" along with "product" and "work," has been "seriously eroded," the constructed polyvalence of these terms radically reduced to a "positivity properly belonging to the narrow or scientific (economic) sense" (p. 70). Preferring to "take up these concepts once more, to try and restore their value and to render them dialectical" he returns to Hegel for whom production, he says, plays a "cardinal role" (p. 70). Emphasizing Marx's critique of Hegel, namely that production in practical terms transcends the opposition between subject and object, and other such philosophical conceits, Lefebvre insists that production is a "concrete universal": production as a social practice suffuses all societies but its concrete form is radically differentiated from one to the other. The inherent contradictoriness of "production," its ambivalence in Marx, yet ultimately its concreteness, represents a potent weapon which, according to Lefebvre, has been "somewhat obscured and watered down," generalized, abstracted and flattened, in a process which in the extreme case – he has Baudrillard in mind – indulges a "willfull dalliance with nihilism" (p. 72).

Lefebvre's own distinction between production and creation, between a product and a work – the one routinized the other irreproducible – is itself dubious, not least because it springs from a strangely traditional assumption of nature as creative, assaulted on all sides, "murdered by 'anti-nature'" (p. 71). Nature for Lefebvre is in the end the uncomplicated ground on which the social complications of space are played out; the subtlety he applies to the production of space is lost on nature, which he sees only as defeated, obliterated, steamrolled by late capitalist production.[4] Entertaining for a fleeting moment the possibility that nature is "in a sense produced," he immediately shuts off this avenue by insisting that the creativity of nature inherently frustrates any possibility of its production. For all that he draws back to a pre-Einsteinian insistence on the philosophical priority of space over nature – an indication of the limits of a Hegelian approach – Lefebvre's analysis of the production of space nevertheless provides a crucial glimpse of the sophisticated theoretical and political possibilities of "production." Transposed to the question of nature, Lefebvre's exploration of production offers a viable way of thinking about alternative productions of nature.

THE RE-ENCHANTMENT OF NATURE?

Waging the science wars in defense of a political lever into the social relations with science; working out a powerful conception of the production of nature; sharpening the analytical tools that can be translated from theory into environmental practice – all of these are important political pursuits in the effort to forge a revolutionary environmentalism. Yet much of this work is strangely detached from its supposed political moorings. The cathartic ridicule aimed at *Social Text* following the Sokal hoax – much of it from the left itself – suggests strongly that however academic and theoretical such issues might appear, they cut to the quick of individual and social identity and insecurity at the end of the millennium. For all our sophisticated analysis and conceptual desquamatory moves on nature, the broad left has completely failed to produce a viable alternative to "establishment environmentalism."

Whether under the rubric of constructionism or the production of nature, socialist ecology or feminist environmentalism, environmental justice or a broader red–green politics, the left has pursued the question of nature in a predominantly analytical mode. The purpose has largely been to peel apart layer upon layer of encrusted ideological meaning crystallized into the discursive body of "nature," and to reconstruct a usable environmental politics. The ability to label some social practice or behavior as natural – and just as important, the correlated ability to label other practices and behaviors unnatural – is a powerful centerpiece of contemporary social ideologies concerning class and gender, race and sexuality, economic organization and political struggle. Popular romantic and sentimental assumptions concerning what we call nature are powerful bearers of such oppressive social ideologies.

As a sometime participant in these critiques of ideologies of nature, however, I am increasingly convinced that for all the merit of our deconstructive erudition, we are left with a rather antiseptic nature which has little if any political appeal. For all that our dismantling of nature idolatry is a vital, always ongoing part of any critical politics of social reconstruction, it has not prevented defeat. Previously outrageous claims from the environmental left have been co-opted now as orthodox environmental wisdom (and strategy) in such a way that the romanticization of nature is intensified. In most people's minds, a four-lane highway has been built in the last two decades connecting environmental awareness with "saving" nature. Where the analytical practicality of left critiques is at all implemented, it is largely subordinated to a science-driven, policy-oriented, bureaucracy centred project of environmental management.

There are exceptions of course. There is a certain left romancing of nature, whether in the form of the Gaia principle, the adoption of nature worship from

279

indigenous peoples, certain eco-feminist appeals to the "earth mother," and so forth. And the environmental justice movement spearheads a very practical concern with environment as the place of daily life, a sphere of involuntary consumption, which draws directly on the analytical dexterity of left critiques combined with a judicious deployment of science and skills of political organizing.

But it remains true that the strengths of left analytical antisepticism have not left much room for the reconstruction of a powerful anti-ideology of nature that takes seriously the realities of the production of nature interwoven with deep emotional significance. Left romanticism of nature abounds but it either shares awkwardly the conservative assumptions of saviour environmentalism or else co-exists in uneasy juxtaposition with our analytical sophistication. The emotional appeal that guides, scripts and emanates from experiences with nature is deep, intense, intuitive and as spontaneous as it is learned, and a new politics of nature will not succeed if it does not rewrite the rich memory banks of experience that are displaced by the critique of ideology. Or as Haraway (1991) puts it, nature is that which we cannot not desire. This, of course, is the emotional terrain so efficiently exploited by saviour environmentalism.

It is obviously unrealistic to expect that such an alternative can fully colonize the psychic main line into people's insecurities and identities, paved and patented by Walt Disney nature idolatry – think of the deeply retrograde environmentalism of child-oriented propaganda movies like *Pocahontas* and *The Lion King* – but even the most modest beginnings of such a project are barely discernible (King 1995). Our analyses may be right as rain but they have little or no ability to move people about such a deeply resonant array of experiences as are implied in "the relation to nature." I want to insist that the re-enchantment of nature not be left to the right, or even to a sentimental liberalism, and that such deep feelings of connectedness to nature somehow be mobilized against establishment environmentalism.

Having said this I am immediately hesitant about where such a project starts, how it can be pursued, and where it might lead. The most obvious difficulty is that such a project takes place at what Kristin Kroptiuch (1988: 9) brilliantly refers to as the "desolate junction of poetics and political economy." The intellectual not to mention the political scars of a Faustian modernity, nurtured by the grotesqueries of capitalist unevenness, run deep. The raw materials for a revolutionary re-enchantment of nature are simultaneously scarce, yet all around. And yet day-to-day experience suffuses the political economic and poetic appropriations of nature in a relentless triptych of sensations and events to be sorted out into ordered categories: access to "poetic" nature is thoroughly policed by political economic power.

But there may be a glimmer of possibility. In a very interesting book, John McClure (1994) revisits Max Weber's lament that the "fate of our times is characterized by rationalization and intellectualization and, above all, by the 'dis-enchantment of the world.'" This represents a highly sentimentalized assessment of an entire era but, as McClure insists, there is a clear "politics of romance" in the literary work of Anglo-American modernism that has in various ways grappled with this "disenchantment." There is, in fact, a "battle of romances," and McClure identifies in the novels of Don DeLillo and Thomas Pynchon the glimmerings of "Late Imperial Romance" which counter the conservative laments of more traditional imperial romance novels – Conrad to Kipling – from earlier in the century. For Pynchon in particular, the novel provides a way of telling certain "counter-stories" which offer "not the usual late imperial consolations of transcendent re-enchantment and world-rejecting resignation, but an eco-spiritual vision that *combines* re-enchantment with political resistance" (p. 154).

There are clearly limits to what the late imperial romance achieves insofar as DeLillo and Pynchon and others remain preservationist in outlook – they seek to "preserve the geography of romance," McClure suggests – but insofar as they initiate this "battle of romances" they also expose a certain "politics of re-enchantment" (p. 8). McClure does not say so explicitly, but there is a strong sense in which this is a politics of re-enchantment which cannot not be engaged, and the same holds true as regards nature. We cannot not engage in the political battle over the re-enchantment of nature.

We can approach the argument by a different route. Fredric Jameson once remarked that a certain "existential bewilderment" characterized the "new postmodern space" (Stephanson 1987: 33). A similar existential bewilderment, more explicitly deployed I suspect, now drapes the dramatic co-option of environmental politics to the simultaneously global and local agendas of First World capital. In not much more than a decade – since Ronald Reagan's curious upholding of a modern anti-environmentalism with the announcement that trees too cause pollution – the USA has reinvented itself, led by such environmental pariahs as Mobil, International Paper, and Dow Chemical ("the environmental people") and abetted by the generalized nature-consumptive ethic of The Nature Company and the Body Shop, as an environmental society. As many of the contributions to this book eloquently attest, the modern language of a wondrous and awe-inspiring nature has been resuscitated and remade as a discursive shield for the intensification of eco-destruction, a jolt for a new round of First World consumerism, and a means of oppression and exploitation of indigenous peoples on an unprecedented scale. Through all of this establishment environmentalism assists in the continued reworking and reinvention of class, gender and race asymmetries. We cannot not engage in the political battle over the re-enchantment of nature.

This project is as dangerous as it is inevitable. Any thought of romancing nature or re-enchanting nature has first to unearth itself from the ideological mucus with which such ideas and concepts come to us. The critique of ideologies of nature has never been more vital than now, and these have to provide the medium in which a counter re-enchantment can maintain critical, vigilant distance from the social magnetism of a simultaneously sentimentalized and oppressive nature. How are we to build a re-enchantment of nature too that does not indulge an academic escapism – all is discourse and discourse all – which already prevails among more idealist formulations of cultural studies; an academic escapism charted by unhinged nature imaginaries? How do we follow through on the conviction that a practical connection to a broadly configured environmental justice and political ecology movement provides the best defense and inspiration?

In every sense this is a millenial issue. Back-to-nature sentiment has had periodic revivals in the ebb and flow of modern capitalism, but more often than not these were associated with wider social crises and upheavals. In the USA, the nature revivalism at the end of the last century was clearly connected to combined crises in nationhood, race, class and masculinity as well as capital accumulation. Teddy Roosevelt rode a distinctive brand of nationalism, war-mongering and environmentalism to the national rescue. The murderous ecocide in Iraq in 1991, ordered by the "environmental president," George Bush, represents a farcical repetition of that history. At the end of the millenium, another nature revival coincides with a dramatic restructuring of economy and identity. If it might take decades for social scientists to get their concepts around the shifts now underway, and especially as they reconfigure nature, as McClure's analysis suggests, signs of that shift and the possibility of an oppositional re-enchantment of nature, begin to show up, however dimly, in literary texts.

Let me point to one simple example, Alix Kates Shulman's *Drinking the Rain* (1995). Shulman, a long-time feminist activist and writer, begins this memoir with an explanation of her escape from a New York life in the 1980s. That life was "more clogged than ever with busyness" even as the decline of the feminist movement she had devoted her life to weighed more heavily and the gathering backlash became more intense. And with her kids grown, her marriage was in trouble. She retreats to an isolated island in Maine, and it is a retreat. Thus far the situation is recognizably traditional, the retreat from city to nature, and we might expect a narrative of joyous renewal in the natural playground of the island, alone with the wind, the sea, and the birds. Some of that happens, and Shulman does indeed take us through her island existence and out the other end, back to the mainland, and into the world again. But her story also rewrites the experience with nature in very different terms. Life on the island is anything but easy: there are few inhabitants, no phones or electricity, and one small pitiful store that can only be reached by a two-mile trek lugging a hand cart through dry sand. There is no sense here of spiritual renewal through hardship, only that it's a pain in the arse. And food is scarce so she takes to eating whatever is scavangable in the environment, from mussels and fish to salads made of everything. Again there is barely a romanticism of "living with nature" in all of this, not even a romanticism of the difficulties; the tone is dead-pan rather than celebratory. Aided by an old tattered country cookbook, she is quietly delighted to discover an extraordinary array of salad ingredients growing wild on the windswept island, but is comparatively unmoved by the pragmatism of it all. This is not a text of nature worship. Her major involvement with nature comprises acts of what we might call "the consumptive production of nature": she is eating nature and transforming it at the same time in an array of starkly pragmatic practices.

Shulman weaves a simple story that recovers a certain redemptive nature immersion as the grist for the politics of everyday life. The consumptive production of nature indeed becomes a means of self-knowledge as she struggles to write and rethink her life. Her New York problems are in no way magically resolved by months on the Maine island. If anything they are sharpened, but at least she grasps her predicament more clearly and, largely through cooking, has a keener sense of the social environment she inhabits in the city. Whatever emotional restitution she enjoyed in Maine, the long summer coincidentally tracked to their source some of the commodities that found their way into her New York kitchen and stimulated her imagination about the pleasure and politics of other foods (cf. Goodman and Redclift 1991). A knowledge simultaneously poetic and political economic hints at the possibility of political difference and redirection.

How a progressive re-enchantment of nature might be engineered via less literary, more social scientific means is more difficult to say. At a time when the academic fashionability of "the imaginary" is so often an escape route from the societal mainland, it is certainly a risky strategy. It is entirely possible that the popular imaginary of nature – that is, the ideology of nature – is already so densely colonized that there is little or no access from which an alternative re-enchantment can be fashioned. Equally, the dangers of this re-enchantment might encourage such a policing of the project that it never gets started. On the other hand, if the power that feeds nature idolatry could be redirected in such a way that it enhanced the analytics of the production of nature thesis, it would represent an incredibly powerful means of leveraging a new future.

As Joel Kovel (1996: 28) has felt compelled to argue, "one can play with the imaginary and live to tell the tale." If a viable revolutionary environmentalism is to be part of nature at the millenium, we will have to find ways of proving Kovel right. But the corollary is just as important: we will have to find a way of playing with science and with political economy and of living to tell these tales too. Whether artifactual constructionism or a broader production of nature perspective is brought to bear, the fulcrum of the politics of nature in the new millennium is social and political practice. Despite the dismal political defeats of the last decades of the twentieth century, therefore, this means that we will also have to begin to "play" with – become involved in – organized political movements, knowing that here too, we can live to tell the tale.

NOTES

1 See, for example, Hawking (1990) and Hogan (1996).
2 My physicist friends assure me that the footnotes in this article are genuine enough but that the real paradox lies in the fact that Sokal's text, with minor editing, is not so

unreasonable a rendition of the current malaise in the physical sciences as a result of the ever-receding promise of scientific closure – the completion of physics – and the inability to reconcile abiding contradictions between relativity and quantum theory.
3 See Smith (1984). For an early, resolutely Kantian approach that is indeed discursive all the way down, see Evernden (1992).
4 Although he makes no reference, Lefebvre's distinction between production and creation, later applied to nature, seems to derive from Benjamin's (1969) discussion of art in the mechanical age. For a more elaborate critique of Lefebvre vis-à-vis nature, see Neil Smith (1997).

REFERENCES

Barad, K. (1996) "Meeting the universe halfway: realism and social constructivism without contradiction," in L. Nelson and J. Nelson (eds) Feminism, Science and the Philosophy of Science, Lancaster: Kluwer Press.

Benjamin, W. (1969) "The work of art in the age of mechanical reproduction," in W. Benjamin, Illuminations, New York: Schocken Books.

Escobar, A. (1996) "Constructing nature: elements for a post-structural political ecology," in R. Peet and M. Watts (eds) Liberation Ecology, London: Routledge.

Evernden, N. (1992) The Social Creation of Nature, Baltimore: Johns Hopkins Press.

Goodman, D. and Redclift, M. (1991) Refashioning Food: Food, Ecology and Culture, London: Routledge.

Gross, P. and Levitt, N. (1994) Higher Superstition: The Academic Left and its Quarrels with Science, Baltimore: Johns Hopkins University Press.

Haraway, D. (1991) Simians, Cyborgs and Women: The Reinvention of Nature, New York: Routledge.

Hawking, S. (1990) A Brief History of Time, New York: Bantam Books.

Hogan, J. (1996) The End of Science, Reading, MA: Helix Books.

King, D. L. (1995) Doing their Share to Save the Planet: Children and Environmental Crisis, New Brunswick: Rutgers University Press.

Kovel, J. (1996) "Negating Bookchin," Capitalism, Nature, Socialism, Pamphlet 5.

Kroptiuch, K. (1988) A Poetics of Petty Commodity Production in Egypt, Minneapolis: University of Minnesota Press.

Latour, B. (1987) Science in Action: How to Follow Scientists and Engineers through Society, Cambridge: Harvard University Press.

Lefebvre, H. (1991) The Production of Space, Oxford: Blackwell.

McLure, J. (1994) Late Imperial Romance, London: Verso.

Marx, K. (1973) Grundrisse, Harmondsworth: Penguin.

Shulman, A. K. (1995) Drinking the Rain, New York: Farrar, Straus, Giroux.

Sismondo, S. (1993) "Some social constructions," Social Studies of Science 23: 515–53.

Smith, N. (1984) Uneven Development: Nature, Capital and the Production of Space, Oxford: Blackwell.

—— (1997) "Antinomies of space and nature in Henri Lefebvre's The Production of Space," Philosophy and Geography 2: 46–69.

Sokal, A. (1996a) "Trangressing the boundaries," Social Text 46–7: 217–52.

—— (1996b) "A physicist experiments with cultural studies," Lingua Franca 6: 62–4.

Stephanson, A. (1987) "Regarding postmodernism – a conversation with Fredric Jameson," *Social Text* 17: xx–xx.

Waddington, C. H. (1941) *The Scientific Attitude*, Harmondsworth: Penguin.

INDEX

Note: page numbers in italics refer to illustrations or tables